ASTROPHYSICAL DISKS

ASTROPHYSICS AND SPACE SCIENCE LIBRARY

VOLUME 337

ASTROPHYSICAL DISKS

Collective and Stochastic Phenomena

Edited by

ALEXEI M. FRIDMAN
Institute of Astronomy, Moscow, Russia

MIKHAIL YA. MAROV
Keldysh Institute of Applied Mathematics, Moscow, Russia

and

ILYA G. KOVALENKO
Volgograd State University, Russia

Springer

A C.I.P. Catalogue record for this book is available from the Library of Congress.

ISBN-10 1-4020-4347-3 (HB)
ISBN-13 978-1-4020-4347-5 (HB)
ISBN-10 1-4020-4348-1 (e-book)
ISBN-13 978-1-4020-4348-2 (e-book)

Published by Springer,
P.O. Box 17, 3300 AA Dordrecht, The Netherlands.

www.springer.com

Printed on acid-free paper

Cover picture: "Work", I, Michael L. Evans, Business Manager of the Astrophysical Research Consortium (ARC), performer of the SDSS, "Astrophysical Research Consortium (ARC) and the Sloan Digital Sky Survey (SDSS) Collaboration, http://www.sdss.org"

Printed in the Netherlands.

Contents

Part III   Posters

Contributing Authors

Victor Afanasiev, Special Astrophysical Observatory, Nizhnij Arkhyz, Russia

Maxim Barkov, Space Research Institute, Moscow, Russia

Vasily Beskin, Lebedev Physical Institute, Moscow, Russia

Dmitri Bisikalo, Institute of Astronomy RAS, Moscow, Russia

Gennady Bisnovatyi-Kogan, Space Research Institute, Moscow, Russia

Anatoly Cherepashchuk, Sternberg State Astronomical Institute of MSU, Moscow, Russia

Chi Yuan, Institute of Astronomy & Astrophysics, Academia Sinica Taipei, Taiwan, ROC

Jim Collett, Division of Physics & Astronomy, University of Hertfordshire, Hatfield, United Kingdom

George Contopoulos, Academy of Athens, Athens, Greece

Mikhail Eremin, Volgograd State University, Volgograd, Russia

Alexei Fridman, Institute of Astronomy RAS, Moscow, Russia

Preben Grosbøl, European Southern Observatory, Garching, Germany

Sergei Kapitza, P.L.Kapitza Institute for Physical Problems, Moscow, Russia

Pavel Kaygorodov, Institute of Astronomy RAS, Moscow, Russia

Alexander Khoperskov, Volgograd State University, Volgograd, Russia

Oleg Khoruzhii, Institute of Astronomy RAS, Moscow, Russia

Sergei Khrapov, Volgograd State University, Volgograd, Russia

Alexander Kilpio, Institute of Astronomy RAS, Moscow, Russia

Elena Kilpio, Institute of Astronomy RAS, Moscow, Russia

Johan Knapen, Department of Physical Sciences, University of Hertfordshire, Hatfield, United Kingdom

Vitalij Korolev, Volgograd State University, Volgograd, Russia

Ilya Kovalenko, Volgograd State University, Volgograd, Russia

Oleg Kuznetsov, Keldysh Institute of Applied Mathematics, Moscow, Russia

Mikhail Marov, Keldysh Institute of Applied Mathematics, Moscow, Russia

Takuya Matsuda, Department of Earth and Planetary Sciences, Kobe University, Kobe, Japan

Evgenij Matvienko, Rostov State University, Rostov, Russia

Elena Mikhailova, Volgograd State University, Volgograd, Russia

Alexei Moiseev, Special Astrophysical Observatory, Nizhnij Arkhyz, Russia

Victor Mustsevoi, Volgograd State University, Volgograd, Russia

Toru Okuda, Hakodate College, Hokkaido University of Education, Hakodate, Japan

Anna Sidorova, Volgograd State University, Volgograd, Russia

Olga Sil'chenko, Sternberg State Astronomical Institute of MSU, Moscow, Russia

Yuri Torgashin, Institute of Astronomy RAS, Moscow, Russia

Nataly Tyurina, Sternberg State Astronomical Institute of MSU, Moscow, Russia

Nikos Voglis, Research Center for Astronomy, Academy of Athens, Athens, Greece

Marina Zabolotskih, Sternberg State Astronomical Institute of MSU, Moscow, Russia

Anatoly Zasov, Sternberg State Astronomical Institute of MSU, Moscow, Russia

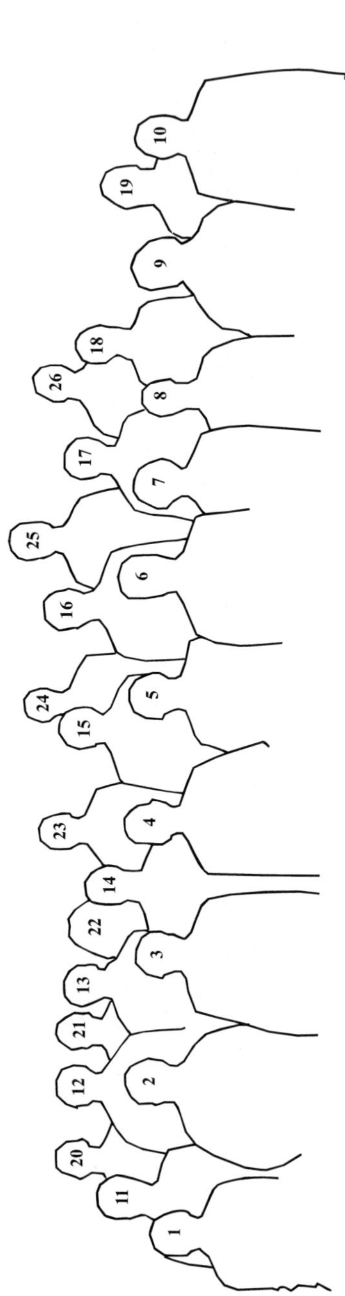

1. Yuri Torgashin
2. Mikhail Marov
3. Alexei Fridman
4. George Contopoulos
5. Gennady Bisnovatyi-Kogan
6. Anatoly Cherepashchuk
7. Elena Mikhailova
8. Olga Sil'chenko
9. Nikos Voglis
10. Oleg Kuznetsov
11. Toru Okuda
12. Preben Grosbøl
13. Anatoly Zasov
14. Sergei Kapitza
15. Dmitri Bisikalo
16. Alexander Khoperskov
17. Chi Yuan
18. Anna Sidorova
19. Pavel Kaygorodov
20. Vasily Beskin
21. Alexander Kilpio
22. Elena Kilpio
23. Ilya Kovalenko
24. Mikhail Eremin
25. Vitalij Korolev
26. Maxim Barkov

Preface

In the recent decades the theory and observations of disk systems became one of the fast progressing branches of astrophysics. This was stimulated by expansion of astronomical facilities including space-born and ground based instruments, data storage with growing resolution, the development of theoretical treatment of the processes involved and their modeling with the use of powerful computers. Among these processes, of particular interest and importance is the study of disks structure, their dynamics and evolution resulting in different configurations, including self-organization, which are basically relevant to the general problems of non-linear dynamic systems.

Discussion of this topic was the main focus of the Colloquium "Progress in the Study of Astrophysical Disks: Collective and Stochastic Phenomena and Computational Tools" that was held from September 9 to 11, 2003 in Volgograd, south-east of European Russia. This book is the Proceedings of a conference attended by the leading scientists in the field from around the world, as well as by young scholars, mostly from the Volgograd State University which hosted the event. Altogether, 28 papers were presented, 22 oral ones and 6 posters, followed by valuable discussions.

The book emphasises models of the disk galaxies, the nature of galactic vortices, dynamics of accretion disks with density waves, chaotic and ordered structures including turbulence in the accretion disks, numerical modelling of galactic and accretion disks, and the results of their observation.

The contents of the book are a good summary of collective and stochastic processes in the astrophysical disks, their structure and evolution, and in-depth study of different approaches to their modelling with the use of modern computational tools. The main topics include: development of the orbital theory leading to a separation of ordered and chaotic regions in the galactic disks; revealing of a spiral-vortex structure (which is a density wave) by numerical simulation of the dynamics of an accretion disk in close binary stars supported by observations; development of a phenomenological model of non-equilibrium turbulence in the compressible fluid with the account of collective non-linear processes; advancement in accretion disk theory with the involvement of turbulent viscosity and in the numerical simulation of structure and dynamics of a gaseous disk in a spiral galaxy, as well as in the study of mass distribution of luminous matter in disk galaxies involving spiral patterns/perturbations and of the internal structure of the thin accretion disk around a black hole. Specially addressed are spiral-vortex structure in galactic and accretion disks, stochastic and ordered structures in the developed turbulence, sources of turbulence in the accretion disks, numerical modelling of Be envelopes in binaries, gaseous disks in spiral galaxies with shock waves formation.

Also, such exciting results as the first experimental detection of an over-reflection instability impact on the rotating shallow water found independently in 1957 for the simplest case of the plane geometry in the pioneer theoretical works of Miles and Ribner, are thoroughly described and discussed in the book.

The authors adeptly brought together collective and stochastic phenomena in the modern field of astrophysical disks, their formation, structure, and evolution involving relevant methodology, the results of observation and modelling, thereby advancing the study in this important branch of astrophysics.

This is a book that belongs in all science libraries. It is intended for professional researchers in the field and graduate students and may be also useful for undergraduates who are going to specialize in astrophysics.

The Editors greatly appreciate the support given by the Volgograd State University in holding this international Colloquium and would like to thank all authors who submitted the refined versions of their papers, the reviewers who generously responded and greatly contributed to the quality of the Proceedings, and Springer-Kluwer Publishers who offered to put out this book.

ALEXEI M. FRIDMAN

MIKHAIL YA. MAROV

ILYA G. KOVALENKO

I

INVITED PAPERS

THE OVER-REFLECTION INSTABILITY: MYTH OR REALITY?

A. M. Fridman,[1] E. V. Polyachenko,[1] Yu. M. Torgashin,[1] S. G. Yanchenko,[1] and E. N. Snezhkin[2]

[1]*Institute of Astronomy RAS, 48 Pyatnitskaya st.*
Moscow 117019 Russia

[2]*Russian Research Center "Kurchatov Institute"*
1, Kurchatov Sq., Moscow, 123182, Russia

Abstract   A scheme of model experiments on differentially rotating free-surface shallow water is proposed for discovering and studying the over-reflection instability. The scheme allows suppressing two stronger instabilities possible in the simulations, i.e. Kelvin-Helmholtz and centrifugal instabilities which the authors have studied earlier in rotating shallow water both theoretically and experimentally. Distinctive features of coherent structures to be generated by the over-reflection instability are figured out. Being observed in the laboratory simulations, the features will evidence that this very instability does exist.

Keywords: hydrodynamic instabilities, over-reflection instability

1.   Introduction

Miles (1957) and Ribner (1957) were the first to research the amplification effect theoretically. They were considering a simple problem about the reflection of a monochrome sound wave from a plane-parallel vortex velocity sheet. It turned out that when the speed of the moving medium was high enough (the incident wave approached the vortex sheet from the motionless medium) there was a possibility of reflection with amplification. In this process the amplitude of the reflected wave becomes higher than that of the incident wave. It is because of the wave of negative energy (more precisely, quasi-energy, McIntaier, 1981) which propagates into the moving medium, while the wave of positive energy propagates into the motionless medium. The energy of the sound wave

3

Alexei M. Fridman, et al. (eds.), Astrophysical Disks; Collective and Stochastic Phenomena 3–22
© 2006 *Springer. Printed in the Netherlands*

radiated into the motionless medium during this process is supplied by the moving medium (Stepanyantz & Fabrikant, 1989). For the development of an instability it is enough to add an acoustic feedback, for example a wall, which would force the wave reflected from the vortex sheet to return and amplify again.

The other source of amplification of the wave during its reflection is a resonant amplification of a sound wave which arises if we smooth the vortex sheet to some finite width (Blumen et al., 1975). Then a thin critical layer appears inside of the vortex sheet and the energy of the sound wave increases in this layer. Thus, an additional branch of unstable oscillations may appear only because of the interaction with resonant particles (Stepanyantz & Fabrikant, 1989).

We mentioned above that Miles and Ribner had investigated the process of wave reflection from the vortex velocity sheet. When there is a reflective wall an over-reflection instability may arise. But simultaneously two other instabilities can exist in the same system where the tangential velocity discontinuity is available. We mean the Kelvin–Helmholtz (vortex sheet) and centrifugal instabilities, which are generally more powerful than the over-reflection instability so the over-reflection instability may be revealed only in case when these both instabilities are absent.

The hydrodynamical medium in which we are going to carry out our investigation of the over-reflection instability is a rotating shallow water. It is known (Landau & Lifshitz, 1984) that dynamics of this medium may be described by two-dimensional equations equivalent to the appropriate dynamic equations of a two-dimensional compressible medium. Exactly for this case Landau (1944) obtained the stabilization criterion for the vortex sheet instability. According to this criterion, if there is a velocity jump only, the vortex sheet instability is absent if the Mach number

$$M \geq 2\sqrt{2}. \tag{1}$$

The correctness of this criterion for the rotating shallow water layer was proved in the experiments (Antipov *et al.*, 1983). When condition (1) was fulfilled in the experiments at the set-up "Spiral" with rotating shallow water (Fridman *et al.*, 1985; Nezlin & Snezhkin, 1993) another type of instability was observed, when the inner part of the fluid rotated faster than the outer one. This instability was called a centrifugal one (Fridman *et al.*, 1985).

In the opposite case

$$\Omega_1 < \Omega_2, \tag{2}$$

where Ω_1, Ω_2 are angular velocities of the inner and outer part of the shallow water, respectively, and when (1) is fulfilled both instabilities – the vortex sheet and the centrifugal ones – are suppressed.

The physics of excitation and stabilization of these two instabilities is qualitatively discussed below. The main part of the article is devoted to the description of the excitation of the over-reflection instability in the rotating shallow

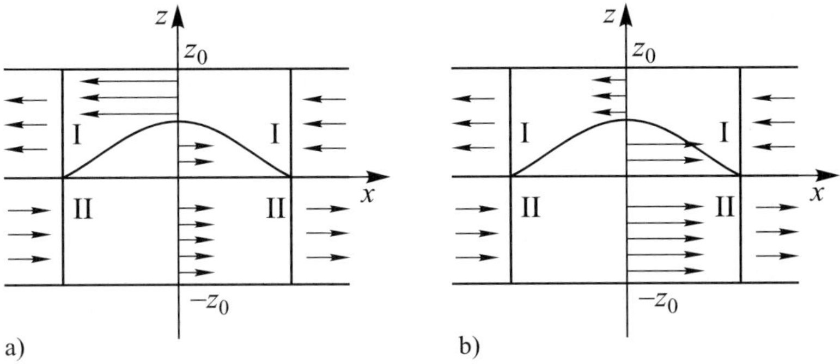

Figure 1. The Bernoulli integral of motion: a) the instability, $M < 2\sqrt{2}$, $V_2 < V_1$, $p_2 > p_1$; b) the stabilization, $M > 2\sqrt{2}$, $V_2 > V_1$, $p_2 < p_1$. Here $M \equiv \Delta V/C_S$, $\Delta V \equiv \left| \vec{V_2} - \vec{V_1} \right|$.

water when the conditions (1) and (2) are fulfilled, i.e. in the absence of "rival" instabilities.

A derivation of the dynamic equations of the shallow water using the Pedlosky method (1982) generalized in this paper for the differentially rotating shallow water is taken out into the Appendix.

Construction peculiarities of the set-up with the rotating shallow water where we expect to discover and research the over-reflection instability are formulated in the concluding part of the paper.

Before describing main results on over-reflection instability we would like to mention main features of other two hydrodynamical instabilities which can arise in rotating shallow water with the jump on the angular rotation velocity $\Omega(r)$ – Kelvin–Helmholtz instability and centrifugal one. There are at least two reasons for such mentioning. First of all in general case these instabilities are much stronger than the over-reflection instability, so, for observing the last one in the course of an experiment we need to choose carefully main parameters of the set-up to avoid generation of two rival instabilities. The second reason is the following: to be convinced that just the over-reflection instability is observed in our experiments, we need to compare it's peculiarities with those of other instabilities mentioned above.

1.1　Kelvin–Helmholtz instability

Let us explain the nature of supersonic stabilization of the Kelvin–Helmholtz instability. When a flat compressible vortex sheet is slightly perturbed, see (Fig. 1), velocity and pressure would behave themselves in different ways in sub- and supersonic flows.

Perturbation amplitudes on both sides of the disturbed discontinuity surface fall exponentially with distance from velocity jump, $\exp\left(-\frac{|z|}{z_0}\right)$, (Landau, 1944). Therefore, it is enough to restrict ourselves to the area $|z| < z_0$.

The region I (over the "hump" of perturbed surface) in both figures (Fig. 1a) and (Fig. 1b) can be considered as the region of motion inside a nozzle (more precisely, in its longitudinal half). It is well known that the behaviour of the flow is principally different in sub- and supersonic nozzles (Loitzyanskii, 1973; Landau and Lifshitz, 1984). In the most narrow (critical) cross-section of a subsonic nozzle, the velocity is maximal, just as it is in the narrowest place of a river. In the critical cross-section of a supersonic diffuser, in contrast, the velocity is minimal. From this comes the different dynamics for two tangential discontinuities of velocity: sub- and supersonic flows. These differences are based on the constancy of the Bernoulli integral of motion: $W(p) + V^2/2 =$ const, where W is enthalpy. In a subsonic flow the velocity under the hump is less than that over the hump ($V_2 < V_1$), hence, $W_2 > W_1$. As in the ordinary situations, the pressure grows as does the enthalpy, $p_2 > p_1$, i.e. the hump will continue to grow.

In case of a supersonic velocity jump (Fig. 1b), the opposite inequality is correct: $p_2 < p_1$, i.e. the hump is suppressed by the pressure gradient – the instability is stabilized.

All observed velocity jumps in gaseous galactic disks are supersonic, therefore, the Landau criterion of stabilization is fulfilled for such disks: the vortex-sheet instability, or Kelvin-Helmholtz instability, is absent in them.

According to the experimental results by Antipov *et al.* (1983) the structures generated by the Kelvin–Helmholtz instability have the form of nearly radially directed spiral waves with m arms, their amplitudes being decaying monotonically on both sides from the velocity jump. Number of the spirals is increasing with the grows of the velocity jump amplitude.

1.2   Centrifugal instability

There is another instability – centrifugal, which develops at any large Mach numbers, under the condition that angular rotation velocity of the central part, Ω_1 (internal in relation to the velocity jump) is larger than that on the periphery, Ω_2: $q = \Omega_2/\Omega_1 < 1$.

Such velocity jump (Fig. 2) is observed in a half of spiral galactic disks (Afanasiev *et al.*, 1988a, 1988b, 1991a, 1992).

The physics of the centrifugal instability is similar to that of the Rayleigh–Taylor instability describing the situation when a heavy liquid lies over a light one. In the case of the centrifugal instability the centrifugal force of the center predominates over that of periphery as well as the gravity force of the heavy liquid predominates over that of the light one in case of Rayleigh–Taylor instability.

The centrifugal instability generates the trailing spiral waves, which rotate with the ends of the spiral pointing backward and so have a good "aerody-

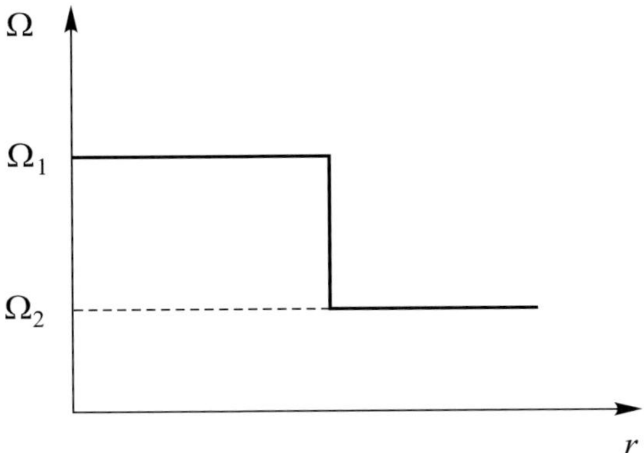

Figure 2.   The rotation curve in the case of centrifugal instability.

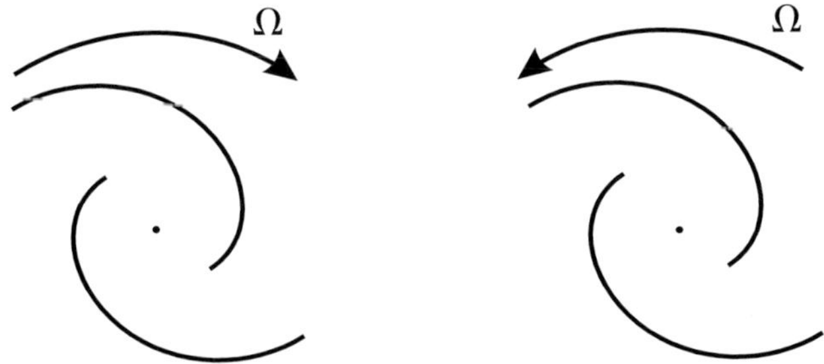

Figure 3.   A schematic representation of the trailing (a) and leading (b) spiral waves. The arrows indicate the directions of the shallow water rotation.

namic" form (Fig. 3) The larger a velocity jump $\Delta\Omega$ the smaller number m of trailing spiral waves are excited by the centrifugal instability.

2.   Over-reflection instability

In two previous subsections we show that a shear layer is the source of hydrodynamical instabilities. The Kelvin–Helmholtz and centrifugal instabilities are the strongest ones. However, they can be suppressed if appropriate parameters of the set-up are chosen.

For the sake of simplicity, we mainly consider a flow with only one discontinuity in the rotation curve $\Omega(r)$: $\Omega(r) = \Omega_1$ within the radius $r = R$, and $\Omega(r) = \Omega_2$ when $r > R$. We assume that the ratio

$$\bar{q} = \frac{\Omega_1}{\Omega_2} \tag{3}$$

is positive and less than unity. In this case the centrifugal instability is suppressed.

Generally speaking, thickness of the critical layer is of the order of the unperturbed shallow water depth H_0. The latter is assumed to be small comparing to all characteristic scales of the problem. Thus, qualitatively, situation can be understood already by analyzing sharp discontinuity of the rotation curve. In the end of this section, however, we address to the effects of finite thickness of the layer.

The velocity jump at the discontinuity is characterized by the Mach number:

$$M = R\frac{|\Omega_2 - \Omega_1|}{C_s}, \tag{4}$$

where C_s is sound speed. By analogy with flat shear layer, for rotating shallow water, Kelvin–Helmholtz instability is suppressed, if $M \geq 2.8$. Numerical analysis show that unlike in the flat shear layer, exact stability boundary slightly varies on the azimuthal number m. Below we consider the Mach number to be sufficiently large to suppress Kelvin-Helmholtz instability.

The dispersion equation for studying the shear layer instability of free surface shallow water can be obtained from inviscid hydrodynamical equations, the derivation on which is available in Appendix A:

$$\frac{\partial v_r}{\partial t} + \Omega\frac{\partial}{\partial \varphi}v_r - 2\Omega v_\varphi = -\frac{\partial}{\partial r}(C_s^2\eta),$$

$$\frac{\partial v_\varphi}{\partial t} + \Omega\frac{\partial}{\partial \varphi}v_\varphi + \frac{\kappa^2}{2\Omega}v_r = -\frac{\partial}{r\partial \varphi}(C_s^2\eta), \tag{5}$$

$$\frac{\partial \eta}{\partial t} + \frac{1}{rH_0}\frac{\partial}{\partial r}(rH_0 v_r) + \frac{1}{r}\frac{\partial}{\partial \varphi}(\eta\Omega r + v_\varphi) = 0.$$

Here v_r, v_φ are the radial and azimuthal velocity disturbances, respectively, H_0 is the shallow water depth (assumed constant), $\eta = h/H_0$ is the normalized depth perturbation, $\Omega(r)$ is the angular rotation velocity, κ is the epicyclic frequency, $\kappa(r) = (4\Omega^2 + rd\Omega^2/dr)^{1/2}$. The role of sound speed C_s is played by the speed of gravity wave propagation: $C_s = (gH_0)^{1/2}$, where g is the gravitational acceleration.

Despite the infinitely thin layer was thoroughly considered in Fridman (1990), in Appendix B we repeat the derivation of the dispersion equation in terms, more suitable for studying over-reflection instability. In case of arbitrary rotation curve and sound speed profile, the system of differential equations (5) can be reduced to the following equations:

$$\frac{d}{dr}(C_s^2\eta) = \frac{2m\Omega}{r\hat{\omega}}(C_s^2\eta) - (\kappa^2 - \hat{\omega}^2)\xi,$$

$$\frac{d}{dr}(rH_0\xi) = -rH_0\left[\left(1 - \frac{m^2C_s^2}{r^2\omega^2}\right)\eta + \frac{2m\Omega}{r\hat{\omega}}\xi\right] \tag{6}$$

with appropriate boundary conditions.

For sharp discontinuity of the rotation curve, the relative perturbation of the shallow water depth is a linear combination of the Bessel functions:

$$\eta(r) = C_1 H_m^{(1)}(kr) + C_2 H_m^{(2)}(kr). \tag{7}$$

Here $H_m^{(1,2)}$ denote the Hankel functions of the first and second kind. The radial wavenumber k depends on the frequency of the perturbation ω:

$$k^2 = \frac{\hat{\omega}^2 - 4\Omega^2}{C_s^2}, \qquad \hat{\omega} \equiv \omega - m\Omega(r). \tag{8}$$

Since the rotation curve is piece-wise constant function, wavenumbers on both sides of the jump are constants (let us denote them as k_1 and k_2 for inner and outer regions, correspondingly). If a wavenumber is real, (7) describes a wave-like solution, otherwise a solution is exponentially increasing or decreasing.

Introducing a dimensionless frequency $x = \omega/(m\Omega_2)$, one can find that k_1 is real when

$$\operatorname{Re} x > \bar{q}\left(1 + \frac{2}{m}\right), \qquad \bar{q}\left(1 - \frac{2}{m}\right), \tag{9}$$

and k_2 is real when

$$\operatorname{Re} x < 1 - \frac{2}{m}. \tag{10}$$

For definiteness, let us assume that the left wall is at $r = 0$, right wall is at $r = \infty$. Depending on whether k_1 and k_2 are real or imaginary, we discern three different cases.

I. Both k_1, k_2 are real. In this case solutions on both sides of the jump are wave-like. It is possible if the azimuthal number

$$m > \frac{2}{1 - \operatorname{Re} x} \tag{11}$$

The inequality shows that minimum number of arms in unstable patterns of this type is $m = 3$. Below in this item and the next one we consider $\bar{q} = 0$.

The wave in the inner part is propagating between the left wall (the origin $r = 0$) and the layer. Under some conditions the wave is reflected with an amplitude greater than that of the incident wave (over-reflection) in full analogy with Miles–Ribner effect (see Fig. 4).

In the presence of feedback, this leads to unstable solutions, described by Kolyhalov (1984) in flat geometry. Number of unstable modes can be large. Real parts of eigen-frequencies can be determined from the equation:

$$J'_m(k_1 R) = 0, \qquad (12)$$

$J'_m(z)$ denotes the derivative of the Bessel function $J_m(z)$ with respect to z. Since $k_1 R$ is of the order of a large parameter Mm, the number of modes of this type is of the order of Mm/π.

If $x_n^{(0)}$ are roots of equation (12), the eigen-frequencies are given by the approximate formula

$$\frac{x_n}{x_n^{(0)}} = 1 - \frac{1}{M^2} + i\frac{m}{M^3}[(1 - x_n^{(0)})^2 - 4/m^2]^{1/2}. \qquad (13)$$

II. k_1 is real, k_2 is imaginary. In the inner region a wave-like solution exists, in the outer region – exponential solution. It happens when $\operatorname{Re} x > 1 - 2/m$. For these frequencies, however, the inner wave cannot be amplified at the layer, as it follows from Fig. 4.

Nevertheless, numerical calculations show that one mode of this type exists for any $m \geq 1$. With gradual change of Mach number, it alternatively becomes neutral or unstable.

The real part of the frequency is roughly equal to

$$x_0 = 1 - \frac{1}{M} + \frac{2.25}{M^2} + ... \qquad (14)$$

Maximum imaginary part is of the order of

$$\max(\operatorname{Im} x) \sim \frac{1 - x_0}{M}\sqrt{2x_0} \sim \frac{1}{M^2}\sqrt{2x_0}. \qquad (15)$$

III. k_1 is imaginary, k_2 is real. In the inner region the exponential solution exists; in the outer region – the wave-like solution. This situation is possible only if $\bar{q} > 0$ and $\operatorname{Re} x < \bar{q}(1 + 2/m)$.

One can show analytically, that one mode of this type exists, if azimuthal number $m \geq 3$. Approximate expressions for real and imaginary parts of x are rather cumbersome. For that reason we do not give their explicit expressions. The real part is close to the right limit, $x_0 = \bar{q}(1 + 2/m)$. The imaginary part in the limit $M \gg m \gg 1$ is of the order of

$$\operatorname{Im} x \sim \bar{q}\frac{2}{M^3}\sqrt{m^2 - 4}. \qquad (16)$$

Unstable mode of this type has the radiative nature: its growth is due to the radiation of the wave of the negative energy to infinity.

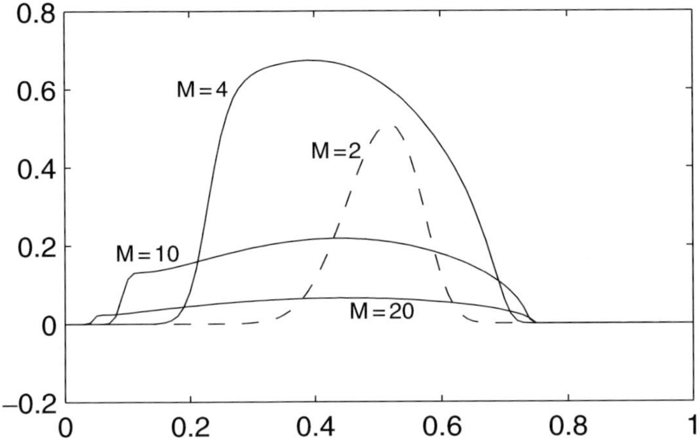

Figure 4. Log of absolute value of the reflection coefficient $\ln|A(x)|$ vs. $\mathrm{Re}\,x$ for $m = 8$, $\bar{q} = 0$ and different Mach numbers: dashed line – $M = 2$, solid lines – $M = 4, 10, 20$. If $M > 4.2$, maximum of these curves decrease with M for any azimuthal number m. Note that in circular case the amplification occurs even for $M < 2$.

Analytical and numerical calculations for moderate values of M and m,

$$2.5 < M \leq 6, \qquad m < 10 \tag{17}$$

show that growth rates of modes of type I and II are comparable, while the growth rate of type III mode is much lower. For that reason, the latter mode can hardly be determined in the experiment.

3. Modelling the over-reflection instability at the experimental set-up

In this section we are going to consider a realistic model of a rotating shear layer. Parameters of this model can be used for constructing an experimental set-up for studying over-reflection instability.

The proposed set-up is shown in Fig. 5. It consists of two parts, one is inside of another. The inner part is a motionless horizontal plate with radius $R = 12\,\mathrm{cm}$. A pivot with radius $R_a = 4.8\,\mathrm{mm}$ is attached to the center of the plate. The outer part is a conical surface with inner radius $R = 12\,\mathrm{cm}$ and outer radius $R_b = 20\,\mathrm{cm}$.

Small amount of water is poured into the set-up so the fluid covers the surface of both parts (when the outer one is rotating) with a thin layer. The depth of the water layer is several millimeters.

The bottom profile of the outer part should be chosen in such a way to make the shallow water depth as uniform as possible. In reality, unperturbed free

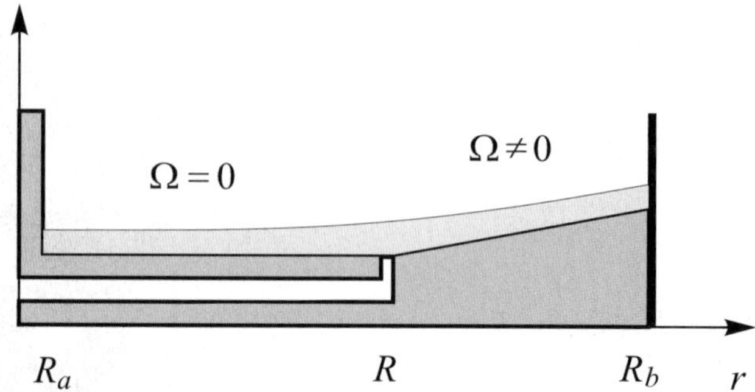

Figure 5.　　The scheme of the experimental setup.

surface is close to the paraboloid shape. However, the paraboloid shape depends on the angular velocity of the outer part. Thus, the condition of constant shallow depth can be strictly fulfilled only for a certain value of the angular velocity. If the velocity jump is fixed, the only way to change the Mach number is to change the shallow water depth. However, this method is less convenient than to change the rotation speed. For this reason we suggest a constant inclination of the surface (about $\alpha_0 = 15°$), which allows us to consider the shallow depth as roughly constant one in the range of Mach number $2.5 < M < 6$.

The vortex velocity sheet is not infinitesimally thin. That is why it is necessary to consider the influence of smoothing at the mechanism of generation of the unstable modes. Therefore instead of the sharp discontinuity we will consider a slightly smoothed jump

$$\Omega\left(r\right) = \frac{\Omega}{2}\left[1 + \tanh\left(\frac{r - R}{L}\right)\right], \tag{18}$$

where L characterizes the width of smoothing of the vortex sheet.

This dynamical system can be considered by the numerical analysis. Instead of solving the dispersion equation we should numerically solve the system (6).

The main result of these calculations is that in comparison with an infinitesimally thin vortex sheet considered in previous section, smoothing of this vortex sheet decreases the increments of unstable solutions, thereby weakening the over-reflection instability. However, the growth rates are still high enough to expect the over-reflection instability to manifest itself in the experiment. For a fixed Mach number, the growth rates for modes of different types and azimuthal numbers can be close to each other. Nevertheless, monotonic growth of the number of arms of the most unstable mode with increasing of Mach number can be observed.

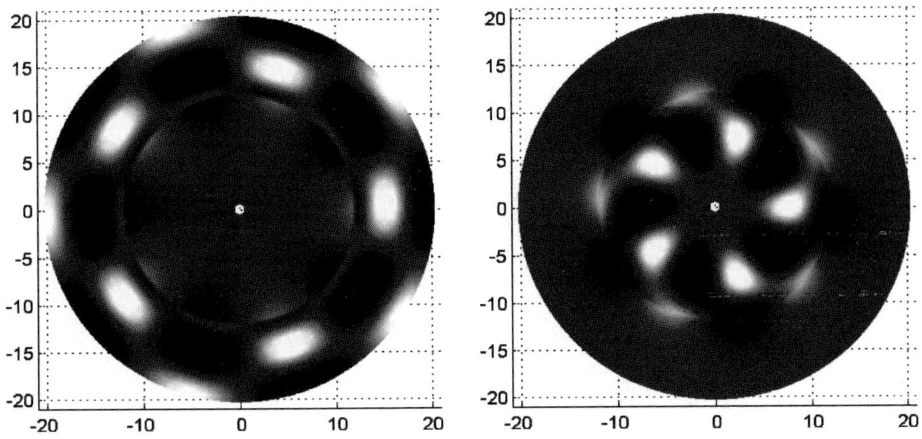

Figure 6.   Type I (left panel) and type II (right panel) five-armed unstable modes, obtained by numerical solution of the eigen-value problem (6). Mach number $M = 2.8$. Rotation is counter-clockwise.

The mode patterns obtained in our numerical calculations using the angular rotation curve (18) are shown in Figs. 6–8. In all cases $L = H_0/2$. Rotation of the patterns is counter-clockwise.

Fig. 6 shows two patterns with five arms (left panel – type I mode, right panel – type II mode), which can develop in the course of the over-reflection instability at Mach number $M = 2.8$.

Fig. 7 shows patterns of unstable modes with three and four arms, which can develop in the course of the over-reflection instability at Mach number $M = 2.8$.

Fig. 8 shows patterns of unstable modes with three arms. On the left panel the mode is unstable due to the Kelvin–Helmholtz instability ($M = 2.0$), on the right panel type II unstable mode is shown ($M = 4.2$).

4.   Conclusion

Thus, our investigation shows that the over-reflection instability generates structures with wave patterns which look like the ones shown in figures Fig. 6, 7, and right panel in Fig. 8 and contain leading spirals. With increasing the Mach number M we obtain the weak growth of sequence of azimuthal numbers m.

The wave patterns differ from those generated by the Kelvin–Helmholtz instability (left panel in Fig. 8). The latters are almost radial structures decreasing monotonically on the both sides from the angular velocity discontinuity.

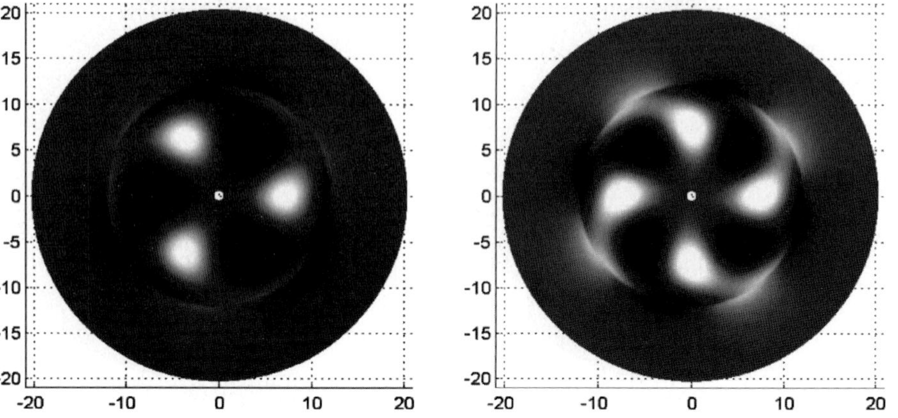

Figure 7.   Type II three- and four-armed unstable modes. Mach number $M = 2.8$. Rotation is counter-clockwise.

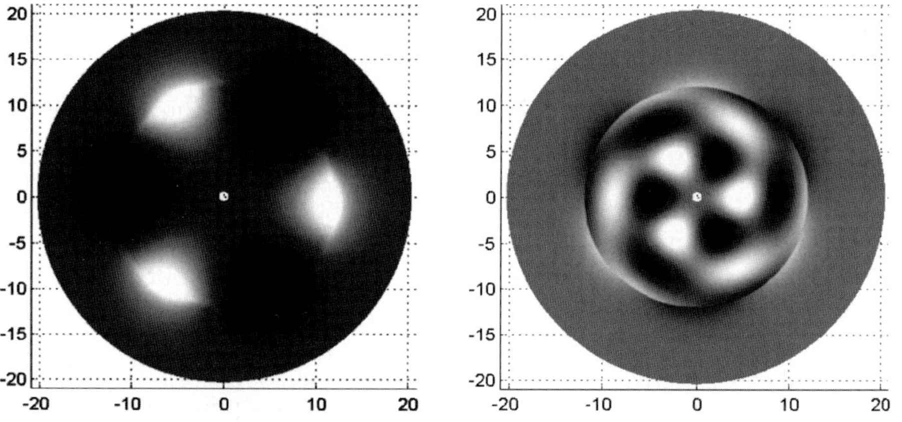

Figure 8.   Unstable mode due to the Kelvin–Helmholtz instability for Mach number $M = 2$, and type II three-armed unstable mode for Mach number $M = 4.2$. Rotation is counter-clockwise.

Also the wave patterns excited by the over-reflection instability differ from those caused by the centrifugal instability. The latter generates the trailing spirals. And the number of observed mode for centrifugal instability decreases with the Mach number growth.

Appendix: A. Derivation of the dynamic equations for rotating shallow water by the generalized Pedlosky method

Pedlosky (1982) has derived the set of dynamic equations for shallow water assuming the availability of two characteristic length scales: the first one lies in the horizontal plane (x, y), the second – along the vertical z-axis. We are forced to take into account both four characteristic length scales (see item (A2)) and additionally the rotation of the shallow water.

A1. Original equations for the "volume" functions.

The initial dynamic equations for incompressible ($\rho = $ const) inviscid fluid have the form

$$\hat{L}_1 V_r - \frac{V_\varphi^2}{r} = -\frac{1}{\rho}\frac{\partial}{\partial r}(P + \rho g z) \equiv -\frac{\partial \chi}{\partial r}, \tag{A.1}$$

$$\hat{L}_1 V_\varphi + \frac{V_r V_\varphi}{r} = -\frac{1}{\rho r}\frac{\partial}{\partial \varphi}(P + \rho g z) \equiv -\frac{\partial \chi}{r \partial \varphi}, \tag{A.2}$$

$$\hat{L}_1 V_z = -\frac{1}{\rho}\frac{\partial}{\partial z}(P + \rho g z) \equiv -\frac{\partial \chi}{\partial z}, \tag{A.3}$$

$$\frac{1}{r}\frac{\partial}{\partial r}(r V_r) + \frac{1}{r}\frac{\partial V_\varphi}{\partial \varphi} + \frac{\partial V_z}{\partial z} = 0, \tag{A.4}$$

$$\hat{L}_1 \equiv \frac{\partial}{\partial t} + V_r\frac{\partial}{\partial r} + \frac{V_\varphi}{r}\frac{\partial}{\partial \varphi} + V_z\frac{\partial}{\partial z}, \tag{A.5}$$

$$\chi \equiv \frac{P}{\rho} + g z. \tag{A.6}$$

For the four unknown functions

$$V_r, \quad V_\varphi, \quad V_z, \quad \chi \tag{A.7}$$

we have so far written down four equations (A.1 – A.4).

A2. Order-of-magnitude estimates of the terms in the initial equations

We shall give some "order-of-magnitude" estimates of the functions (A.7) and derivatives:

$$|V_r| \sim U; \ |V_\varphi| \sim V; \ |V_z| \sim W; \ |\chi| \sim \chi;$$

$$\left|\frac{\partial}{\partial r}\right| \sim L^{-1}, R^{-1}, D^{-1}; \ \left|\frac{1}{r}\frac{\partial}{\partial \varphi}\right| \sim \zeta^{-1}; \ \left|\frac{\partial}{\partial z}\right| \sim D^{-1}; \ \left|\frac{\partial}{\partial t}\right| \sim T^{-1}. \tag{A.8}$$

Unlike Pedlosky (1982) who took into account two characteristic length scales we introduce here four characteristic length scales: D is the characteristic thickness of the shallow water along the rotational z-axis; L, R, D are the scales of the perturbations, equilibrium parameters and the jump of Ω, respectively, along the radius r; ζ is the scale of the perturbations along the azimuth.

Substituting (A.8) into continuity equation (A.4) we find as to order of magnitude[1]:

$$W \lesssim (\varepsilon_1 U, \varepsilon_2 V)_{\max}, \tag{A.9}$$

---

[1] If v_r and v_φ have the same sign, then $W \sim (\varepsilon_1 U, \varepsilon_2 V)_{\max}$, if their signs are different $W < (\varepsilon_1 U, \varepsilon_2 V)_{\max}$.

where we must take the largest of the terms within the rounded brackets and where

$$\varepsilon_1 \equiv \frac{D}{L}, \; \varepsilon_2 \equiv \frac{D}{\zeta}. \tag{A.10}$$

Substituting (A.8) into (A.1) and (A.2) and dropping terms containing derivatives with respect to z for reasons which will become clear in what follows we find the χ function of the order of magnitude from the left hand side of equations (A.1), (A.2)

$$|\chi| \sim \left(\frac{L}{T}U, \; U^2, \; \frac{L}{\zeta}UV, \frac{L}{R}V^2, \; \frac{\zeta}{T}V, \; \frac{\zeta}{D}UV, \; V^2, \; \frac{\zeta}{R}UV \right)_{\max}. \tag{A.11}$$

The dimensionless quantities $\frac{L}{R}$, $\frac{\zeta}{R}$ and $\frac{L}{\zeta}$ which occur within the brackets appear due to the difference in scale over which χ and the velocities change. Using (A.8) we determine the order of magnitude of the vertical gradient of χ from (A.3)

$$D \left| \frac{\partial \chi}{\partial z} \right| \sim \left(\frac{D}{T}W, \; \frac{D}{L}UW, \; \frac{D}{\zeta}VW, \; W^2 \right)_{\max}. \tag{A.12}$$

Let us now find the conditions under which all terms inside the brackets in (A.12) turn out to be much smaller than magnitude of $|\chi|$ following from (A.11). To obtain those conditions we substitute in (A.12) instead of W successively its expressions from (A.9). Substituting $W \sim \varepsilon_1 U, \; \varepsilon_2 V$ leads, respectively, to the following estimates:

$$D \left| \frac{\partial \chi}{\partial z} \right| \sim \left(\frac{D}{T}\varepsilon_1 U, \; \varepsilon_1^2 U^2, \; \varepsilon_1 U \varepsilon_2 V \right)_{\max}, \tag{A.13}$$

$$D \left| \frac{\partial \chi}{\partial z} \right| \sim \left(\frac{D}{T}\varepsilon_2 V, \; \varepsilon_1 U \varepsilon_2 V, \; \varepsilon_2^2 V^2 \right)_{\max}. \tag{A.14}$$

Taking the ratios of all three terms within the brackets in (A.13) to the first three terms within the brackets in (A.11), respectively, we find (taking (A.10) into account)

$$\frac{D}{|\chi|} \left| \frac{\partial \chi}{\partial z} \right| \sim \varepsilon_1^2, \; \varepsilon_1^2, \; \varepsilon_1 \varepsilon_2 \frac{\zeta}{L} = \varepsilon_1^2. \tag{A.15}$$

Taking the ratios of the last term within the brackets in (A.13) to the sixth and last terms within the brackets in (A.11), respectively, we obtain

$$\varepsilon_1 \varepsilon_2 \frac{D}{\zeta} = \varepsilon_1 \varepsilon_2^2, \; \varepsilon_1 \varepsilon_2 \frac{R}{\zeta} = \varepsilon_2^2 \frac{R}{L}. \tag{A.16}$$

Taking the ratios of all three terms within the brackets in (A.14) to the fifth, third and fourth terms within the brackets in (A.11), respectively, we find

$$\frac{D}{|\chi|} \left| \frac{\partial \chi}{\partial z} \right| \sim \left(\frac{D}{\zeta}\varepsilon_2 = \varepsilon_2^2, \; \frac{\zeta}{L}\varepsilon_1 \varepsilon_2 = \varepsilon_1^2, \; \frac{R}{L}\varepsilon_2^2 \right)_{\max}. \tag{A.17}$$

At last, taking the ratios of the second term within the brackets in (A.14) to the sixth and last terms within the brackets in (A.11), respectively, we obtain the ratio (A.16).

The comparison of the corresponding terms of (A.11) and (A.12) can be presented in the following form

$$\frac{D}{|\chi|} \cdot \left| \frac{\partial \chi}{\partial z} \right| \sim \left\{ \varepsilon_1^2, \varepsilon_2^2, \varepsilon_1 \varepsilon_2^2, \varepsilon_2^2 \frac{R}{L} \right\}. \tag{A.18}$$

Let us assume that in our future experiments with rotating shallow water the condition $\varepsilon_1^2 \ll 1$ will be fulfilled. On the one hand, from (A.8) it follows that we always have $\frac{R}{L} > 1$. [2] On the other hand, from $\varepsilon_1^2 \ll 1$ it follows that $\varepsilon_1 \varepsilon_2^2 < \varepsilon_2^2$. Hence under the conditions (Fridman, Khoruzhii, 1999)

$$\varepsilon_1^2 \ll 1, \qquad \varepsilon_2^2 \frac{R}{L} \ll 1 \tag{A.19}$$

it follows that

$$\left| \frac{\partial \chi}{\partial z} \right| \ll \frac{|\chi|}{D}.$$

i.e. the χ function does not depend on z with the accuracy of $\mathrm{O}\left\{ \varepsilon_1^2, \varepsilon_2^2 \frac{R}{L} \right\}$

$$\frac{D}{|\chi|} \cdot \left| \frac{\partial \chi}{\partial z} \right| \sim \mathrm{O}\left\{ \varepsilon_1^2, \varepsilon_2^2 \frac{R}{L} \right\} \tag{A.20}$$

A3. Dynamic equations for rotating shallow water.

From (A.20) and (A.6) we find that

$$\frac{\partial P}{\partial z} = -g\rho \cdot \left\{ 1 + \mathrm{O}\left(\varepsilon_1^2, \varepsilon_2^2 \frac{R}{L} \right) \right\}, \tag{A.21}$$

i.e. $\left| \frac{\partial \tilde{p}}{\partial z} \right|$ is negligibly small, where $\tilde{p}$ is the deviation of the full pressure P from the hydrostatic one. So the hydrostatic approximation is fulfilled.

Independence of the χ function from z enables us to present this function in the form

$$\chi = \chi\left(r, \varphi, t \right) \tag{A.22}$$

with the accuracy mentioned above. Thus with the same accuracy the right hand sides of equations (A.1) and (A.2) are independent from z, the left hand sides of these equations turn out to be z-independent as well. From this it follows that

$$V_r \approx V_r\left(r, \varphi, t \right), \qquad V_\varphi \approx V_\varphi\left(r, \varphi, t \right). \tag{A.23}$$

Integrating (A.4) over z we find

$$V_z\left(r, \varphi, z, t \right) \approx -z \cdot \left[\frac{1}{r} \frac{\partial}{\partial r}\left(r V_r \right) + \frac{\partial V_\varphi}{r \partial \varphi} \right] + A\left(r, \varphi, t \right), \tag{A.24}$$

where A is an arbitrary function.

Let the bottom surface be described by function $z = h_B(r)$ (Fig. A.1).

From Fig. A.1 it follows that in an arbitrary point C on the bottom $z = h_B(r)$ we can write[3]

$$\frac{dh_B(r)}{dr} = \tan \alpha(r) = \frac{dz}{dr} = \frac{dz/dt}{dr/dt} = \frac{V_z}{V_r},$$

whence

$$V_z\left(r, \varphi, h_B, t \right) = V_r\left(r, \varphi, h_B, t \right) \cdot \frac{dh_B(r)}{dr}. \tag{A.25}$$

---

[2] Indeed, according to (A.8) $\frac{1}{L} \sim \left| \frac{\partial}{\partial r} \right| \sim |k_r|$ where k_r is the radial component of the wave vector. Hence it follows that $k_r R \sim \frac{2\pi R}{\lambda_r} > 1$, where λ_r is the wave length along radius.
[3] Only in the case, if the bottom surface $z = h_B(r)$ is described by a smooth function.

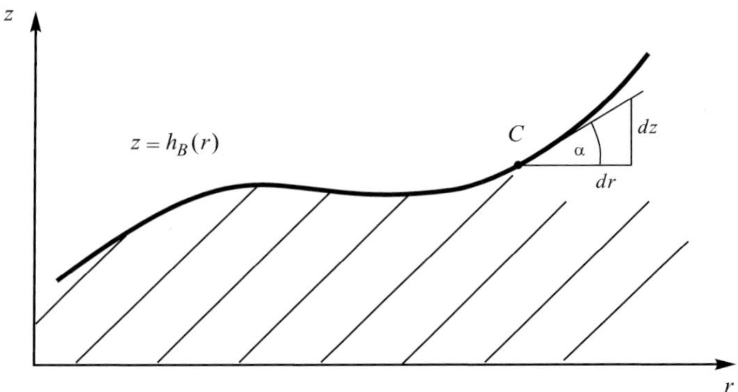

Figure A.1.   The profile of the bottom in a shallow water set-up is described by a smooth function $z = h_B(r)$.

On the surface $z = h_B(r)$ equation (A.24) is reduced to the following relationship

$$V_z\left(r, \varphi, h_B, t\right) \approx -h_B \cdot \left[\frac{1}{r}\frac{\partial}{\partial r}\left(rV_r\right) + \frac{\partial V_\varphi}{r\partial\varphi}\right] + A\left(r, \varphi, t\right),$$

whence we find $A\left(r, \varphi, t\right)$ and substitute this and (A.25) in (A.24). As a result we obtain

$$V_z\left(r, \varphi, z, t\right) \approx \left(h_B - z\right) \cdot \left[\frac{1}{r}\frac{\partial}{\partial r}\left(rV_r\right) + \frac{\partial V_\varphi}{r\partial\varphi}\right] + V_r \cdot \frac{dh_B\left(r\right)}{dr}. \tag{A.26}$$

On the free surface of the shallow water $z = h_S\left(r, \varphi, t\right)$ we have an evident kinematic condition

$$V_z\left(r, \varphi, h_S, t\right) = \frac{dh_S}{dt} = \frac{\partial h_S}{\partial t} + V_r\frac{\partial h_S}{\partial r} + V_\varphi\frac{\partial h_S}{r\partial\varphi}. \tag{A.27}$$

As a result of substitution (A.26) in (A.27) we obtain

$$\frac{\partial h_S}{\partial t} + V_r\frac{\partial}{\partial r}\left(h_S - h_B\right) + V_\varphi\frac{\partial h_S}{r\partial\varphi} + \left(h_S - h_B\right) \cdot \left[\frac{1}{r}\frac{\partial}{\partial r}\left(rV_r\right) + \frac{\partial V_\varphi}{r\partial\varphi}\right] = 0.$$

Introducing the full depth H of the shallow water

$$H\left(r, \varphi, t\right) = h_S\left(r, \varphi, t\right) - h_B\left(r\right), \tag{A.28}$$

we come to the continuity equation

$$\frac{\partial H}{\partial t} + \frac{1}{r}\frac{\partial}{\partial r}\left(rV_rH\right) + \frac{\partial}{r\partial\varphi}\left(V_\varphi H\right) = 0. \tag{A.29}$$

On the free surface of the shallow water the pressure $P(r, \varphi, z, t)$ should be equal to the atmospheric pressure P_a:

$$P\left(r, \varphi, h_s, t\right) = P_a = \mathrm{const}. \tag{A.30}$$

Integrating equation (A.21) over z we find

$$P\left(r, \varphi, z, t\right) = -g\rho z + B\left(r, \varphi, t\right). \tag{A.31}$$

In the point $z = h_s(r, \varphi, t)$ we have

$$P(r, \varphi, h_s, t) = -g\rho h_s + B(r, \varphi, t) = P_a,$$

where

$$B(r, \varphi, t) = P_a + g\rho h_s. \tag{A.32}$$

Substituting (A.32) in (A.31) we obtain

$$P(r, \varphi, z, t) = g\rho(h_s - z) + P_a. \tag{A.33}$$

By (A.6) we find

$$\chi(r, \varphi, t) = \frac{P_a}{\rho} + gh_s(r, \varphi, t). \tag{A.34}$$

Substituting (A.34) in (A.1) and (A.2) and taking into account (A.23), we obtain the following equations of motion for rotating shallow water

$$\hat{L}_2 V_r - \frac{V_\varphi^2}{r} = -g\frac{\partial h_s}{\partial r}, \tag{A.35}$$

$$\hat{L}_2 V_\varphi + \frac{V_r V_\varphi}{r} = -g\frac{1}{r}\frac{\partial h_s}{\partial \varphi}, \tag{A.36}$$

where

$$\hat{L}_2 \equiv \frac{\partial}{\partial t} + V_r\frac{\partial}{\partial r} + \frac{V_\varphi}{r}\frac{\partial}{\partial \varphi}. \tag{A.37}$$

The set of equations (A.28), (A.29), (A.35) and (A.36) is a closed system of four equations for four unknown functions h_s, H, v_r and v_φ, describing the shallow water dynamics.

A4. Linearized system of equations for small perturbations.

Suppose we have a stationary rotating shallow water layer, parameters of which are noted by the index "0"

$$\begin{cases} V_{\varphi 0}(r) = r \cdot \Omega(r), \\ V_{r0} \equiv 0, \\ H_0(r) = h_{S_0}(r) - h_B(r), \end{cases} \tag{A.38}$$

where $h_{S_0}(r)$ is an unperturbed surface of the shallow water.

According to (A.35), (A.36) the equilibrium conditions are the following:

$$\frac{V_{\varphi 0}^2}{r} = g \cdot \frac{dh_{S_0}}{dr},$$

$$0 \le \tan \alpha_0(r) \ll 1, \tag{A.39}$$

$\alpha_0(r)$ is the slope angle of the rotating shallow water in equilibrium, which is small;

$$\tan \alpha_0(r) = \frac{dh_{S_0}}{dr} = \frac{r \cdot \Omega^2(r)}{g} \tag{A.40}$$

Considering small perturbations of the stationary state from (A.28), (A.29) and (A.35), (A.36) we obtain a set of the linearized equations:

$$\widehat{L}v_r - 2\Omega \cdot v_\varphi = -\frac{\partial}{\partial r}\left(C_s^2 \cdot \eta\right), \tag{A.41}$$

$$\widehat{L}v_\varphi + \frac{\kappa^2}{2\Omega} \cdot v_r = -\frac{\partial}{r\partial \varphi}\left(C_s^2 \cdot \eta\right), \tag{A.42}$$

$$\widehat{L}\eta + \frac{1}{r}\cdot\frac{\partial}{\partial r}(r\cdot v_r) + \frac{\partial v_\varphi}{r\partial\varphi} + \frac{1}{H_0}\cdot\frac{dH_0}{dr}\cdot v_r = 0, \tag{A.43}$$

where

$$\widehat{L} \equiv \frac{\partial}{\partial t} + \Omega(r)\frac{\partial}{\partial\varphi}, \qquad \kappa^2 = 2\Omega\left(2\Omega + r\frac{d\Omega}{dr}\right), \tag{A.44}$$

$$H = H_0(r) + h(r,\varphi,t), \qquad \eta(r,\varphi,t) \equiv \frac{h(r,\varphi,t)}{H_0(r)},$$

$$V_r = v_r(r,\varphi,t), \qquad V_\varphi = V_{\varphi 0}(r) + v_\varphi(r,\varphi,t), \tag{A.45}$$

$$C_s^2(r) \equiv g\cdot H_0(r),$$

where C_s is the speed of the wave propagation.

Appendix: B. Derivation of the dispersion equation for the rotation curve with a jump

All coefficients in the set of linear homogeneous equations (A.41)–(A.43) are independent of t and φ. So perturbations may be presented as the superposition of individual harmonics of the form

$$f(r,\varphi,t) = \tilde{f}(r)e^{i(m\varphi-\omega t)}. \tag{B.1}$$

Let us introduce in ordinary terms the radial Lagrangian displacement ξ

$$v_r = \frac{d\xi}{dt} = \frac{\partial\xi}{\partial t} + \Omega\frac{\partial\xi}{\partial\varphi} = -i\hat{\omega}\xi, \tag{B.2}$$

where $\hat{\omega} \equiv \omega - m\Omega(r)$. Using dependencies (B.1) we may find from (A.41)–(A.43) and (B.2) for "radial" functions $\eta(r)$ and $\xi(r)$ the following system of linear ordinary homogeneous differential equations:

$$\frac{d}{dr}(C_s^2\eta) = \frac{2m\Omega}{r\hat{\omega}}(C_s^2\eta) - (\kappa^2 - \hat{\omega}^2)\xi,$$

$$\frac{d}{dr}(rH_0\xi) = -rH_0\left[\left(1 - \frac{m^2 C_s^2}{r^2\omega^2}\right)\eta + \frac{2m\Omega}{r\hat{\omega}}\xi\right]. \tag{B.3}$$

System (B.3) is used in numerical simulations of dynamics of small perturbations on the rotating shallow water. Eigen functions calculated must obey boundary conditions at inner $r = R_a$ and outer $r = R_b$ rigid borders of the shallow water region. It gives us the eigen-value problem for finding ω and eigen functions $\eta(r)$ and $\xi(r)$.

For the sharp discontinuities, the dispersion equation was previously derived by Fridman (1990) for studying the Kelvin–Helmholtz instability. In this Appendix we give this derivation in slightly modified way, which is more convenient for description of over-reflection.

B1. Sharp discontinuity.   For the sharp discontinuity, the radial part of perturbations of the shallow water depth H obey the Bessel equation:

$$\frac{d^2\eta(r)}{dr^2} + \frac{1}{r}\frac{d\eta(r)}{dt} + \left(k^2 - \frac{m^2}{r^2}\right)\eta(r) = 0, \tag{B.4}$$

where k is the radial wavenumber depending on the eigen-frequency ω, m is the azimuthal number.

The solution must obey boundary conditions and conditions at $r = R$. Assuming that the inner boundary is $r = 0$, and the outer boundary is $r = \infty$, the former conditions give

$$\eta(r) \;=\; C_1 J_m(k_1 r), \quad r < R, \tag{B.5}$$

$$\eta(r) \;=\; C_2 H_m^{(1)}(k_2 r), \quad r > R, \tag{B.6}$$

where $H_m^{(1)}$ denotes the Hankel function of the first kind, $k_{1,2}$ are the wavenumbers:

$$k_1^2(x) = k_0^2((x - \bar{q})^2 - 4\bar{q}^2/m^2); \qquad k_2^2(x) = k_0^2((x - 1)^2 - 4/m^2), \tag{B.7}$$

x is the normalized frequency, $x \equiv \omega/(m\Omega_2)$,

$$k_0^2 \equiv \frac{M^2 m^2}{R^2(1 - \bar{q})^2}. \tag{B.8}$$

The constants $C_{1,2}$ are determined from conditions that functions

$$\xi, \qquad C_s^2 \eta(r) + r\xi\Omega^2(r) \tag{B.9}$$

are continuous at $r = R$. This leads to the dispersion equation:

$$\alpha_1[(x - 1)^2 - 4/m^2] \quad \alpha_2[(x - \bar{q})^2 - 4\bar{q}^2/m^2] + \alpha_1\alpha_2(\bar{q}^2 - 1)/m^2 = 0, \tag{B.10}$$

where

$$\alpha_1 = \frac{2\bar{q}}{x - \bar{q}} - k_1 R \frac{J_m'(k_1 R)}{J_m(k_1 R)}, \qquad \alpha_2 = \frac{2}{x - 1} - k_2 R \frac{H_m'^{(1)}(k_2 R)}{H_m^{(1)}(k_2 R)}. \tag{B.11}$$

Functions $J_m'(z)$ and $H_m'^{(1)}(z)$ denote derivatives of functions $J_m(z)$ and $H_m^{(1)}(z)$ with respect to z.

B2. Smeared critical layer.

In reality, the discontinuity of the rotation curve cannot be abrupt. In fact, there is a narrow transitional region with characteristic length L of the order of shallow water depth H_0. Generally, however, equations (B.3) cannon be reduced to an analytical equation like (B.10), (B.11), and unstable solutions are obtained from numerical integration.

Assume that beyond the critical layer the rotation curve $\Omega(r)$ is constant. Thus, in those regions the solutions are combinations of Bessel functions, which take into account boundary conditions. Starting from these analytical solutions, we integrate equations (B.3) in some point R_s. The left-hand and right-hand solutions are linearly dependent, if the determinant

$$W(\omega) = \begin{vmatrix} \lim_{r=R_s-0} \xi(r) & \lim_{r=R_s+0} \xi(r) \\ \lim_{r=R_s-0} \eta(r) & \lim_{r=R_s+0} \eta(r) \end{vmatrix} \tag{B.12}$$

is zero. Thus, finding of unstable eigen-frequencies is equivalent to determining of the roots of the equation:

$$W(\omega) = 0. \tag{B.13}$$

Acknowledgments

EVP thanks Dr. I.G. Shukhman for many fruitful discussions. The work was supported in part by Russian Science Support Foundation, RFBR grant No. 05-02-17874, grant "Leading Scientific Schools" No. 925.2003.2, provided by Ministry of Industry, Science, and Technology, and grant of Division of Physical Sciences, RAS "Elongated objects in the Universe".

References

Afanasiev, V.L., A.N. Burenkov, A.V. Zasov, and O.K. Sil'chenko, Astrofizika **28**, 243, 1988.

Afanasiev, V.L., A.N. Burenkov, A.V. Zasov, and O.K. Sil'chenko, Astrofizika **29**, 155, 1988.

Afanasiev, V.L., A.N. Burenkov, A.V. Zasov, and O.K. Sil'chenko, Sov. Astron. **35**, 569, 1991.

Afanasiev, V.L., A.N. Burenkov, A.V. Zasov, and O.K. Sil'chenko, Astron. Zhurn. **69**, 19, 1992.

Antipov, S.V., M.V. Nezlin, V.K. Radianov, E.N. Snezhkin, and A.S. Trubnikov. Stabilization of tangential shear instability in shallow water with supersonic fluid flow. JETP, **37** (7), 378, 1983.

Blumen, W., P.G. Drazin, and D.F. Billings. J. Fluid Mech., **71**(2), 305, 1975.

Fridman, A.M., A.G. Morozov, M.V. Nezlin, and E.N. Snezhkin. *Phys. Lett.*, **109A**, 228, 1985.

Fridman, A.M. ZhETP, **98**, No. 4, 1121, 1990.

Fridman, A.M., and O.V. Khoruzhii, Appendix I To A.M. Fridman, N.N. Gor'kavyi, Physics of Planetary Rings, Springer, New-York etc, 1999.

Kolykhalov, P.I. Mech. Zhid. i Gaza, **3**, 145, 1984.

Landau, L.D. "On the stability of tangential gaps in contracted fluid". Dokl. Akad. Nauk USSR, **44**, 151, 1944.

Landau, L.D., and E.M. Lifshitz, Fluid Mechanics, Pergamon Press, 1984.

Loitzyanskii, L.G., Mechanics of Liquid and Gas (Mekhanika Zhidkosti i Gaza, in Russian), Nauka, Moscow, 1973.

McIntaier, M.E., J. Fluid Mech. **106**, 454, 1981.

Miles, J.W. J. Acoust. Soc. Am., **29**, 226, 1957.

Nezlin, M.V., and E.N. Snezhkin. Rossby vortices, spiral structure, solitons. Astrophysics and plasma physics in shallow water experiments. Springer–Verlag. New York, etc., 1993.

Pedlosky, J. "Geophysical Fluid Dynamics: Springer Study Edition", Springer, 1982.

Ribner, H.S.J. Acoust. Soc. Am., **29**, 435, 1957.

Stepanyantz, Y.A., A.L. Fabrikant. Uspekhi Fiz. Nauk, **159**, vyp. 1, 83, 1989.

CHAOTIC AND ORDERED STRUCTURES IN THE DEVELOPED TURBULENCE

Mikhail Ya. Marov and Aleksander V. Kolesnichenko

M.V. Keldysh Institute of Applied Mathematics
Miusskaya sq 4, Moscow 125047, Russia
marov@spp.keldysh.ru

Abstract   In an open hydrodynamic turbulent system new relationships between different regions set up because of coherent interactions of the flow subsystems involved. This increases internal ordering of the system compared to random fluctuations on molecular level. Multiple spatial-temporal scales of the developed shear turbulence are ensured by the collective behavior of numerous particles in the system resulting to an appearance of dissipative organized structure. Hence order of chaos set up that is closely related to the synergetics problem. Using the methods of irreversible thermodynamics, the defining relationships for turbulent flows and forces, which describe most comprehensively the transport and ordering processes in quasi-stationary case, are derived. Phenomenological macroscopic 3D model of the stationary non-equilibrium turbulence in the compessible fluid is developed with the account for collective non-linear processes. Internal parameters of the medium are introduced into the model, which characterizes the excitation of macroscopic degrees of freedom. It made possible to generalize Onsager formalism over turbulent pulsations relative to the average motion and describe thermodynamically the Kolmogorov cascade process. Various kinetic equations of the Fokker-Planck type in the configuration space for the functions of distribution of small-scale turbulence characteristics are deduced including the unsteady kinetic equation for the distribution of probability of dissipation of turbulent energy. Synergetic approach to the developed turbulence study open new horizons in the theory of dynamic systems and improvement of astrophysical models.

Keywords: Irreversible thermodynamics, turbulence in compessible fluids

Introduction

Turbulence is regarded as the most complicated natural phenomena caused by liquid flow in an essentially non-equilibrium open system. Various multi-

Alexei M. Fridman, et al. (eds.), Astrophysical Disks; Collective and Stochastic Phenomena 23–54
© 2006 *Springer. Printed in the Netherlands*

scales coherent structures[1] set up in chaotic turbulent flows, in particular those in our space environment and in the Universe. Since natural gaseous objects used to be non-homogeneous in their structure and composition, they exhibit very complicated patterns of turbulent motions of different temporal-spatial scales. Numerous evidence in support were brought in the field of astrophysics and space physics, specifically when processes in the multicomponent reactive media are involved and turbulence strongly influences properties and thermodynamics of local flow regions and overall system (Kolesnichenko and Marov, 1997; Marov and Kolesnichenko, 2001). In-depth study of turbulence as interdisciplinary subject is closely related with the theory of dynamic systems incorporating fundamentals of statistical mechanics, hydrodynamics, and thermodynamics[2].

Turbulent fluid continuum can be represented as thermodynamic complex consisting of two interacting subsystems (continua): subsystem of the mean motion and subsystem of turbulent chaos. The latter is considered as conglomerate of eddies of different scales in dissipative medium evolving in accordance with Kolmogorov's cascade process and thus responding to incremental mass, momentum, and heat transfer in the turbulent flow. In the recent years, alongside with some progress in the direct numerical modeling of turbulent motions based on the exact (instantaneous) hydrodynamic equations and attempts to resolve eddies structure at subgrid level, interest to the development of new phenomenological (macroscopic) approaches to the description of turbulent small-scale dynamics using different closure theories grows. Such approaches require introduction of averaged parameters of the medium and finding of universal and partial relations for their evaluation, in addition to the known relations based on mass, momentum, and energy conservation laws. The ultimate goal is the development of macroscopic theory of the turbulent gas motion as closely related to the natural media modeling and practical needs as possible. This is a challenge though because of many difficulties creation of such a theory seems problematic. Indeed, in the review of the above referred book after Marov and Kolesnichenko (2001) written by the known US scientist A. Buckingham it is emphasized "the need for... a model useful for evaluating the average influence of turbulence on the evolution of a concomitant physical processes, rather than focusing on a deeper understanding of the nature

---

[1]It should be stressed, however, that "the number of macroscopic spatial and temporal scales present in turbulent chaos is so great that the behavior of the system appears to be chaotic" (Prigogine and Stengers, 1984).

[2]Let us note that in the case of plasma, along with magneto hydrodynamic turbulence, wave turbulence also occurs caused by charged particles fluxes that excite oscillations and waves in the plasma. Such turbulence is usually weak and used to be referred to as Langmuir or ionic-sonic turbulence. It is generated by a swing of the broad spectrum of plasma waves and resembles waves on the water surface rather than eddy motions in the turbulent liquid flow.

of turbulence, a pursuit which while having academic appeal is fraught with potential disappointment" (Buckingham, 2003).

The self-organization processes, which occur on the background of turbulent motion, represent the most important mechanism of formation astrophysical and geophysical objects at different stages of their evolution. These include the emergence of galaxies and galactic clusters (Fridman and Choruzhy, 2003), the birth of stars in a diffuse medium of gas-dust clouds, the formation of protoplanetary disks and subsequent accumulation of planetary systems (Marov and Kolesnichenko, 2001; Makalkin, 2003; Ruskol, 2003), the formation of gaseous envelopes (atmospheres) of the planets (Marov and Grinspoon, 1998), flows of various scales in the atmospheres and planetary plasma environment, and so on. It is worth to note that astrophysical objects are open dissipative structures and in an ideal case, they can be addressed as isolated (conservative) only on relatively small time intervals.

When transition from laminar to turbulent motion in an open hydrodynamic system occurs, new macro-dependencies between different flow regions set up due to collective interaction of different subsystems. This increases the internal ordering in the system that results in the coherent behavior of a huge number of particles, which manifests itself in the form of supermolecular organization. In particular, the cascade process of fragmentation of eddies, which takes place in the fully developed turbulence, can be treated as an unlimited sequence of self-organization processes[3]. Granulation in the solar photosphere at the background of smaller scale turbulent motion serves as an example of numerous coherent structures in the turbulent flow. Thus, we come to the (at the first sight paradoxical conclusion that the advanced turbulent motion, despite its great complexity, corresponds to a higher ordering (in some regions) than the laminar motion.

The phenomenon giving rise to ordering from chaos in turbulent fluid is a part of general self-organization problem (synergetics). Self-organization of dissipative structures in such a system set up due to stabilization of spatially non-uniform instabilities in an open non-linear dynamic system. In other words, coherence accompanying the process of bifurcations results in "ordering through fluctuations" (Prigogine and Stengers, 1984) and probability becomes an objective property intrinsically generated within dynamics, which is fundamental structure of the dynamic system. Hence investigation of an open non-linear hydrodynamic system behavior, capable to create macroscopic dissipative spatial-temporal structures by an internal self-organization rather than by an external influence, is the great appeal for the contemporary science.

---

[3]Let us note that emerged vortex structures sometimes alternate with zones of laminar stream (the intermittence phenomenon).

Extension of the formalism of irreversible thermodynamics (see, e.g., Prigogine, 1955) to the media with excited macroscopic degrees of freedom (serving as internal parameters to describe macrostructure of the medium) presumably allows one to apply this approach to the macroscopic description of the cascade process of transfer of turbulent energy by eddies of different sizes. Thus our goal is to take advantage of the known methods of irreversible thermodynamics, which proved their efficiency in the study of various non-equilibrium dissipative structures well outside the thermodynamic equilibrium (see, e.g., Ebeling et al., 1990), in order to develop phenomenological macro-scopic model of the stationary non-equilibrium turbulence taking into account nonlinear cooperative processes. As the first step along this track, the closed set of averaged hydrodynamic equations which describe self-consistently 3D turbulence, should be formulated.

Some Introductory Remarks on Phenomenological Models

Let us recall that phenomenology of the developed turbulence can be studied on either micro or macro levels. Because the basic ideas of macroscopic model will be further used, we first briefly review the phenomenological approach to the small-scale eddy turbulence. We shall proceed from the model of one component fluid for which state vector characterizes a simple hydrodynamic system represented by the set of random extensive variables ρ, $\rho\mathbf{u}$, U, where ρ is density of the medium, $\mathbf{u}$ is its velocity vector, and U is the density of internal energy.

Classical approach to the phenomenological description of the developed turbulence is based on the Reynolds idea of time and/or space averaging of hydrodynamic equations for variables ρ, $\rho\mathbf{u}$, U. Another equivalent approach that is used, for example, in the statistical hydrodynamics (Monin and Yaglom, 1975), is theoretical-probabilistic procedure of averaging over an ensemble of possible realization, i.e. set of identical systems in some external conditions suitable for their compact description. All these averaging provided statistical physics ergodicity concept is valid, filters out motion modes having scales lower than temporal-spatial averaging scale. The latter are pulsations of physical parameters relative to the respective averaged values.

Evidently, the separation of motion of a liquid into the mean and turbulent depends entirely on the choice of the space-time region G for which the mean and regular functions of macro coordinates $\mathbf{x}, t$ of the hydrodynamic variables are established. The procedure is generally depending on the specific modeling problem. The scale of non-uniformity of regular flow called Obukhov's scale of observation $L \geq L_0$ (Obukhov, 1941) determined by the size $d\mathbf{x} \sim \Lambda^3$ of the G region, is usually assumed small compared to the characteristic scale L_0 of the whole system. In such a case, all larger scale eddies contribute to the regular motion described by the synchronous averaged values (mathematical expectation) $\overline{f(\mathbf{n}(\mathbf{x},t))} = \int f(\mathbf{n}(\mathbf{x},t))W_1(\mathbf{n},\mathbf{x},t)\,d\mathbf{n}$ of the random hydrodynamic variables f which are the functions of vector-valued stochastic process $\mathbf{n}(\mathbf{x},t)$. Here the set of random hydrodynamic variables ρ, $\rho\mathbf{u}$, U are represented as the state vector $\mathbf{n}$. In turn, eddies of lower scales filtered out in due course of averaging, contribute to the turbulent motion defined by corresponding pulsations $f'(\mathbf{x},t) = f(\mathbf{x},t) - \bar{f}(\mathbf{x},t)$ of the same variables used in the averaging procedure.

When dealing with physical ensemble, two functions are sufficient to describe completely the stochastic processes $\mathbf{n}(\mathbf{x},t)$. These are the function $W_1(\mathbf{n},\mathbf{x},t)$ – probability density to find out $\mathbf{n}$ in the $(\mathbf{n},\mathbf{n}+d\mathbf{n})$ interval of temporal-spatial point $(\mathbf{x},t)$, and two-point function $W_2(\mathbf{n}^0,\mathbf{x}^0,t^0;\mathbf{n},\mathbf{x},t)$ – combined probability density distribution. The processes described by these functions are known to be the Markov's processes. One also needs to utilize two-point density of the conditional probability $P_2(\mathbf{n}^0,\mathbf{x}^0,t^0|\mathbf{n},\mathbf{x},t)$, i.e. the probability to find the value $\mathbf{n}$ in the point $(\mathbf{x},t)$ provided the probability of $\mathbf{n} = \mathbf{n}^0$ in the point $(\mathbf{x}^0,t^0)$ equals to unity. The expression $P_2(\mathbf{n}^0,\mathbf{x}^0,t^0| \mathbf{n},\mathbf{x},t) = W_2(\mathbf{n}^0,\mathbf{x}^0,t^0;\mathbf{n},\mathbf{x},t)/W_1(\mathbf{n}^0,\mathbf{x}^0,t^0)$ contains implicitly the relationship between mean values over the conditional and the whole assembles and it introduces the transition probability P_2. It is worth to note that only conditional mean values $\overline{f(\mathbf{n}(t))}^0 = \int f(\mathbf{n})P_2(\mathbf{n}^0,t^0|\mathbf{n},t)\,d\mathbf{n}$ for the stationary assemble depends on time and this is just the values which are related with the macroscopic transfer equations.

In the case of isothermal fluid ($\rho = \text{const}, \nu = \text{const}$) regarded to as the continuous medium, small-scale structure of turbulence is determined according to Kolmogorov (see, e.g., Landau and Lifshitz, 1988) by the cascade energy transfer through eddy spectrum of the different temporal-spatial scales. Qualitatively, Richardson cascade scheme looks as follows: Small eddies take energy due to progressive splitting of the large ones, the process being controlled by the Reynolds number $\mathbf{Re} = L_0 u_0/\nu$ that corresponds to large-scale motion of the flow (L_0 – characteristic size of large-scale eddies; ν – kinematic molecular viscosity). Specific kinetic energy $\sim u_0^3/L_0$ redistribution in the stream continuously occurs until the smallest eddies with size of an order of the internal turbulence scale $\eta = (\nu^3/\bar{\varepsilon})^{1/4}$ appear, the latter characterizing the viscous effects on the structure of small-scale turbulence.

Formula for the dissipative length scale η follows from the first Kolmogorov (1941) similarity hypothesis which states that statistical regime of the small-scale isotropic and homogeneous turbulence is strictly defined by the two dimensional parameters: the mean (over ensemble of possible realization of the medium flow) rate of the energy dissipation $\bar{\varepsilon}$ (which serves as the key characteristic of the local isotropic turbulence) and the viscosity ν. At the same time, under very large $\mathbf{Re}$ quasi-stationary regime is established in the so called inertial interval of eddies scale ($\eta < \lambda_k < L_1$) where neither production nor dissipation of the kinetic energy occurs. Here λ_k is an intermediate linear scale of eddy structures. Following the second Kolmogorov (1941) similarity hypothesis, in the inertial interval statistical regime of turbulence is defined by the parameter $\bar{\varepsilon}$ only.

The quantitative description of the local small-scale isotropic turbulence in G-region (supposing $\Lambda \lesssim L_1$) is based on utilization of the structure functions

$$D_{ij}(\mathbf{r}) = \overline{[u_i'(\mathbf{x}) - u_i'(\mathbf{x} + \mathbf{r})][u_j'(\mathbf{x}) - u_j'(\mathbf{x} + \mathbf{r})]}$$

and their spectra

$$E_{ij}(\mathbf{k}) = \int D_{ij}(\mathbf{r}) \exp\left(i\mathbf{k} \cdot \mathbf{r}\right) d\mathbf{r},$$

where $i = \sqrt{-1}$ and $\mathbf{k}$ is the wave vector (Kolmogorov, 1941; Obukhov, 1941). From the first Kolmogorov similarity hypothesis and an assumption that parameters of the large-scale turbulence slow vary at $r = |\mathbf{r}|$ (if $r < L_1 \ll L_0$) follows:

$$D_{ij}(r) = (\bar{\varepsilon}r)^{2/3}[f(r/\eta)r_i r_j r^{-2} + g(r/\eta)\delta_{ij}],$$

where f and g are random functions of the dimensionless argument r/η. According to the second Kolmogorov similarity hypothesis, structure function in the inertial interval $L_1 \gg r \gg \eta \gtrsim L\mathbf{Re}^{-3/4}$ does not depend on the viscosity ν, i.e. $f = \mathrm{const}$ and $g = \mathrm{const}$ under $r/\eta \gg 1$. These hypotheses underlie the two-third law regarded as one of the most important laws of small-scale turbulent motions. It states that in any turbulent flow when Reynolds number $\mathbf{Re}$ is sufficiently large the mean square of velocities residual in two points of the flow located at a distance r apart (r assuming to be comparable with the scale Λ of the averaged flow) should be proportional to $r^{2/3}$:

$$D_{11}(r) = C(\bar{\varepsilon}r)^{2/3}, \tag{1}$$

where C is the universal constant[4]. One may also write the statement equivalent to this law in terms of the structure function of the velocity field: $E(k) = C^* \bar{\varepsilon}^{2/3} k^{-5/3}$ under $1/\eta \gg k \gg 1/L_0$ (Obukhov, 1941)[5]. The two-third law is nicely confirmed experimentally for the diverse turbulent motions (see Monin and Yaglom, 1996).

One should note, however, that for the structure functions of n-order the expression $\overline{V^n} \gtrsim (\bar{\varepsilon}r)^{n/3}$ (where $V \equiv u'(\mathbf{x}) - u'(\mathbf{x} + \mathbf{r})$), following from the

---

[4]Let us note that in the original work of Kolmogorov (1941) the two-third law follows from the condition $\bar{\varepsilon}(r) = \mathrm{const}$, where $\eta < r < L_0$ occurs in the inertial interval.

[5]Energy spectrum $E(k)$, along with the function of energy dissipation spectrum $2\nu k^2 E(k)$ and the function of energy re-distribution over spectrum $T(k)$, describe time variations of the spectral distribution of the turbulent energy.

Kolmogorov theory, is poorly supported in experiments, especially for $n \gg 1$. This discrepancy is known to be caused by the fact that similarity hypotheses for the local isotropic turbulence in their original form suggested entropy input to small-scale disturbances in the inertial interval to be constant.

It means that Kolmogorov parameter $\bar{\varepsilon}$ precisely defined as

$$\bar{\varepsilon} = \bar{\varepsilon}(\mathbf{x}, t) = \overline{\frac{\nu}{2} \left(\frac{\partial u_i'}{\partial x_j} + \frac{\partial u_j'}{\partial x_i} \right)^2} \tag{2}$$

is constant (hereafter summing is taken by repeated indexes) and also that joint multivariate functions of the probability density distribution $W_1(\mathbf{u}'; \mathbf{x}, t)$ for velocity field pulsations within relatively small temporal-spatial region G having diameter $\Lambda \gg \eta$ depend only on the parameter $\bar{\varepsilon} = \mathrm{const}$.

Kolmogorov's ideas on the random cascade were more accurately addressed by Obukhov (1962), to overcome these difficulties. Instead of the original condition $\bar{\varepsilon}(\mathbf{x}, t) = \mathrm{const}$ in the G-region with characteristic scale $\Lambda \ll L_0$ and point $\mathbf{x}$ as a center) he suggested to proceed from the idea that statistical characteristics of small-scale motions (for instance, structure functions) are dependent on the energy dissipation $\varepsilon_r(\mathbf{x}, t)$ averaged over some volume V_r of characteristic scale r (small compared to the typical inhomogeneity scale of the mean motion $r \ll \Lambda$), rather than on theoretical probabilistic mean value $\bar{\varepsilon}(\mathbf{x}, t)$ of random ε. It turned out that if one select as an averaged region within G a sphere of r radius, then the result is only negligibly dependent on a shape of the averaged region and

$$\varepsilon_r(\mathbf{x}, t) = \frac{6}{\pi r^3} \int_{|\mathbf{r}|^* \leq r} \varepsilon(\mathbf{x} + \mathbf{r}^*, t) \, \mathrm{d}\mathbf{r}^* . \tag{3}$$

Statistical variability of $\varepsilon_r(\mathbf{x}, t)$ required to clarify the concept of physical ensemble introduced for theoretical-probabilistic averaging. Indeed, statistical ensemble depending parametrically on r and ε_r ("pure ensemble") allows us to evaluate only conditional mean values of any statistical characteristics of small-scale motions which are defined under fixed parameter $\varepsilon_r = \mathrm{const}$. However, in practice one deals with a mixed assemble where ε_r change according to some general statistical law. Hence it is necessary to average conditional value of the respective moment over possible ε_r values in order to figure out unconditional mean value of a small-scale turbulence characteristic. In particular, to evaluate Reynolds tensor $R_{ij}(\mathbf{x}, t) = -\rho \overline{u_i'(\mathbf{x}, t) u_j'(\mathbf{x}, t)}$ as one of such characteristics it is necessary to calculate the integral $R_{ij}(\mathbf{x}, t) = \int_0^\infty R_{ij}(\mathbf{x}, t, \varepsilon_r) W_1(\varepsilon_r; \mathbf{x}, t) \, \mathrm{d}\varepsilon_r$. Here $W_1(\varepsilon_r; \mathbf{x}, t)$ is the density of probability distribution for ε_r, which is defined by large-scale regular motions; $R_{ij}(\mathbf{x}, t, \varepsilon_r) = -\rho \int u_i'(\mathbf{x}, t) u_j'(\mathbf{x}, t) P(\varepsilon_r | \mathbf{u}', \mathbf{x}, t) \, \mathrm{d}\mathbf{u}'$; $P(\varepsilon_r | \mathbf{u}', \mathbf{x}, t)$ is the conditional function of probability density of the velocity fluctuations, i.e.

probability to find $\mathbf{u}'$ within an interval from $\mathbf{u}'$ to $\mathbf{u}' + \mathrm{d}\mathbf{u}'$ in a sub-ensemble where $\varepsilon_r = \mathrm{const}$ under all $\mathbf{x}, t$. Generally speaking, the values of any integrals can be different for the different kind of flows and depend, in particular, on $\mathbf{Re}$ number.

With the use of $\varepsilon_r(\mathbf{x}, t)$, Kolmogorov (1962) reformulated his first and second similarity hypotheses to make them more accurate. He introduced $\varepsilon_r(\mathbf{x}, t)$ instead of $\bar{\varepsilon}(\mathbf{x}, t)$ and complemented these two original hypotheses with the third one which postulates (for large scale ratio $L_0 : r$) the normal probability density distribution of $\ln \varepsilon_r$ and the linearity of the dispersion $\sigma^2_{\ln \varepsilon_r}$ dependence on $\ln L_0/r$:

$$W_1(\varepsilon_r) = \frac{1}{\sqrt{2\pi}\varepsilon_r \sigma_{\ln \varepsilon_r}} \exp - \left[\frac{\ln^2 \varepsilon_r / m_{\ln \varepsilon_r}}{2\sigma^2_{\ln \varepsilon_r}} \right], \qquad (4)$$

$$\ln m_{\ln \varepsilon_r} = -\sigma^2_{\ln \varepsilon_r}/2 + \ln \bar{\varepsilon}, \qquad \bar{\varepsilon}_r = \bar{\varepsilon}, \qquad \sigma^2_{\ln \varepsilon_r} = \mu \ln \frac{L_0}{r} + B(\mathbf{x}, t),$$

where $\sigma^2_{\ln \varepsilon_r}$ is the dispersion of $\ln \varepsilon_r$; $m_{\ln \varepsilon_r} = \exp(\overline{\ln \varepsilon_r})$ is the median of distribution; μ is the parameter of alteration (≈ 0.5); and $B(\mathbf{x}, t)$ is the addend depending on characteristics of the averaged regular motion. The distribution allows one to evaluate the important characteristics of the stationary small-scale turbulence, in particular the moments $\overline{\varepsilon_r^n}$:

$$\overline{\varepsilon_r^n} = \int\limits_0^\infty \varepsilon_r^n W_1(\varepsilon_r)\mathrm{d}\varepsilon_r = \bar{\varepsilon}_r^n \exp\left[\tfrac{1}{2}n(n-1)\sigma^2_{\ln \varepsilon_r} \right] =$$

$$F_n(\mathbf{x}, t)\bar{\varepsilon}_r^n (r/L_0)^{-\mu n(n-1)/2}, \qquad (5)$$

where the coefficients $F_n(\mathbf{x}, t)$ depend on macrostructure of the turbulent motion. Hence for the structure function of n-order $\overline{V^n} \propto \overline{\varepsilon^{n/3}} r^{n/3}$ the expression follows $\overline{V^n} \propto F_n(\mathbf{x}, t)(\bar{\varepsilon})^{n/3}(r/L_0)^{-\mu n(n-1)/6(r)^{n/3}}$. It turns out in accord with experimental data at the extreme accuracy available for $n = 2$ (formula (1)) and also for structure functions of higher order, though in the latter case it is in discrepancy with the dependence suggested by the original Kolmogorov theory for the isotropic turbulence. Similar approach was used to reduce the theory of the temperature and chemically active admixture local pulsation field in the turbulent flow (Monin and Yaglom 1975).

For the phenomenological turbulence model of our particular interest utilization of the parameter ε_r is especially important because the original Kolmogorov's cascade theory do not take into account explicitly small-scale coherent dissipative structures in the turbulent flow. Meanwhile, an existence and dynamics of such dissipative structures, as well as their irregular temporal-spatial distribution in the chaotic flow should be accounted for any adequate turbulence theory (Krow and Champagne, 1971; Brown and Rozhko, 1974).

In turn, chaotic kinetic energy transfer in cascade caused by instability of dissipative structures is intrinsically related with the phenomenon of hydrodynamic internal alteration. Here the region occupied by the so called turbulent spots where intense pulsations of the velocity gradients, $\varepsilon_r > 0$ are observed, are tightly linked with the regions of weakly turbulent or fully irrotational flow where there are no pulsations and $\varepsilon_r \cong 0$. Alteration appears when $\mathbf{Re}$ number is small and the power of permanently acting mechanism of the turbulence set up is insufficient for the formation of fully developed turbulence in the whole fluid region D. Thus the values of $\varepsilon_r(\mathbf{x}, t)$ or similar quadratic by the velocity gradient quantities may serve as an indicator of alteration.

Macroscopic Model of the Developed Turbulence

The results of phenomenological modeling of the Kolmogorov-Obukhov small-scale turbulent structure allow us to advance the macroscopic description of the developed turbulence, in particular, thermodynamic approaches to the construction of semi-empirical models for one-point moments. Defining relations in such models bear a local sense. For example, Reynolds stress tensor $\mathbf{R}(\mathbf{x}, t)$ depends only on the kinetic energy of turbulence, dissipation, and gradient of the mean velocity in the same temporal-spatial point $(\mathbf{x}, t)$. Introduction in the hydrodynamic model of the internal parameters of the medium responsible for the exited macroscopic degrees of freedom, allows us to describe the Kolmogorov's (stationary non-equilibrium) cascade process using statistic irreversible thermodynamics approach. Moreover, with involvement of the Kolmogorov's parameter $\langle \varepsilon_r \rangle$ depending on the global Reynolds number $\mathbf{Re}$ as the alteration indicator, we may derive different kinetic equations of Fokker-Planck type in the respective configuration space for the distribution functions of small-scale turbulence characteristics.

In order to compose the set of averaged equation of the turbulent fluid motion we shall assume that it is characterized by the two linear scales: external scale L_0 of the motion at large, and internal scale L_1 representing a turbulent mole size. Then we introduce two coordinate systems for micro-scale $\delta x_i' \gtrsim \lambda$ and macro-scale $\mathrm{d}x_j \gtrsim \Lambda \gg \lambda$, with $L_0 \gg \Lambda \geq L_1$ and $L_1 \gg \lambda \gg l_0$ where l_0 is the molecular micro-scale. Basically, this makes possible to formulate the Navier-Stokes equations in the macro-scale x_j taking advantage of the known Navier-Stokes equations in the micro-scale x_j'. Keeping in mind astrophysical application of the model where the ratio of averaged velocity to the mean sound speed is much more than unity (and hence, density fluctuations play an important role), we shall also assume variability of the mass density.

The set of the "regular" hydrodynamic equations of the mean motion scale and equitable also in the micro-scale of the instantaneous equations of motion has the form:

$$\bar{\rho}\frac{\mathrm{d}\langle v \rangle}{\mathrm{d}t} = \nabla \cdot \langle \mathbf{u} \rangle , \quad (\langle \mathbf{v} \rangle = 1/\bar{\rho}) , \tag{6}$$

$$\bar{\rho}\frac{\mathrm{d}\langle \mathbf{u} \rangle}{\mathrm{d}t} = -\nabla\bar{p} + \nabla \cdot \mathbf{\Pi}^{\Sigma} + \bar{\rho}\mathbf{F} , \tag{7}$$

$$\bar{\rho}\frac{\mathrm{d}\langle U \rangle}{\mathrm{d}t} = -\nabla\cdot\mathbf{q}^{\Sigma} - \bar{p}\nabla\cdot\langle \mathbf{u} \rangle + \overset{\circ}{\overline{\mathbf{I}}}\cdot\cdot\overset{\circ}{\mathbf{E}} - \frac{\mathrm{d}t}{-p'\nabla \cdot \mathbf{u}''} + \mathbf{J}_{\mathrm{v}}^{\mathrm{turb}}\cdot\nabla\bar{p} + \bar{\rho}\langle \varepsilon_r \rangle. \tag{8}$$

It was obtained by means of the theoretical-probabilistic averaging by ensemble of realizations (Kolesnichenko and Marov, 1997; Marov and Kolesnichenko, 2001). Here $\mathbf{u}$, p, v, U are the instantaneous values of the velocity, pressure, specific volume ($v = 1/\rho$) and specific internal energy of liquid particle, respectively; $\langle \mathbf{u} \rangle(\mathbf{x}, t) \equiv \overline{\rho\mathbf{u}}/\bar{\rho}$ is the averaged (mean-weight) hydrodynamic velocity of the medium; $\mathrm{d}(\ldots)/\mathrm{d}t \equiv \partial(\ldots)/\partial t + \langle \mathbf{u} \rangle \cdot \nabla(\ldots)$ is the total derivative with respect to the time relative to the averaged velocity field; $\bar{p}(= R\bar{\rho}\langle T \rangle)$ is the averaged pressure; $\mathbf{F}$ is an external force acting upon a mass unit (for the sake of simplicity, we shall neglect here the mass force pulsations); $\mathbf{R}(\mathbf{x}, t) \equiv -\overline{\rho\mathbf{u}''\mathbf{u}''}$ is turbulent (Reynolds' stress) tensor; $\mathbf{\Pi}^{\Sigma}(\mathbf{x}, t) \equiv \overline{\mathbf{\Pi}} + \mathbf{R}$ is the total stress tensor in the turbulent flow; $\mathbf{J}_{v}^{\mathrm{turb}}(\mathbf{x}, t) \equiv -\overline{\rho'\mathbf{u}''}/\bar{\rho}$ and $\mathbf{q}_{v}^{\mathrm{turb}}(\mathbf{x}, t) \equiv \overline{\rho i''\mathbf{u}''}$ are the turbulent fluxes of specific volume and heat, respectively (where $i \equiv U + p/\rho$ is the instantaneous value of specific enthalpy of the medium); $\mathbf{q}^{\Sigma}(\mathbf{x}, t) \equiv (\overline{\mathbf{q}} + \mathbf{q}^{\mathrm{turb}} - \overline{p'\mathbf{u}''})$ is the total heat flux in the subsystem of the averaged motion (where $\overline{\mathbf{q}}(\mathbf{x}, t)$ is the averaged molecular heat flux); $\langle \varepsilon_r \rangle(\mathbf{x}, t) = \langle \varepsilon \rangle(\mathbf{x}, t) \equiv \overline{\rho\varepsilon}/\bar{\rho} = \overline{\mathbf{\Pi}\cdot\cdot\nabla\mathbf{u}''}/\bar{\rho}$ is the mean-weight value of the specific rate of turbulent kinetic energy dissipation into heat due to molecular viscosity ν (this relation generalizes Obukhov's formula (3) over compressible fluid); $\mathbf{I}$ is the unit tensor; $\overline{\mathbf{\Pi}}(\mathbf{x}, t)$, $\mathbf{E}(\mathbf{x}, t) \equiv 1/2(\nabla\langle \mathbf{u} \rangle + \nabla^*\langle \mathbf{u} \rangle)$; and $\overset{\circ}{\overline{\mathbf{\Pi}}}(\mathbf{x}, t)$, $\overset{\circ}{\mathbf{E}}(\mathbf{x}, t)$ are the averaged viscosity stress tensor and deformation rate tensor for the averaged continuum, respectively; and their parts with zero trace defined by the following relations:

$$\overset{\circ}{\mathbf{E}} \equiv \mathbf{E} - \frac{1}{3}(\mathbf{E}\cdot\cdot\mathbf{I})\mathbf{I} = \mathbf{E} - \frac{1}{3}(\nabla\cdot\langle \mathbf{u} \rangle)\mathbf{I} , \quad \overset{\circ}{\overline{\mathbf{\Pi}}} \equiv \overline{\mathbf{\Pi}} - \frac{1}{3}(\overline{\mathbf{\Pi}}\cdot\cdot\mathbf{I})\mathbf{I} = \overline{\mathbf{\Pi}} - \pi\mathbf{I} , \tag{9}$$

where $\pi(\mathbf{x}, t) \equiv (1/3\overline{\mathbf{\Pi}}\cdot\cdot\mathbf{I})$; ∇ is the Hamilton operator; symbols $\mathbf{A}\cdot\cdot\mathbf{B}$ and $\mathbf{AB}$ mean the internal product of two tensors and the external product of two vectors (dyad), respectively; and symbol $\nabla \cdot \mathbf{A}$ means the generalized divergence since $\mathbf{A}$ is not always a vector.

The set of equations (6)-(8) describes the large-scale structure of turbulent field in the framework of complete macroscopic turbulent model. It involves firstly, the averaged molecular thermodynamic fluxes $\overline{\mathbf{q}}(\mathbf{x}, t)$ and $\overline{\mathbf{\Pi}}(\mathbf{x}, t)$ for which the defined relations are additionally required (see Kolesnichenko, 1985); and secondly, uncertain mixed one-point correlations – second order moments

$\mathbf{R}(\mathbf{x}, t)$, $\mathbf{q}^{\mathrm{turb}}(\mathbf{x}, t)$ and $\mathbf{J}_v^{\mathrm{turb}}(\mathbf{x}, t)$ that represent the hydrodynamic characteristics transfer with turbulent pulsations. These fluxes and moments, as well as correlation terms containing pressure pulsations ($\overline{p'\nabla \cdot \mathbf{u}''}$ and $\overline{\nabla \cdot (p'\mathbf{u}'')}$) and the averaged value of viscous turbulent energy dissipation $\langle \varepsilon_r \rangle$ should be found. Choice of the closed linear defining relations can be performed in accordance with the continuum mechanics rheology rules using the Onsager procedure. It is based on the idea that certain thermodynamic forces are responsible for the relaxation of averaged variables to a stable stationary state (see Prigogine, 1955). Moreover, in our model the turbulent continuum is considered to posses some internal structure and the developed thermodynamic approach allows us to generalize the Onsager procedure over the description also turbulent pulsations relative to the mean values.

Thermodynamics of the Subsystems of Averaged Motion and Turbulent Chaos

As it was already noted, turbulent gaseous continuum can be represented as thermo-hydrodynamic complex composed of two mutually interacting subsystems: averaged motion and turbulent chaos. Every subsystem in a physical infinitely small volume $\mathrm{d}\mathbf{x}$ is considered open hydrodynamic system exchanging with other system(s) with energy and entropy (but not with a matter). Velocity fields within such subsystems supposed to be congruent since in the process of turbulent motion no separation of the respective Lagrangian volumes (diffusion) occurs, in other words, the subsystem of turbulent chaos has no hydrodynamic velocity relative to the subsystem of averaged motion. Following the concept on turbulence as macroscopically highly organized motion, one may consider subsystem of turbulent chaos in terms of continuum. Its internal structure is formed by eddies of various temporal-spatial scales that occupy equilibrium interval (total of the inertial and viscous intervals of the turbulence spectrum) and fulfill continuously the whole space at every cascade step. We shall further assume that generalized state parameters characterizing turbulent chaos are connected with the ordinary in the local-equilibrium thermodynamics relations of Gibbs, Gibbs-Dugem, etc. type which remains valid outside local thermodynamic equilibrium provided that an appropriate stationary-non-equilibrium state is selected as the point of reference (Prigogine, 1955). This assumption is regarded as an axiom introduced in the thermodynamic approach for evaluation of the developed turbulence.

Let us first address thermodynamics of the averaged motion beginning from analysis of the balance equations for the averaged entropy $\langle S \rangle$ of the turbulent medium. Fundamental Gibbs identity written along an averaged trajectory of the center mass of an unit volume has the form (Kolesnichenko and Marov,

1985; Kolesnichenko, 1998):

$$\frac{\mathrm{d}\langle S\rangle}{\mathrm{d}t} = \frac{1}{\langle T\rangle}\frac{\mathrm{d}\langle U\rangle}{\mathrm{d}t} + \frac{\bar{p}}{\langle T\rangle}\frac{\mathrm{d}}{\mathrm{d}t}\left(\frac{1}{\bar{\rho}}\right), \tag{10}$$

where $\langle S\rangle \equiv \overline{\rho S}/\bar{\rho}$ and $\langle U\rangle \equiv \overline{\rho U}/\bar{\rho}$ are, respectively, the specific entropy and internal energy of the averaged motion; $\langle \mathbf{u}\rangle \equiv \overline{\rho\mathbf{u}}/\bar{\rho}$ is the averaged hydrodynamic velocity of the medium; and $\mathrm{d}()/\mathrm{d}t \equiv \partial()/\partial t + \langle \mathbf{u}\rangle \cdot \nabla()$ is the substantive derivative with respect to the averaged velocity field. Equation (10) can be rewritten as an equation of local balance of the entropy $\langle S\rangle$ by eliminating from it the substantive time derivatives of the parameters $\langle U\rangle(\mathbf{x}, t)$ and $\langle 1/\bar{\rho}\rangle(\mathbf{x}, t)$ using the averaged hydrodynamic equations (6) and (8) (see, e.g., Kolesnichenko and Marov, 1997). This results in:

$$\frac{\partial}{\partial t}(\bar{\rho}\langle S\rangle) + \nabla \cdot \left(\bar{\rho}\langle S\rangle\langle \mathbf{u}\rangle + \frac{\mathbf{q}^{\Sigma}}{\langle T\rangle}\right) = \sigma^{i}_{\langle S\rangle} + \sigma^{e}_{\langle S\rangle}, \tag{11}$$

where

$$0 \leq \sigma^{i}_{\langle S\rangle} \equiv \frac{1}{\langle T\rangle}\left(-\mathbf{q}^{\Sigma}\cdot\frac{\nabla\langle T\rangle}{\langle T\rangle} + \pi\nabla\cdot\langle\mathbf{u}\rangle + \overset{\circ}{\overline{\mathbf{\Pi}}}\cdot\cdot\overset{\circ}{\mathbf{E}}\right), \tag{12}$$

$$\sigma^{e}_{\langle S\rangle} \equiv \frac{1}{\langle T\rangle}\left(-\overline{p'\nabla\cdot\mathbf{u}''} + \mathbf{J}^{\mathrm{turb}}_{v}\cdot\nabla\bar{p} + \bar{\rho}\langle\varepsilon_r\rangle\right) = \frac{\Xi}{\langle T\rangle}, \tag{13}$$

Here, $\mathbf{J}^{\mathrm{turb}}_{v}(\mathbf{x}, t) \equiv \overline{\rho v''\mathbf{u}''}$ and $\mathbf{q}^{\mathrm{turb}}_{v}(\mathbf{x}, t) \equiv \overline{\rho i''\mathbf{u}''}$ are, respectively, the turbulent flows of the specific volume $v(\equiv 1/\rho)$ and heat; $i(= U + p/\rho)$ is the instantaneous value of the specific enthalpy of the medium; $\mathbf{q}^{\Sigma}(\mathbf{x}, t) \equiv (\bar{\mathbf{q}}+\mathbf{q}^{\mathrm{turb}} - \overline{p'\mathbf{u}''})$ is the total heat flow in the mean-motion subsystem; $(\bar{\mathbf{q}}(\mathbf{x}, t)$ is the averaged molecular heat flow; $\langle\varepsilon_r\rangle(\mathbf{x}, t) = \overline{\rho\varepsilon_r}/\bar{\rho} \equiv \overline{\mathbf{\Pi}\cdot\cdot\nabla\mathbf{u}''}/\bar{\rho}$ is the mean specific rate ε_r of dissipation of turbulent kinetic energy into heat due to molecular viscosity ν; $\mathbf{I}$ is the unit tensor; $\overline{\mathbf{\Pi}}(\mathbf{x}, t)$, $\mathbf{E}(\mathbf{x}, t) \equiv 1/2(\nabla\langle\mathbf{u}\rangle + \nabla^{*}\langle\mathbf{u}\rangle)$, $\overset{\circ}{\overline{\mathbf{\Pi}}}$, and $\overset{\circ}{\mathbf{E}}$ are, respectively, the averaged tensor of viscous stresses, the tensor of deformation rates (for the averaged continuum) and their zero-trace parts. These parameters are defined by the relationships:

$$\overset{\circ}{\mathbf{E}} \equiv \mathbf{E} - \frac{1}{3}(\mathbf{E}\cdot\cdot\mathbf{I})\mathbf{I} = \mathbf{E} - \frac{1}{3}(\nabla\cdot\langle\mathbf{u}\rangle)\mathbf{I}, \qquad \overset{\circ}{\overline{\mathbf{\Pi}}} \equiv \overline{\mathbf{\Pi}} - \frac{1}{3}(\overline{\mathbf{\Pi}}\cdot\cdot\mathbf{I})\mathbf{I} = \overline{\mathbf{\Pi}} - \pi\mathbf{I},$$

where $\pi(\mathbf{x}, t) \equiv (1/3\overline{\mathbf{\Pi}}\cdot\cdot\mathbf{I})$.

The positive quantity $\sigma^{i}_{\langle S\rangle}(\mathbf{x}, t)$ defines the rate of local entropy production $\langle S\rangle$ due to irreversible processes within the averaged-motion subsystem. In turn, $\sigma^{e}_{\langle S\rangle}(\mathbf{x}, t) \equiv \Xi/\langle T\rangle$ defines the entropy sink or influx and corresponds to the entropy exchange between the subsystems of turbulent chaos and mean motion. This quantity can be positive or negative, depending on the particular regime of turbulence. Indeed, the rate of turbulent-energy dissipation $\bar{\rho}\langle\varepsilon_r\rangle$ is a strictly positive quantity. However, the rate of energy transfer $\overline{p'\nabla\cdot\mathbf{u}''}$, caused by the ambient medium on turbulent eddies per unit time in a unit volume as the result of pressure pulsations, i.e. expansion $(\nabla\cdot\mathbf{u}'' > 0)$ or compression $(\nabla\cdot\mathbf{u}'' < 0)$ of eddies, may be either positive or negative. As for the quantity $\mathbf{J}_v^{\mathrm{turb}}\cdot\nabla\bar{p}$ is concerned, it is positive in the case of small-scale turbulence and negative in the case of large eddies (Kolesnichenko and Marov, 1997). Thus, it follows from (11) that, in the general case, the averaged entropy $\langle S\rangle$ of the mean-motion subsystem can either rise or decline, which is the characteristic of the open physical systems.

One may see that the averaged entropy $\langle S\rangle$ only is not sufficient to describe adequately all peculiarities of the turbulent medium in an elementary volume $d\mathbf{x}$ since it is not related with parameters responsible for internal structure of the turbulent chaos subsystem, in particular with such key parameter as turbulent energy

$$\langle e\rangle(\mathbf{x},t) = -(\mathbf{R}\cdot\cdot\,\mathbf{I})/2\bar{\rho} \equiv \overline{\rho(\mathbf{u}')^2}/2\bar{\rho}\,. \tag{14}$$

Thus, it is necessary, for a complete description, to introduce the elements of thermodynamics of a "turbulent superstructure" that consists of the ensemble of eddies of different sizes.

Let us assume that the structure of "turbulent chaos" is determined by the following local variables of state: the "internal energy" of turbulization $U_{\mathrm{turb}}(\mathbf{x},t)$; specific volume of the medium $\langle v\rangle$; and the local turbulent entropy $S_{\mathrm{turb}}(\mathbf{x},t)$. The generalized entropy U_{turb} is intrinsically related with fluctuations and variations of the medium in quasi-stationary state, similar to local equilibrium entropy in the quasi-equilibrium state. Because the baseline of the adopted conception is the presence of a great number $(\propto N)$ of excited macroscopic degrees of freedom in the developed turbulent motion, we shall further introduce, in addition to $U_{\mathrm{turb}}(\mathbf{x},t)$ and $\langle v\rangle$, a set of internal variables $n(\xi_k,\mathbf{x},t)$, $(\equiv n_k,\ k = 1,2,\ldots,N)$ which suppose to fluctuate relative to a stable stationary state ξ_k^{st}. The set of ξ_k represents the vector-valued stochastic process that characterizes the structure of small-scale turbulence within Lagrange volume $d\mathbf{x}$. It is sufficient to describe the state of turbulent field, such as eddies in the cascade process of interaction of turbulent motions of different scales; ring vortexes of deforming torus type; eddy tubes and layers, etc. As parameters ξ_k (assumed to be non-correlated) could be taken pulsating velocity $\xi_u = u''$; residual square of the velocities (temperatures) in two neighbor points of space; eddies kinetic energy $\xi_e = \overline{\rho(\mathbf{u}'')^2}/2$; the energy dissipation rate in eddies $\xi_k = \varepsilon$; the rate of decay of the temperature dispersion $\xi_N = N \equiv \overline{\chi(\nabla T'')^2}\,.$

Following the Gibbs' basic thermodynamic relationship, we choose (as *a priory* specified) local characteristic function of the subsystem of turbulent chaos in the stationary state:

$$U_{\text{turb}}(\mathbf{x}, t) = U_{\text{turb}}(1/\bar{\rho}(\mathbf{x}, t), S_{\text{turb}}(\mathbf{x}, t), n(\xi_k, \mathbf{x}, t)). \tag{15}$$

Here $n(\xi)\mathrm{d}\xi$ is the number of eddies in a unit volume having characteristic internal coordinates in the ξ to $\xi + \mathrm{d}\xi$ range, $n(\xi)$ being considered as continuous function of coordinates and time. We also use, as it is typically done in the irreversible thermodynamics, the following definitions of the quantities T_{turb}, p_{turb}, and $\mu_{\text{turb,k}}$, assuming that all derivatives shown below are positive:

$$T_{\text{turb}} = (\partial U_{\text{turb}}/\partial S_{\text{turb}})_{1/\bar{\rho}, n_k}, \quad p_{\text{turb}} = (\partial U_{\text{turb}}/\partial \langle v \rangle_{\text{turb}})_{S_{\text{turb}}, n_k},$$

$$\mu_{\text{turb}}(\xi_k)/n_{\Sigma} = (\partial U_{\text{turb}}/\partial n_k)_{1/\bar{\rho}, S_{\text{turb}}, n_k}.$$

Here, T_{turb} and p_{turb} have the sense of generalized temperature and pressure of turbulization; $\mu_{\text{turb}}(\mathbf{x}, t, \xi_k)(\equiv \mu_{\text{turb,k}})$ is the so-called "chemical potential" (per unit mass) for the configuration ξ_k; $n_{\Sigma}(\mathbf{x}, t) = \Sigma_k n_k(\mathbf{x}, t)$ is the total number of eddies specified by the symbol ξ in a unit volume Λ^3 corresponding to the observation scale; n_k denotes concentrations of eddies of all scales, except for those of the order k.

Then, the respective differentiation form of the fundamental Gibbs identity (15) for the subsystem of turbulent chaos written along the trajectory of unit mass volume motion takes the form (Prigogine and Kondepudi, 2002):

$$\bar{\rho}\frac{\mathrm{d}S_{\text{turb}}}{\mathrm{d}t} = \frac{\bar{\rho}}{T_{\text{turb}}}\frac{\mathrm{d}\langle e \rangle}{\mathrm{d}t} + \frac{\bar{\rho}p_{\text{turb}}}{T_{\text{turb}}}\frac{\mathrm{d}}{\mathrm{d}t}\left(\frac{1}{\bar{\rho}}\right) - \frac{\bar{\rho}}{n_{\Sigma}T_{\text{turb}}}\int_{\xi} \omega(\xi)\frac{\partial \mu_{\text{turb}}(\xi)}{\partial \xi}\,\mathrm{d}\xi.$$

$$\tag{15$'$}$$

Evidently, various relationships between the intensive variables T_{turb}, p_{turb} and $\mu_{\text{turb,k}}$, which can be derived in a usual way from (15'), can be interpreted as "equations of state" for the subsystem under consideration. Then the internal energy U_{turb} is supposed to be identical to the turbulent energy $\langle e \rangle \equiv \overline{\rho(\mathbf{u}'')^2}/2\bar{\rho}$ (the averaged kinetic energy of pulsations in a unit mass of the turbulent medium)

$$U_{\text{turb}}(\mathbf{x}, t) = \langle e \rangle + \text{const} = \overline{\rho(\mathbf{u}'')^2}/2\bar{\rho} + \text{const} \tag{16}$$

and the turbulent-chaos system is regarded in the thermodynamic sense as "a perfect gas" possessing three degrees of freedom with uniform energy distribution.

We shall further take advantage of the basic kinetic equation for the rate of change of eddy moles $n(\xi)$ in the cascade process of interaction of turbulent motions of different scale as a one-dimensional continuity equation in the space of internal coordinate ξ. Since in the developed turbulence (under large but finite **Re** number) a continuum of excited degrees of freedom exist, we assume that ξ takes continuously only discrete values and that in the process of consecutive transitions caused by the loss of stability due to the emergence of each new mode of motion in the medium, only the memory of the last transition is retained (*Markov's process*). In other words, the concentration of eddies $n(\xi_k)$ in the state ξ_k may change only as a result of transition of eddies from the neighboring state ξ_{k-1} (the decay of large eddies of size ξ_{k-1} into smaller eddies with size ξ_k or transition into the neighboring states ξ_{k+1} (the principle of local interaction). In chemical terms, this process can be treated as a series of consecutive chemical reactions expressed by the sequence $\ldots \longrightarrow (k-1) \longrightarrow k \longrightarrow (k+1) \longrightarrow \ldots$.(Prigogine, 1955). Let us note, however, that in an actual cascade process of destruction of large eddies and formation of small eddies in a unit event of interaction, only infinitely small changes of the quantities ξ occur, whereas finite changes arise as a result of the cumulative action of a large number of interactions. Then using ξ as a continuous parameter, the equation for the rate of change of the concentration of eddies can be rewritten in the form of the continuity equation (in both coordinate space and internal ξ-coordinate space) as follows:

$$\frac{\partial n(\xi, \mathbf{x}, t)}{\partial t} + \nabla_x \cdot [n(\xi, \mathbf{x}, t)\langle \mathbf{u}\rangle] = -\nabla_\xi \cdot \mathbf{J}(\xi, \mathbf{x}, t), \qquad (19)$$

where $\mathbf{J}(\xi, \mathbf{x}, t)$ is thermodynamic flux in the internal ξ-coordinate space (or, more precisely, probability flux in the ξ-state). Similarly, for a single stochastic process ξ with the account for (6), one has:

$$\bar{\rho}\frac{\mathrm{d}}{\mathrm{d}t}\left(\frac{n(\xi, \mathbf{x}, t)}{\bar{\rho}}\right) = -\frac{\partial \mathbf{J}(\xi, \mathbf{x}, t)}{\partial \xi}, \qquad (19')$$

where $\mathbf{J}(\xi, \mathbf{x}, t)$ is the thermodynamic flux in the space of internal coordinate ξ (or, more accurately, probability flux into the state ξ).

Using (19'), we can now transform the Gibbs identity (15') by means of integration by parts (and assuming also that the flux $\mathbf{J}(\xi, \mathbf{x}, t)$ becomes zero at the both boundaries ξ_η and ξ_{L_1} of the variable ξ definition due to the condition $\int_{\xi_\eta}^{\xi_{L_1}} \delta n(\xi)\,\mathrm{d}\xi = 0$ into the form:

$$\bar{\rho}\frac{\mathrm{d}S_{\text{turb}}}{\mathrm{d}t} = \frac{\bar{\rho}}{T_{\text{turb}}}\frac{\mathrm{d}\langle e\rangle}{\mathrm{d}t} + \frac{\bar{\rho}p_{\text{turb}}}{T_{\text{turb}}}\frac{\mathrm{d}}{\mathrm{d}t}\left(\frac{1}{\bar{\rho}}\right) - \frac{\bar{\rho}}{n_\Sigma T_{\text{turb}}}\int_\xi \omega(\xi)\frac{\partial \mu_{\text{turb}}(\xi)}{\partial \xi}\,\mathrm{d}\xi.$$

$$(20)$$

In writing this relationship, it was assumed that the flux $\omega(\xi)$ vanishes beyond the limits of the equilibrium range. The last term in (20) describes the net growth of turbulization entropy, which is caused by irreversible processes of formation of eddy structures inside the "turbulent-chaos" subsystem corresponding to the changes of the internal parameter ξ:

$$\frac{\mathrm{d}_i S_{\mathrm{turb}}}{\mathrm{d}t} \equiv \int_\xi \sigma_\xi(S_{\mathrm{turb}})\,\mathrm{d}\xi = -\frac{1}{n_\Sigma T_{\mathrm{turb}}} \int_\xi \omega(\xi)\frac{\partial \mu_{\mathrm{turb}}(\xi)}{\partial \xi}\,\mathrm{d}\xi \geq 0 \,. \quad (21)$$

We can see from this expression that the local production of pulsation entropy $\sigma_\xi(S_{\mathrm{turb}})$, which corresponds to every part of the internal coordinate ξ space has a standard thermodynamic form $T_{\mathrm{turb}}\sigma_\xi(S_{\mathrm{turb}}) = \omega(\xi)A_{\mathrm{turb}}(\xi)/n_\Sigma$ where the term

$$\begin{aligned} A_{\mathrm{turb}}(\xi) &= -\frac{\partial \mu_{\mathrm{turb}}(\xi)}{\partial \xi} = -\frac{RT_{\mathrm{turb}}}{n(\xi)}\frac{\partial n(\xi)}{\partial \xi} - \frac{\partial C(\xi)}{\partial \xi} \\ &= -\frac{RT_{\mathrm{turb}}}{n(\xi)}\exp\left(-\frac{C(\xi)}{RT_{\mathrm{turb}}}\right)\frac{\partial}{\partial \xi}\left[\exp\left(\frac{\mu_{\mathrm{turb}}}{RT_{\mathrm{turb}}}\right)\right] \end{aligned} \quad (22)$$

is the generalized chemical affinity for ξ-configuration (state function of the turbulent chaos subsystem) which was obtained in view of the formula (17) for the chemical potential μ_{turb}.

The evolutionary equation for the turbulence entropy S_{turb} can be deduced from (20) using the same approach that has led us to the (11). Eliminating substantial derivatives of the specific volume $(1/\bar\rho)$ and the turbulent energy $\langle e \rangle$ from (20) with the help of the corresponding equations of motion (Kolesnichenko and Marov, 1997), we obtain

$$\frac{\partial}{\partial t}\left(\bar\rho S_{\mathrm{turb}}\right) + \nabla\cdot\left(\bar\rho S_{\mathrm{turb}}\langle \mathbf{u}\rangle + \frac{\mathbf{J}_e^{\mathrm{turb}}}{T_{\mathrm{turb}}}\right) = \sigma_{S_{\mathrm{turb}}} \equiv \sigma_{S_{\mathrm{turb}}}^i + \sigma_{S_{\mathrm{turb}}}^e \,, \quad (24)$$

where

$$\sigma_{S_{\mathrm{turb}}}^e \equiv \frac{1}{T_{\mathrm{turb}}}\left(\overline{p'\nabla\cdot\mathbf{u}''} - \mathbf{J}_v^{\mathrm{turb}}\cdot\nabla\bar p - \bar\rho\langle \varepsilon_r\rangle\right) = -\frac{\Xi}{T_{\mathrm{turb}}} \,, \quad (25)$$

and

$$\begin{aligned} 0 \leq \sigma_{S_{\mathrm{turb}}}^i &\equiv \frac{1}{T_{\mathrm{turb}}}\left(-\mathbf{J}_e^{\mathrm{turb}}\cdot\frac{\nabla T_{\mathrm{turb}}}{T_{\mathrm{turb}}} + \mathbf{R}\,\raisebox{0.3ex}{\scriptsize$..$}\,\nabla\langle\mathbf{u}\rangle\right. \\ &\qquad \left. + p_{\mathrm{turb}}\nabla\cdot\langle\mathbf{u}\rangle - \frac{\bar\rho}{n_\Sigma}\int_\xi \omega(\xi)\frac{\partial \mu_{\mathrm{turb}}(\xi)}{\partial \xi}\,\mathrm{d}\xi\right) \\ &= \frac{1}{T_{\mathrm{turb}}}\left(-\mathbf{J}_e^{\mathrm{turb}}\cdot\frac{\nabla T_{\mathrm{turb}}}{T_{\mathrm{turb}}} + \overset{\circ}{\mathbf{R}}\,\raisebox{0.3ex}{\scriptsize$..$}\,\overset{\circ}{\mathbf{E}} + \frac{\bar\rho}{n_\Sigma}\int_\xi \omega(\xi)A_{\mathrm{turb}}(\xi)\,\mathrm{d}\xi\right) \,. \end{aligned} \quad (26)$$

Here, $\mathbf{J}_e^{\text{turb}}(\mathbf{x},t) \equiv \overline{\rho(\mathbf{u}''^2/2 + p'/\rho)\mathbf{u}''} - \overline{\boldsymbol{\Pi}\cdot\mathbf{u}''}$ is the diffusion flux of turbulent energy; $\mathbf{R}(\mathbf{x},t) \equiv -\overline{\rho\mathbf{u}''\mathbf{u}''}$, and $\overset{\circ}{\mathbf{R}}$ are, respectively, the symmetric tensor of turbulent (Reynolds) stresses and its zero-trace part determined by the relationship

$$\overset{\circ}{\mathbf{R}} \equiv \mathbf{R} - \frac{1}{3}(\mathbf{R}\cdot\cdot\,\mathbf{I})\mathbf{I} = \mathbf{R} + \frac{2}{3}\bar{\rho}\langle e\rangle\mathbf{I} = \mathbf{R} + p_{\text{turb}}\mathbf{I}, \qquad p_{\text{turb}} = \frac{1}{3}(\mathbf{R}\cdot\cdot\,\mathbf{I}).$$

The quantities $\sigma^i_{S_{\text{turb}}}(\mathbf{x},t)$ and $\sigma^e_{S_{\text{turb}}}(\mathbf{x},t)$ have the sense of the local production and sink of the entropy S_{turb} of turbulent chaos subsystem, respectively. Evidently the work of turbulent stresses $\mathbf{R}\cdot\cdot\,\nabla\langle\mathbf{u}\rangle$ results in the entropy of chaos grow, whereas the viscous dissipation $\bar{\rho}\langle e\rangle$ decreases the turbulence entropy S_{turb}.

Now, summing up the equations (11) and (24) for the entropies $\langle S\rangle$ and S_{turb}, we can find the evolutionary equation for the total entropy $S_\Sigma = (\langle S\rangle + S_{\text{turb}})$ of the system

$$\frac{\partial}{\partial t}(\bar{\rho}S_\Sigma) + \nabla\cdot\left(\bar{\rho}S_\Sigma\langle\mathbf{u}\rangle + \frac{\mathbf{q}^\Sigma}{\langle T\rangle} + \frac{\mathbf{J}_e^{\text{turb}}}{T_{\text{turb}}}\right) = \sigma_\Sigma, \tag{27}$$

where

$$0 \leq \sigma_\Sigma = \sigma^i_{\langle S\rangle} + \sigma^e_{\langle S\rangle} + \sigma^i_{S_{\text{turb}}} + \sigma^e_{S_{\text{turb}}} = \sigma^i_{\langle S\rangle} + \sigma^i_{S_{\text{turb}}} + \frac{T_{\text{turb}} - \langle T\rangle}{T_{\text{turb}}\langle T\rangle}\Xi \tag{28}$$

is the local production of the total entropy due to irreversible processes inside the "closed" thermodynamic complex. In view of the formulas (12), (13), (25), and (26), the quantity σ_Σ has a structure of the bilinear form $\sigma_\Sigma = \sum_\alpha \text{Im}_\alpha(\mathbf{x},t)X_\alpha(\mathbf{x},t)$:

$$0 \leq \sigma_\Sigma \equiv \mathbf{q}^\Sigma\cdot\nabla\left(\frac{1}{\langle T\rangle}\right) + \frac{1}{\langle T\rangle}\pi\nabla\cdot\langle\mathbf{u}\rangle + \frac{1}{\langle T\rangle}(\overset{\circ}{\boldsymbol{\Pi}}\cdot\cdot\,\overset{\circ}{\mathbf{E}}) + \mathbf{J}_e^{\text{turb}}\cdot\nabla\left(\frac{1}{T_{\text{turb}}}\right)$$

$$+ \frac{1}{T_{\text{turb}}}\overset{\circ}{\mathbf{R}}\cdot\cdot\,\overset{\circ}{\mathbf{E}} + \frac{\bar{\rho}}{n_\Sigma T_{\text{turb}}}\int_\xi J(\xi)A_{\text{turb}}(\xi)\,\mathrm{d}\xi + \frac{T_{\text{turb}} - \langle T\rangle}{T_{\text{turb}}\langle T\rangle}\Xi. \tag{29}$$

According to the basic postulate of irreversible thermodynamics, fluxes can be expressed in terms of linear functions of thermodynamic forces $\text{Im}_{\alpha i} = \sum_\beta \Lambda^{ij}_{\alpha\beta}X_{\beta j}$, $(\alpha, \beta = 1, 2, \ldots)$ (see De Groot and Masur, 1962). Thus, the formula (29) enables us to find the defining (rheologic) relationships between the thermodynamic fluxes and forces for the three major regions of the turbulent flow: the laminar region; zone of the developed turbulent flow where turbulent fluxes are much more effective than the corresponding averaged molecular fluxes ($\mathbf{R} \gg \overline{\overline{\Pi}}$, $\mathbf{q}^{\mathrm{turb}} \gg \overline{\mathbf{q}}$, etc); and the buffer layer – an intermediate region where effects of molecular and turbulent transfer are comparable in magnitude (Kolesnichenko, 1997; Marov and Kolesnichenko, 2001).

It is worth to emphasize that in the case of a turbulent continuum the matrix $\Lambda_{\alpha\beta}$ of the Onsager phenomenological coefficients depends not only on the averaged parameters of state of the medium (temperature, density, etc.), but also on the characteristics of the turbulent superstructure itself, i.e., on the parameters $\langle \varepsilon_r \rangle$, T_{turb}, L_0, etc. Such a situation (the dependence of $\Lambda_{\alpha\beta}$ on the thermodynamic fluxes Im_α, for example on the dissipation rate ε_r) is typical of the self-organizing systems (Haken, 1983, 1988). Basically, it may result in the presence of some non-positively defined terms $\mathrm{Im}_\alpha(\mathbf{x}, t) X_\alpha(\mathbf{x}, t)$ in the sum σ_Σ though generally $\sigma_\Sigma \geq 0$. In such a case, the superposition of various fluxes may lead, in principle, to the appearance of some negative diagonal elements of the matrix $\Lambda_{\alpha\beta}$. The latter probably accounts for the known effect of negative viscosity in some turbulent flows.

As follows from (29), in the case of a turbulent system the spectrum of possible crossing effects is expanded compared to the laminar regime. For instance, the total heat flux $\mathbf{q}^\Sigma$ in the turbulent continuum is affected by the thermodynamic forces $\nabla(1/T_{\mathrm{turb}})$ conjugated with the flux $\mathbf{J}_e^{\mathrm{turb}}$ which describes the "diffusion" transfer of the turbulent kinetic energy. For the present, however, there is no experimental data that would support and/or describe quantitatively the crossing effects. Moreover, their contribution to the general rate of the process under study is an order of magnitude less compared to the direct effects (De Groot and Masur, 1962). Therefore, the requirement of the positive values of the intensities $\sigma^i_{\langle S \rangle}$, $\sigma^i_{S_{\mathrm{turb}}}$ and σ_Σ can be taken irrespective of each other and hence, without special reservations, crossing effects in rheologic relations can be omitted.

It is noteworthy that the last term on the right-hand side of (29), which describes the entropy production inside the entire thermodynamic system due to the irreversible transfer of heat between its subsystems, is equal to the product of the thermodynamic flux $\Xi(\mathbf{x}, t)$ and the corresponding function of state $(1/\langle T \rangle - 1/T_{\mathrm{turb}})$. By virtue of the second law of thermodynamics, this quantity is always positive:

$$\sigma_{\langle S \rangle, S_{\mathrm{turb}}}(\mathbf{x}, t) \equiv \Xi \left(\frac{1}{\langle T \rangle} - \frac{1}{T_{\mathrm{turb}}} \right) \geq 0. \tag{30}$$

Thus, the "direction" of the flux $\Xi(\mathbf{x}, t)$ is determined by the sign of the function of state $X_\Xi \equiv (1/\langle T \rangle - 1/T_{\mathrm{turb}})$. Macroscopically, it can be treated as a conjugative thermodynamic force causing this flux. Such an entropy exchange between two mutually open subsystems may be a source of structurized collective behavior, hosting coherent processes of self-organization in one of them (Prigogine and Stengers, 1993). In our case, such relation as the heat flux

$\Xi(\mathbf{x}, t)$ determines possible stationary states and explains the dissipative activity of the turbulent-chaos subsystem, thus representing the necessary condition for self-organization.

Turbulent Chaos in the Stationary Non-equilibrium State

Here we shall focus on linear regime of stationary non-equilibrium turbulence as the most important case of the developed turbulent motion of fluid with application of the thermodynamics of irreversible processes procedure. It allows us to bring evidence that dissipative activity of the turbulent chaos subsystem is just determined by the input of negative entropy from the subsystem of averaged motion.

Evidently to ensure such a regime, some permanently operating mechanism of turbulence set up in the medium should be at work (for example, large-scale shear velocity in the flow or thermal convective large-scale instability) that transfers stored energy to eddies at the large scales and prevents the system to reach the complete thermodynamic equilibrium. The power of such an energy source must be sufficient to compensate for the loss of the turbulent energy caused by molecular viscosity. In this case, nearly all the expended energy, without substantial though existing losses, should be transferred from the energy range to the viscous range through the inertial range, and the process of energy transfer from large-scale to small-scale eddies occurs. Then the non-equilibrium stationary state (Prigogine, 1955) is established in the structure of turbulent chaos, in which the production of the turbulence entropy $\sigma^i_{S_{\text{turb}}}$ inside the system is compensated for by its outflow $\sigma^e_{S_{\text{turb}}}$ so that the total local production of the entropy S_{turb} is negligible, i.e., $\sigma_{S_{\text{turb}}} \equiv \sigma^e_{S_{\text{turb}}} + \sigma^i_{S_{\text{turb}}} \cong 0$, and thus the entropy flux in the stationary case is constant.

Because $\sigma^i_{S_{\text{turb}}} \geq 0$ we have $0 > \sigma^e_{S_{\text{turb}}} \cong -\sigma^i_{S_{\text{turb}}}$. It means that the subsystem of "turbulent chaos" must export entropy to the "external medium" (the subsystem of averaged motion) in order to compensate the entropy production at the expense of internal irreversible processes. In other words, the influx of negative entropy (*negentropy*) $\sigma^e_{S_{\text{turb}}} \equiv -\Xi/T_{\text{turb}} = -\langle T \rangle \sigma^e_{\langle S \rangle}/T_{\text{turb}} < 0$ coming from the external medium is required in order to maintain the stationary state inside the subsystem of turbulent chaos. The arriving negentropy is expended to maintain and improve of the internal structure of the subsystem. Then the relationship $0 \leq \sigma^e_{\langle S \rangle} = -T_{\text{turb}}\sigma^e_{S_{\text{turb}}}/\langle T \rangle \cong (T_{\text{turb}}/\langle T \rangle)\sigma^i_{S_{\text{turb}}}$ is satisfied and the equation (11) for the entropy balance $\langle S \rangle$ takes the form:

$$\bar{\rho}\frac{\partial}{\partial t}\langle S \rangle + \nabla \cdot \left(\frac{\mathbf{q}^{\Sigma}}{\langle T \rangle}\right) = \sigma^i_{\langle S \rangle} + \sigma^e_{\langle S \rangle} \cong \sigma^i_{\langle S \rangle} + (T_{\text{turb}}/\langle T \rangle)\sigma^i_{S_{\text{turb}}} \cong \sigma_{\Sigma} \,, \quad (31)$$

where the local scattering of energy $\langle T \rangle \sigma_\Sigma$ is expressed as follows

$$\langle T \rangle \sigma_\Sigma \equiv -\left(\mathbf{q}^{\text{turb}} - \overline{p'\mathbf{u}''} \right) \frac{\nabla \langle T \rangle}{\langle T \rangle} + \overset{\circ}{\mathbf{R}} \cdot\cdot \overset{\circ}{\mathbf{E}} + \frac{\bar{\rho}}{n_\Sigma} \int_\xi \omega(\xi) A_{\text{turb}}(\xi)\, d\xi \geq 0.$$

$$(32)$$

Here $\mathbf{q}^\Sigma(x,t) \equiv -\left(\mathbf{q}^{\text{turb}} - \overline{p'\mathbf{u}''} \right)$ is the total heat flux in the subsystem of average motion of the developed turbulence.

Since in the non-equilibrium stationary state under consideration the entropy outflow from the subsystem of averaged motion $\sigma^e_{\langle S \rangle} \equiv \Xi/\langle T \rangle$ is positive, the rate Ξ of heat exchange between the averaged and turbulent motion is also positive, $\Xi \geq 0$. Then from (30) follows that, in contrast to Klimontovich (1995), the temperature in the turbulent subsystem T_{turb} is higher than the temperature in the subsystem of averaged motion, $T_{\text{turb}} > \langle T \rangle$. This result is in the full agreement with the synergetic principle of self-organization of dissipative system.

Using (32) and the Curie-Prigogine principle (according to which the relation between tensors of different rank in the isotropic medium is impossible), the generalized rheological relations can be derived for the turbulent fluxes and the corresponding thermodynamic forces in the following form:

$$\mathbf{q}^{\text{turb}}(\mathbf{x},t) = \overline{p'\mathbf{u}''} - k^{\text{turb}} \nabla \left(\ln \langle T \rangle \right) , \tag{33}$$

$$\mathbf{R}(\mathbf{x},t) = -\frac{2}{3} \bar{\rho} \langle e \rangle \mathbf{I} + \bar{\rho} v^{\text{turb}} \left(\frac{1}{2} \left(\nabla \langle \mathbf{u} \rangle + \nabla^* \langle \mathbf{u} \rangle \right) - \frac{1}{3} \left(\nabla \cdot \langle \mathbf{u} \rangle \right) \mathbf{I} \right), \tag{34}$$

$$\omega(\mathbf{x},t,\xi) = -\int_\xi L(\mathbf{x},t,\xi,\tilde{\xi}) \frac{\partial \mu(\tilde{\xi})}{\partial \tilde{\xi}}\, d\tilde{\xi} , \tag{35}$$

which correspond to the stationary non-equilibrium state of the turbulent field. Since strong turbulence is locally uniform and isotropic, the phenomenological coefficients (turbulent-exchange coefficients) $k^{\text{turb}}(\mathbf{x},t)$, $v^{\text{turb}}(\mathbf{x},t)$ and $L(\mathbf{x},t,\xi,\tilde{\xi})$ in these relationships are scalar quantities. Clearly, these quantities, unlike the molecular-exchange coefficients, are not material constants. This is accounted for by the fact that, for the turbulent continuum, all processes of transfer (of matter, momentum, and energy) from one to another region of the system are determined by the collective motions of molecules (vortex structures). Therefore, they depend on the parameters which specify the intensity of the turbulent field (in particular, on $\langle \varepsilon_r \rangle$ and L_0). For example, in the inertial scale range $\eta < l < L_0$ the turbulent-viscosity coefficient v^{turb} (which corresponds to the empirical Richardson-Obukhov "four-thirds law" emerging from the dimensional and similarity theory), has the form: $v^{\text{turb}} \propto \langle \varepsilon_r \rangle^{1/3} l^{4/3}$. It means that in application models dealing with the stationary non-equilibrium turbulence when energy dissipation processes are important, the equation of heat transfer for the averaged motion in the form (31) should be brought and complemented with the defining relations (33)-(35).

According to the formula (35), the phenomenological relationship for the rate $\omega(\xi)$ of destruction of eddies of sort ξ and for the corresponding "instantaneous affinity" $A_{\text{turb}}(\xi)$ [see formula (22)] has generally an integral form. Then taking advantage of the (Prigogine, 1955) principle we can assume that in every part of the coordinate space the processes direction ensures only entropy increment. It means that both integral (21) and the quantity

$$\omega(\xi)A_{\text{turb}}(\xi) = -\omega(\xi)\frac{\partial \mu_{\text{turb}}(\xi)}{\partial \xi} \geq 0$$

which specifies the "energy scattering" in a unit volume of the internal configuration space, are positive. In this case, the defining relation between the rate $\omega(\xi)$ and affinity $A_{\text{turb}}(\xi)$, being regarded as an equivalent of the process of eddies decay, has a simple form:

$$\omega(\xi) = L_\xi A_{\text{turb}}(\xi) = -L_\xi \frac{\partial \mu_{\text{turb}}(\xi)}{\partial \xi} \, , \tag{36}$$

where L_ξ is the Onsager's coefficient.

Thus the approach defined by the formula (36) allows us to find out the nonlinear defining relations for the total fluxes which are connected to the internal parameter. It also constitute the basis for obtaining evolutionary equations in partial derivatives (of the Fokker-Planck type in the configuration space ξ for various statistical characteristics of the low-scale turbulence provided some hypotheses on the distribution function of these characteristics in the stationary non-equilibrium state are adopted. One should keep in mind, however, that all hypotheses of such kind are not entirely rigorous, representing idealization or simplification of the real processes (Monin and Yaglom, 1975).

We now apply the principle (36) to derive the *kinetic equation* that would describe the continuous eddies interaction in terms of time evolution of the conditional probability density of eddy velocities $P_2(0|u'', \mathbf{x}, t)$ ($\equiv n(u'', \mathbf{x}, t)/n_\Sigma$). This function is considered as an internal variable of the turbulent chaos subsystem, while the pulsation eddy velocity u'' as an internal coordinate ξ. From the physical kinetics follows that time evolution of such distribution function in the approximation of continuous interaction between eddies should be described by the Fokker-Planck equation in the velocities configuration space. Let us deduce it here using the thermodynamic approach.

It is known that in the developed turbulence the probability distribution function of pulsation velocity is not a universal one because it depends on the mechanism that generates the turbulent field. Nevertheless, following

Millionschikov (1941) we use the hypothesis on normal distribution of the locally-isotropic pulsation velocity field in the steady-state case

$$W_1(u'') \equiv n^{\text{stat}}(u'')/n_\Sigma = (\beta/\sqrt{\pi}) \exp\left(-\beta^2 u''^2\right). \tag{37}$$

Then, following the kinetic gas theory, we can relate the β constant in (37) with the turbulent temperature of the subsystem of turbulent chaos. Using further (17) and (37) it is easy to find that $\beta^2 = (2RT_{\text{turb}})^{-1}$ from where another equivalent expression for the $n^{\text{stat}}(u'')$ follows:

$$n^{\text{stat}}(u'') = n_\Sigma (1/2\pi RT_{\text{turb}})^{1/2} \exp\left(-u''^2/2RT_{\text{turb}}\right)$$
$$= \text{const} \cdot \exp\left(-u''^2/2RT_{\text{turb}}\right). \tag{37*}$$

Substituting now this distribution in (17) we obtain the following expression for the chemical potential μ_{turb}:

$$\mu_{\text{turb}}(u,t) = u'^2/2 + RT_{\text{turb}} \ln\left[n(u',t)\right] + \text{const}. \tag{38}$$

In view of this formula, the phenomenological equation (36) for the probability flow $J(u'', \mathbf{x}, t)$ takes the form

$$J(u'') = -\frac{\bar{\rho}}{n_\Sigma} L_{u''} \left(u'' + \frac{RT_{\text{turb}}}{n(u'')} \frac{\partial n(u'')}{\partial u''}\right)$$
$$= -\alpha \left[u'' n(u'') + RT_{\text{turb}} \frac{\partial n(u'')}{\partial u''}\right], \tag{39}$$

where $R(= n_\Sigma k/\bar{\rho})$. Here we introduced the coefficient $\alpha \equiv L_{u'}/n(u',t)$ which can be interpreted as "*mobility*" in the space of internal coordinate u'' per unit volume. This coefficient is considered in the first approximation independent of both $n(u'')$ and u''. Substituting (38) into (19) yields the desired Fokker-Planck equation for the function of conditional probability density of pulsation velocity u'':

$$\frac{\partial P_2(0|u'', \mathbf{x}, t)}{\partial t} + \nabla_x \cdot \left(P_2(0|u'', \mathbf{x}, t)\langle \mathbf{u}(\mathbf{x}, t)\rangle\right) \tag{40}$$

$$-\alpha \left(\frac{\partial}{\partial u''}[u'' \cdot P_2(0|u'', \mathbf{x}, t)] + \frac{\partial}{\partial u''}\left[RT_{\text{turb}}(\mathbf{x}, t)\frac{\partial P_2(0|u'', \mathbf{x}, t)}{\partial u''}\right]\right) = 0.$$

This dynamic equation being complemented with the initial condition $P_2(0|u'', \mathbf{x}, 0) = \delta(u'' - 0)$ where the right side is δ-function in 0 point, describes the temporal evolution of the $P_2(0|u'', \mathbf{x}, t)$ function of u'' in particular in the case of the damped (degenerated) turbulence. Let us note that

$K \equiv -\alpha u''$ (that is the friction coefficient in the corresponding Langevin equation) serves as the drift coefficient in the standard notation of the Fokker-Planck equation, while the quantity $Q \equiv 2L_{u''}kT_{\text{turb}}/n(u'') = 2\alpha RT_{\text{turb}} = \alpha\beta^{-2}$ can be treated as the diffusion coefficient.

The normal distribution (37) that is a stationary solution of the one-dimensional in parameter u'' Fokker-Planck equation (40) can be adopted as initial statistical state of the pulsating velocities field for a set of different motions of the degenerated turbulence. Then a non-stationary solution of this equation can be found in the analytical form (Haken, 1985):

$$P_2(0|u'', \mathbf{x}, t) = \{\pi a(\mathbf{x}, t)\}^{-1/2} = \exp\{-[u'' - b(\mathbf{x}, t)]^2/a(\mathbf{x}, t)\}, \quad (41)$$

where

$$a(\mathbf{x}, t) = \frac{Q}{\alpha}\{1 - \exp(-2\alpha t)\} + a_0 \exp(-2\alpha t), \quad b(\mathbf{x}, t) = b_0(\mathbf{x})\exp(-\alpha t);$$
$$(42)$$

$a_0(\mathbf{x})$ and $b_0(\mathbf{x})$ are the initial conditions. This solution allows us to calculate the different n-point moments (correlation functions) of m-order, which describe the statistical relationship between random temporal-spatial velocities. Analysis of eddies evolution results in the relations that are in accord with the assumption on local isotropy of eddy velocities field in the case of the developed turbulence and also with the basic concepts of the statistical turbulence theory.

It should be emphasized that non-negative macroscopic variables (even velocity functions) of eddies kinetic energy, rates of energy dissipation, etc. are addressed as the most suitable internal coordinate to characterize the small-scale turbulence. According to Kolmogorov's hypothesis, similar stochastic characteristics satisfy asymptotically the log-normal law of probabilities distribution. This is accounted for by the fact that the process of consecutive fragmentation of the vortex structures is similar, in particular, to the process of coagulation of solid particles (the latter leads, as is known, to the log-normal size distribution of particles). At the same time, it is important that the log-normal distribution does not describes accurately the edges of an actual distribution of the given characteristics and can be used only with great care for calculating the higher moments.

Another example of using the developed approach is the Richardson-Kolmogorov strong-turbulence scheme (large eddies $\longrightarrow$ small eddies $\longrightarrow$ heat) which can be considered as a process of above mentioned consecutive chemical reactions. If we choose the specific kinetic energy of eddies as an internal coordinate, the basic kinetic equation (19) for the function of distribution $P_2(0|\xi; \mathbf{x}, t)$ of eddies in the space of their energies $\xi = \rho u''^2/2$ may be represented as

$$\frac{\partial P_2(0|\xi;\mathbf{x},t)}{\partial t} + \nabla \cdot [P_2(0|\xi;\mathbf{x},t)\langle \mathbf{u}\rangle] = \frac{\partial}{\partial \xi}[P_2(0|\xi;\mathbf{x},t)\varepsilon(\xi,\mathbf{x},t)]\,, \quad (47)$$

where

$$\varepsilon(\xi,\mathbf{x},t) \equiv -J(\xi,\mathbf{x},t)/n(\xi,\mathbf{x},t) \qquad (48)$$

is the reaction rate of transition from state ξ in $\xi + \mathrm{d}\xi$ corresponding to the probability flux $J(\xi)$ in the state ξ. In other words, this relationship defines the parameter $\varepsilon(\xi,\mathbf{x},t)$ which can be formally interpreted as the rate of transfer of the kinetic energy $\rho \mathbf{u}''^2/2$ according to eddies rank along the ξ coordinate. Simultaneously, this value determines the mean kinetic energy dissipation in the eddies of ξ kind.

Indeed, the equation for the first moment of the normalized distribution function $\int \xi P_2(0|\xi;t)\,\mathrm{d}\xi = \overline{\rho \mathbf{u}''^2/2} = \bar{\rho}\langle e\rangle$ deduced from (47) by integrating by parts and assuming that the energy flows $J(\xi,\mathbf{x},t)$ vanish at the integration limits, has the form

$$\bar{\rho}\frac{\mathrm{d}\langle e\rangle}{\mathrm{d}t} \cong -\int \varepsilon(\xi,\mathbf{x},t)P_2(0|\xi;\mathbf{x},t)\,\mathrm{d}\xi = -\overline{\varepsilon(\mathbf{x},t)}^0 \cong -\bar{\rho}\langle \varepsilon_r\rangle\,. \qquad (49)$$

Here the conditional mean value is taken according to the stochastic process ξ and the parameter $\overline{\varepsilon(\mathbf{x},t)}^0$ corresponds to the rate of turbulent energy dissipation in the point $((\mathbf{x},t)$ (see, e.g., Landau and Lifshitz, 1988). Hence, the quantity $\varepsilon_r(\mathbf{x},\xi)$ can be interpreted as a local dissipation of energy of eddies at the point (ξ,t) of the configuration space.

Then for the part of dissipation energy $\langle T\rangle \sigma_\Sigma$ (see formula (32)) which is stipulated by cascade turbulent energy transfer, we have:

$$(\langle T\rangle \sigma_\Sigma)^{Ch} \equiv -\bar{\rho}\int_\xi \varepsilon(\xi,\mathbf{x},t)n(\xi,\mathbf{x},t)A_{\text{turb}}(\xi,\mathbf{x},t)\,\mathrm{d}\xi \geq 0\,. \qquad (50)$$

From this formula the local phenomenological equation of Prigogine's type follows

$$\varepsilon(\xi,\mathbf{x},t) = L_\xi A_{\text{turb}}(\xi,\mathbf{x},t)/n(\xi,\mathbf{x},t) = -\alpha' A_{\text{turb}}(\xi,\mathbf{x},t)\,, \qquad (51)$$

where

$$A_{\text{turb}}(\xi) = -kT_{\text{turb}}\frac{\partial \ln n(\xi)}{\partial \xi} + f(\xi) \qquad (52)$$

is the chemical affinity of the eddies splitting (state function of the subsystem of turbulent chaos); $f(\xi) = -\partial V/\partial \xi$ – is the so called friction force; $\alpha' = -L_\xi/n(\xi)$ is mobility coefficient supposed to be independent on ξ.

If non-equilibrium stationary state is established in the system and the rate of energy transfer over the cascade of eddies is constant, i.e. $\varepsilon(\xi, \mathbf{x}, t) \cong \langle \varepsilon(\mathbf{x}, t) \rangle$ (this assumption was adopted in the classical phenomenology of the Kolmogorov's cascade) then (50) takes the form:

$$\langle T \rangle \sigma_\Sigma = -\bar{\rho} \langle \varepsilon(\mathbf{x}, t) \rangle A^{gl}_{\text{turb}}(\mathbf{x}, t) \geq 0 , \tag{53}$$

where $A^{gl}_{\text{turb}}(\mathbf{x}, t) \equiv \int_\xi n(\xi) A_{\text{turb}}(\xi) \, \mathrm{d}\xi = n_\Sigma \overline{A_{\text{turb}}}$ – the so called global affinity of the process of turbulent structures formation that with the account for (52) can be rewritten as

$$\overline{A_{\text{turb}}}(\mathbf{x}, t) = -kT_{\text{turb}} \int_\xi \frac{\partial P_2(0|\xi; t)}{\partial \xi} \mathrm{d}\xi + \int_\xi P_2(0|\xi; t) f(\xi) \mathrm{d}\xi \tag{54}$$
$$= -kT_{\text{turb}}[P_2(\xi_{L_1}) - P_2(\xi_\eta)] + \overline{f(\mathbf{x}, t)} \cong \overline{f(\mathbf{x}, t)} ,$$

since P_2 becomes zero at the integration limits. Thus, the increment of the averaged entropy σ_Σ in the steady-state process of formation of new turbulent structures (eddies) is expressed as the product of general rate ε_r of the energy transfer along the cascade $\langle \varepsilon \rangle$ and the global affinity A^{gl}_{turb} corresponding to the entire cascade process of fragmentation of large eddies into progressively smaller ones. In this case, the linear phenomenological relation

$$\langle \varepsilon \rangle = -\alpha' A^{gl}_{\text{turb}} \tag{55}$$

is valid, in accordance with the results of irreversible thermodynamics for the consecutive chemical reactions (see Prigogine and Kondepudi, 2002).

On the other hand, the condition of the steady-state (independent on the parameter ξ) thermodynamic flow of the kinetic energy according to eddies rank $J(\xi, \mathbf{x}, t) \cong J(\mathbf{x}, t) \equiv -n_\Sigma(\mathbf{x}, t) \langle \varepsilon(\mathbf{x}, t) \rangle$ jointly with the linear phenomenological relationship (51) leads to the more general than (55) nonlinear relation between ε and characteristic affinity of the overall cascade process given by the formula

$$\tilde{A}(\mathbf{x}, t) \equiv \int_\xi A(\xi) \mathrm{d}\xi = \mu(\xi_{L_1}, \mathbf{x}, t) - \mu(\xi_\eta, \mathbf{x}, t) .$$

This nonlinear defining relation can be easily obtained using the formula (22) for the "local affinity" $A_{\text{turb}}(\xi, \mathbf{x}, t)$. As a result, we have

$$\langle \varepsilon \rangle \cong \gamma[1 - \exp(-\tilde{A}/kT_{\text{turb}})] , \tag{56}$$

where

$$\gamma = \frac{(\alpha' kT_{\text{turb}}/n_\Sigma) \exp(\mu(\xi_\eta)/kT_{\text{turb}})}{\int_{\xi_\eta}^{\xi_{L_1}} \exp[V(\xi)/kT_{\text{turb}}] \mathrm{d}\xi}$$

and η is the local Kolmogorov's scale.

Thus, we can use two interpretations of the Kolmogorov parameter ε: as the quantity that describes the rate of energy dissipation into heat, and also as the rate of turbulent energy transfer along the cascade of eddies. This allowed us to obtain the defining relations (55) and (56) for the dissipation rate $\langle \varepsilon \rangle$. These relations close the set of the averaged hydrodynamic equations (6)-(8) and make self-sufficient the thermodynamic approach to the modeling of the developed turbulence.

It is worthwhile to emphasize the following. At the first sight, by analogy to the liquid laminar motion, certain constraints on the coefficient of turbulent transfer in the defining relationships (33), (34), and (55), place the condition of the entropy increase for the aggregate continuum (32) as well. It is known that positiveness of the direct coefficients molecular exchange follows precisely from the similar conditions though the cross-coefficients may be different in sign (Landau and Lifshitz, 1988). Substituting (33), (34), and (55) into expression (32) for the production of the total entropy of the turbulent system yields

$$0 \leq \sigma_{\Sigma} \equiv \frac{1}{\langle T \rangle} \left\{ \lambda^{\text{turb}} \left(\frac{\nabla \langle T \rangle}{\langle T \rangle} \right)^2 + \bar{\rho} v^{\text{turb}} \left(\mathbf{E} - \frac{1}{3} (\nabla \cdot \langle \mathbf{u} \rangle) \mathbf{I} \right)^2 \right.$$
$$\left. + \bar{\rho} \alpha' \left(A^{gl}_{\text{turb}} \right)^2 \right\} . \tag{57}$$

As we pointed out earlier, specific interaction between various dissipative processes in the turbulent continuum (related to the functional dependence of the turbulent exchange coefficients on the parameters $\langle \varepsilon_r \rangle$ and $\langle e \rangle$ occurs in such a way that an absence of one of the thermodynamic forces (for example, the affinity A^{gl}_{turb} may impact on some processes (for example, viscous effects). This means that the second law of thermodynamics which requires the entire sum (57) to be positive, cannot generally be applied to its individual terms. For example, the inequality $v^{\text{turb}} \left(\mathbf{E} - \frac{1}{3} (\nabla \cdot \langle \mathbf{u} \rangle) \mathbf{I} \right)^2 < 0$ may appear only provided the sum $\sigma_{\Sigma} \geq 0$. This proves, in particular, the possibility of turbulent flow regimes for which the turbulent viscosity coefficient is negative, $v^{\text{turb}} < 0$.

Let us now consider in more detail the most interesting small-scale variable – the rate of energy dissipation $\varepsilon_r(\xi)$. We have earlier supposed that, in the case of the developed turbulence, the rate of energy transfer along the eddies cascade within the Lagrange volume is constant and this is why we did not distinguish between ε, ε_r and $\langle \varepsilon_r \rangle$. However, in the real turbulent motion averaged over small (within $d\mathbf{x}$) volume V_r the rate of energy dissipation $\varepsilon_r(\mathbf{x}, t) = (1/V_r) \int_{V_r} \varepsilon(\mathbf{x} + \mathbf{x}', t) \, d\mathbf{x}'$ is a random function which pulsates jointly with the gradient of velocity field $\mathbf{u}''(\mathbf{x}, t)$. As it was already noticed, this Obuchov's parameter defined by formula (3) serves as the most important small-scale variable responding to the refined Kolmogorov's similarity criteria. Therefore, this averaged characteristic of eddies can be used as an internal coordinate ξ of the subsystem of "turbulent chaos", $\xi = \varepsilon_r$. In this case, the quantity $n(\mathbf{x}, t, \varepsilon_r)\mathrm{d}\varepsilon_r$ represents the number of eddies (in a unit volume of the medium), whose dissipation rates fall in the range from ε_r to $\varepsilon_r + \mathrm{d}\varepsilon_r$. We shall apply again the principle (36) to derive the Fokker-Planck kinetic equation describing temporal evolution of the conditional probability distribution $P_2(\varepsilon_r^{st}|\varepsilon_r, \mathbf{x}, t)$ of the turbulent energy dissipation ε_r. Such a probabilistic equation will contain parameters (in particular, Reynolds number) defining the system state.

If a random function ε_r in the stationary state is obeyed to the log-normal probabilities distribution (Kolmogorov, 1962) according to (4) then substituting (4) into (17) we obtain the desired formula for the chemical potential $\mu_{\mathrm{turb}}(\varepsilon_r)$ in the internal coordinate ε_r:

$$\mu_{\mathrm{turb}}(\varepsilon_r) = kT_{\mathrm{turb}} \ln n(\varepsilon_r) + V(\varepsilon_r, T_{\mathrm{turb}}) , \tag{58}$$

where

$$V(\varepsilon_r, T_{\mathrm{turb}}) = \mu_{\mathrm{turb}}^{st}(\varepsilon_r) - kT_{\mathrm{turb}} \ln n^{st}(\varepsilon_r) = \mathrm{const}$$

$$- kT_{\mathrm{turb}} \ln \frac{n_\Sigma}{\sqrt{2\pi}\sigma_{\ln \varepsilon_r}} + kT_{\mathrm{turb}} \ln n(\varepsilon_r) + \frac{kT_{\mathrm{turb}}}{2\sigma_{\ln \varepsilon_r}^2} \left[\ln \left(\frac{\varepsilon_r}{m_{\ln \varepsilon_r}} \right) \right]^2 \tag{59}$$

is the friction force potential appearing also in the stochastic Langevin equation $\varepsilon_r = \alpha' f(\varepsilon_r) + F(t)$ where $F(t)$ is a "random force". Then the local phenomenological equation (36) for the probability flux takes the form:

$$J(\varepsilon_r, \mathbf{x}, t) = -L_\xi \frac{\partial \mu_{\mathrm{turb}}(\varepsilon_r, t)}{\partial \varepsilon_r} = \alpha^* n(\varepsilon_r, t) A_{\mathrm{turb}}(\varepsilon_r, \mathbf{x}, t) . \tag{60}$$

Here $\alpha^*(\varepsilon_r) \equiv L_{\varepsilon_r}/n(\varepsilon_r)$ is the mobility coefficient in the internal coordinate ε_r space (which, in the first approximation, does not depend on $n(\varepsilon_r)$), while the quantity

$$A_{\mathrm{turb}}(\varepsilon_r, \mathbf{x}, t) = -kT_{\mathrm{turb}} \frac{\partial \ln n(\varepsilon_r, \mathbf{x}, t)}{\partial \varepsilon_r} + f(\varepsilon_r, \mathbf{x}, t)$$

$$= -kT_{\mathrm{turb}} \frac{\partial \ln n(\varepsilon_r, \mathbf{x}, t)}{\partial \varepsilon_r} - \frac{kT_{\mathrm{turb}}}{\varepsilon_r} \left(1 + \frac{1}{\sigma_{\ln \varepsilon_r}^2} \ln \frac{\varepsilon_r}{m_{\ln \xi}} \right) \tag{61}$$

defines the generalized chemical affinity of ε_r state.

Using now the Kolmogorov's formula (4) for dispersion $\sigma_{\ln \varepsilon}^2$ as well as the known relations $\bar{\varepsilon}_r = \exp\left(\sigma_{\ln \varepsilon}^2/2 + \overline{\ln \varepsilon_r}\right)$ and $m_{\ln \xi} = \exp\left(\overline{\ln \varepsilon_r}\right)$ for the mean value and median of the log-normal distribution we can derive the explicit expression for the friction force $f(\varepsilon_r)$:

$$
\begin{aligned}
f(\varepsilon_r, t) &= -\frac{\partial U}{\partial \varepsilon_r} = -\frac{kT_{\text{turb}}}{\varepsilon_r}\left[1 + \frac{1}{\sigma_{\ln\varepsilon}^2}\ln\left(\frac{\varepsilon_r}{m_{\ln\varepsilon}}\right)\right] \\
&= -\frac{kT_{\text{turb}}}{\varepsilon_r}\left(1 + \frac{\ln\varepsilon_r - \overline{\ln\varepsilon_r}}{\sigma_{\ln\varepsilon}^2}\right) \\
&= -\frac{kT_{\text{turb}}}{\varepsilon_r}\left(1 + \frac{\ln\varepsilon_r - \ln\bar{\varepsilon}_r + \sigma_{\ln\varepsilon}^2/2}{\sigma_{\ln\varepsilon}^2}\right) \\
&= -\frac{kT_{\text{turb}}}{\varepsilon_r}\left(\frac{3}{2} + \frac{1}{\sigma_{\ln\varepsilon}^2}\ln\left(\frac{\varepsilon_r}{\bar{\varepsilon}_r}\right)\right).
\end{aligned}
\tag{62}
$$

It depends on both $\mathbf{Re}$ number controlling the motion regime and specificity of large-scale motions, in the latter case through Kolmogorov's relation $\sigma_{\ln\varepsilon_r}^2 \approx \frac{3}{4}\mu\ln\mathbf{Re} + B(\mathbf{x}, t)$.

Finally, substituting (60) into (19) we obtain the desired kinetic equation for the probability distribution function $P_2(\varepsilon_r^{st}|\varepsilon_r, t)$ in the form:

$$
\begin{aligned}
\bar{\rho}\frac{\mathrm{d}}{\mathrm{d}t}\left(\frac{P_2(\varepsilon_r^{st}|\varepsilon_r, t)}{\bar{\rho}}\right) &= \frac{\partial}{\partial\varepsilon_r}\left[D(\varepsilon_r)\frac{\partial P_2(\varepsilon_r^{st}|\varepsilon_r, t)}{\partial\varepsilon_r}\right] \\
&\quad - \frac{\partial}{\partial\varepsilon_r}\left[\alpha^*(\varepsilon_r)P_2(\varepsilon_r^{st}|\varepsilon_r, t)f(\varepsilon_r, t)\right].
\end{aligned}
\tag{63}
$$

Here $D(\varepsilon_r) = \alpha^* kT_{\text{turb}}$ is the diffusion coefficient in the configuration space ε_r, which bears an existence of eddy structure of the turbulent chaos subsystem and serves as a source of the so called "natural noise". The equation (63) reflects the fact that both this noise and control parameters (such as the parameter $\sigma_{\ln\varepsilon}^2$) influence on the relative increment of the rate of energy transfer fluctuations along the cascade significantly increases when approaching different critical points (Haken, 1985, 1991). Here one may notice a deep analogy with Landau phase transitions theory.

Let us also emphasize that depending on $\mathbf{Re}$ fluctuations of the variable ε_r (and hence $\langle\varepsilon\rangle$ as well) can reach such critical values that a bifurcation of the turbulent field may occur. This is connected with the formation of new relatively stable dissipative eddy structures corresponding to the diverse attractors set up, the latter being characterized by exponential instability of nearly all their intrinsic trajectories and changing the macroscopic behavior of the whole system. Such a bifurcation is responsible, for example, for the transition from the flow with alteration to the fully turbulent flow.

Conclusion

Modeling of the stationary non-equilibrium turbulence is carried out using synergetic approach to the study of complicated systems. Turbulent continuum is represented in terms of the thermodynamic complex composed of two mutually open subsystems: mean motion and turbulent chaos, the latter considering as an ensemble of eddies of different temporal-spatial scales (with the caveat for the case of the plasma wave turbulence). Macroscopic description of the developed turbulence as self-organizing process in an open system is presented and methods of irreversible thermodynamics is applied to derive defining relations for the turbulent fluxes and forces which describe comprehensively the system structure and transfer processes. It is shown that a deep analogy exist between the consecutive chemical reactions ($A \longrightarrow B \longrightarrow C \longrightarrow$ etc.) and the Richardson-Obukhov cascade process of eddies splitting when assigning them the respective chemical potential and affinity. Continuous internal parameters of the turbulent medium specifying excited macroscopic degrees of freedom are introduced and Prigogine's postulate on irreversible processes localized in the configuration space is used in order to describe thermodynamically the Kolmogorov's cascade process. This allowed us, in particular, to derive the different equations of Fokker-Planck type for the distribution functions of small-scale turbulence characteristics and to find non-linear defining relations for the Kolmogorov parameter that describes the rate of kinetic energy dissipation into heat and simultaneously, the rate of energy transfer through eddies cascade.

Basically, the developed modeling approach is a part of the rapidly progressed non-linear dynamics involving evolution of chaotic motions and formation of ordered dissipative structures. Duality of the irreversible processes demonstrating disordering nearby equilibrium and, in contrast, ordering well outside equilibrium, is observed when dealing with various turbulent objects in the outer space. Cascades of temporal-spatial configurations comprising chaotic processes are of key importance in the origin and evolution of the Universe, in stellar-planetary cosmogony, in formation of accretion discs, to mention a few. This new approach broaden also the entropy concept as an underlying important entity and brings support to the allegation (Prigogine and Stengers, 1984) that understanding of chaos as a category of irreversibility opens up the wide horizons and that without irreversible processes it is impossible to describe our space environment.

References

Batchelor, G.K. (1953) *The theory of homogeneous turbulence.* Cambridge.

Buckingham A.C. (2003) *Appl. Mech. Rev.* vol **56**, no 1, 1R38, B11-B13.

De Groot, S.R., and Mazur, P. (1962) *Non-equilibrium thermodynamics.* North Holland Publishing Company, Amsterdam.

Ebeling, W., Engel, A., and Feistel, R. (1990) *Physic der evolutionsprozesse*. Akademic-Verlag Berlin.

Fridman, A.M., and Choruzhy, O.V. (2003) Progress in the study of galaxies: Structures, Collective phenomena, and Methods. *Space Science Reviews* **105**, 1-284.

Fridman, A.M., Sagdeev, R.Z., Choruzhy, O.V., and Polyachenko, E.V. In: "The Contemporary Problems of Mechanics and Physics of Space", dedicated to the 70th birthday of M.Ya. Marov. Fizmatlit, Nauka, 12-26.

Frisch, U., Sulem, P.L., and Nelkin, M. (1978) A simple dynamical model of intermittent fully developed turbulence. *J. Fluid Mech.* **87**, 719-736.

Frisch, U. (1995) *Turbulence. The legacy of A.N.Kolmogorov*. Cambridge University Press, London.

Haken, H. (1983) *Advanced Synergetics. Springer Ser. Synergetics,* **20**, 2nd ed., Verlag/Berlin/ Heidelberg; Springer/Berlin, Heidelberg.

Haken, H. (1988) *Information and Self-Organization. A macroscopic approach to complex systems*. Springer Verlag, Berlin, Heidelberg.

Klimontovich, Yu.L. (1995) *Statistical Theory of Open System*. Kluwer Academic Publishers, Dordrect.

Kolesnichenko, A.V., and Marov, M.Ya. (1985) Methods of non-equilibrium thermodynamics for description of multicomponent turbulent gas mixtures. *Arch. Mech.* Warszawa. **37**(1-2), 3-19.

Kolesnichenko, A.V. (1995) On the theory of turbulence in the planetary atmospheres: A numerical modeling of structure parameters. *Astronom. Vestnik* **29**(2), 135-155.

Kolesnichenko, A.V., and Marov, M.Ya. (1997) *Turbulence of multicomponent media*. Nauka, Moscow.

Kolesnichenko, A.V. (2001) Hydrodynamic aspects of modeling of mass transfer and coagulation in the turbulent accretion disc. *Astronom. Vestnik* **35** (2), 139-155.

Kolmogorov, A.N. (1941) Local structure of turbulence in incompressible fluid under very large Reynolds numbers. *Doklady AN USSR.* **30**, 299-303.

Landau, L.D., and Lifshitz, E.M. (1988) *Hydrodynamics*. Nauka, Moscow.

Makalkin, A.B. (2003) Problems of the Protoplanetary Discs Evolution. In: "The Contemporary Problems of Mechanics and Physics of Space", dedicated to the 70th birthday of M.Ya. Marov. Fizmatlit, Nauka, 402-446.

Marov, M.Ya., and Grinspoon, D.H. (1998) The Planet Venus. Yale University Press, New Haven & London.

Marov, M.Ya., and Kolesnichenko, A.V. (2001) *Mechanics of turbulence of multicomponent gases*. Kluwer Academic Publishers, Dordrecht/Boston/London.

Millionshchikov, M.D. (1941) On the theory of homogeneous and isotropic turbulence. *Izvestiya AN USSR.* **5**(4-5), 433-446.

Monin, A.S., and Yaglom, A.M. (1975) *Statistical Fluid Mechanics,* vol.**2**.Ed. J. Lumley. MIT Press, Cambridge, MA.

Nevzglyadov, V.G. (1945) On the phenomenological theory of turbulence. *Doklady AN USSR.* **47**(3), 169-172.

Obukhov, A.M. (1941) On the energy distribution in the spectrum of turbulent flow. *USSR Acad. of Sci. Reports, Geography and Geophysics series*, **5** (No. 4), 453-456.

Obukhov, A.M. (1962) Some specific features of atmospheric turbulence. *J. Fluid Mech.* **13** Pt.1, 77-81.

Onsager, L. (1949) Statistical hydrodynamics. *Nuovo Cimento* (Supplement). **6,** 279-287.

Prigogine, I. (1955) *Introduction to thermodynamics of irreversible processes.* Wiley Interscience, New York.

Prigogine, I., and Stengers, I. (1984) *Ordes of Chaos.* N.Y.: Dantam Books.

Prigogine, I., and Stengers, I. (1993) *Das Paradox der Zeit.* Munchen: R. Piper & Co. Verlag.

Prigogine, I., and Kondepudi, D. (2002) *Contemporary Thermodynamics. From Heat Engines to Dissipative Structures.* MIR P.H., Moscow.

Serrin, J. (1959) *Mathematical principles of classical fluid mechanics. Handbuch der physik Band VIII/1.* Stromungsmechanic I. Berlin-Gottingen-Heidelberg.

Ruskol, E.L. (2003) Formation of planets and satellites. In: "The Contemporary Problems of Mechanics and Physics of Spacc", dedicated to the 70th birthday of M.Ya. Marov. Fizmatlit, Nauka, 353-368.

INTERNAL STRUCTURE OF THIN ACCRETION DISKS

Vasily Beskin
Lebedev Physical Institute, Moscow
beskin@lpi.ru

Alexander Tchekhovskoy
Moscow Institute of Physics and Technology
chekhovs@lpi.ru

Abstract   We present analytical results available for a large class of axisymmetric stationary flows in the vicinity of compact astrophysical objects. First, the most general case is formulated corresponding to axisymmetric stationary MHD flows in the Kerr metric. Then, we discuss the hydrodynamical version of the Grad-Shafranov equation. Although not so well-known as the full MHD one, it allows us to clarify the nontrivial structure of the Grad-Shafranov approach as well as to discuss the simplest version of the 3+1-split language – the most convenient one for the description of ideal flows in the vicinity of spinning black holes. Finally, we consider several examples that demonstrate how this approach can be used to obtain the quantitative description of real transonic flows in the vicinity of black holes.

Keywords: accretion: hydrodynamics, accretion disks, black hole physics

1.   Introduction

Many astrophysical sources are axisymmetric and stationary to a good accuracy. These include both accreting neutron stars and black holes, axisymmetric stellar (solar) winds, jets from young stellar objects, and ejection of particles from magnetospheres of rotating neutron stars. It cannot be ruled out that such magnetohydrodynamic flows also play an important role in other galactic sources, e.g. microquasars. The latter ones are regarded as candidates for black holes not to say about active galactic nuclei where the electrodynamical

Alexei M. Fridman, et al. (eds.), Astrophysical Disks; Collective and Stochastic Phenomena 55–74
© 2006 *Springer. Printed in the Netherlands*

processes in the vicinity of spinning supermassive black holes are considered as the most reasonable model of their central engine (Shapiro & Teukolsky, 1983; Thorne et al., 1986). So, it is not surprising that ideal magnetohydrodynamics, which allows sufficiently simple formalization of the problem, is actively applied when describing these flows.

The point is that due to axial symmetry and stationarity (as well as the ideal freezing-in condition), in the most general case it is possible to introduce five integrals of motion which are constant at axisymmetric magnetic surfaces. This remarkable fact allows us to separate the problem of finding the poloidal field structure (the poloidal flow structure in the hydrodynamics) from the problem of particle acceleration and the structure of electric currents. The solution of the latter task for a given poloidal field can be obtained in terms of quite simple algebraic relations. It is important that such an approach can be straightforwardly generalized to the flows in the vicinity of the spinning black holes, as the Kerr metric is also axially symmetric and stationary.

On the other hand, it is much more difficult to find the two-dimensional poloidal magnetic field structure (the hydrodynamical flow structure). First of all, this is due to the complex structure of the equation describing axisymmetric stationary flows. In the general case, it is a nonlinear equation of the mixed type which changes from elliptical to hyperbolical at singular surfaces and in addition contains integrals of motion in the form of free functions. Generally speaking, similar equations, which stem from the classical Tricomi equation, have been discussed since the beginning of the last century in connection with transonic hydrodynamic flows (Guderley, 1962; von Mises, 1958). Later on, the equations describing axially symmetric stationary flows were called Grad-Shafranov equations after the authors who formulated in the late 1950s an equation of such a type in connection with controlled thermonuclear fusion (Shafranov, 1958; Grad, 1960). This equation, however, was originally related to equilibrium static configuration only and required strong revision when it was generalized to the transonic case. The full version of such an equation was formulated by L.S. Soloviev in 1963 in the third volume of *Problems of Plasma Theory* (Soloviev, 1967) and was well-known to physicists. However, as it often occurs, the full version of the Grad-Shafranov equation was little known in the astrophysical literature, so it was later 'rediscovered' scores of times (Okamoto, 1975; Heinemann & Olbert, 1978).

As it turned out, the difficulty lay in the fact that the very formulation of the direct problem within the Grad-Shafranov approach proved to be nontrivial. For example, in the hydrodynamical limit, when there are only three integrals of motion, the problem requires four boundary conditions for the transonic flow regime. This implies that, for instance, two thermodynamic functions and two velocity components either should be specified at some surface. However, to determine the Bernoulli integral, which, naturally, should be known in order to solve the Grad-Shafranov equation, all three components of the velocity must be specified, which is impossible since the third velocity component itself is to be obtained from the solution. This is in fact one of the main difficulties of the approach under consideration.

Nevertheless, there are several ways that allow us to construct analytical solutions of direct problems within the framework of the Grad-Shafranov method. For example, this is possible when the exact solution of this equation is known and we explore the flows weakly diverging from the known one. Spherically symmetric accretion (ejection) of matter could be such an exact solution. The known structure of the flow in the zeroth approximation enables us to determine (with the required accuracy) both the location of singular surfaces and all the integrals of motion directly from boundary conditions, thus making it possible to solve the Grad-Shafranov equation within the direct formulation of a problem.

2. Grad-Shafranov equation

Let us consider the axisymmetric stationary plasma flow in the vicinity of a spinning black hole, i.e., in the Kerr metric (Thorne et al., 1986):

$$ds^2 = -\alpha^2 dt^2 + g_{ik}(dx^i + \beta^i dt)(dx^k + \beta^k dt), \tag{1}$$

where

$$\alpha = \frac{\rho}{\Sigma}\sqrt{\Delta}, \qquad \beta^r = \beta^\theta = 0, \qquad \beta^\varphi = -\omega = -\frac{2aMr}{\Sigma^2},$$

$$g_{rr} = \frac{\rho^2}{\Delta}, \qquad g_{\theta\theta} = \rho^2, \qquad g_{\varphi\varphi} = \varpi^2. \tag{2}$$

Here α is the lapse function (gravitational red shift) vanishing on the horizon

$$r_\mathrm{g} = M + \sqrt{M^2 - a^2}, \tag{3}$$

ω is the angular velocity of local nonrotating observers (the so-called Lense-Thirring angular velocity), and

$$\Delta = r^2 + a^2 - 2Mr, \qquad \rho^2 = r^2 + a^2 \cos^2\theta,$$

$$\Sigma^2 = (r^2 + a^2)^2 - a^2\Delta\sin^2\theta, \qquad \varpi = \frac{\Sigma}{\rho}\sin\theta. \tag{4}$$

As usual, M and a are the black hole mass and angular momentum per unit mass ($a = J/M$) respectively. Here indices without hats denote components of vectors with respect to the coordinate basis $\partial/\partial r$, $\partial/\partial\theta$, and $\partial/\partial\varphi$, and indices with hats correspond to their physical components.

Below we use the system of units with $c = G = 1$. We shall also use the $3 + 1$ split language (Thorne et al., 1986). Within this approach, the physical quantities are expressed in terms of three-dimensional vectors which would be measured by observers moving around the spinning black hole with the angular velocity ω (the so-called ZAMOs – zero angular momentum observers). The convenience of the $3 + 1$ split language is connected with the fact that the representation of many expressions has the same form as in the flat space. On the other hand, all the thermodynamic quantities are determined in the comoving reference frame.

Now, we shall demonstrate how the five 'integrals of motion', which are constant at the magnetic surfaces, can be derived in the general case of axisymmetric stationary flows. It is convenient to introduce a scalar function $\Psi(r,\theta)$, which has the meaning of the magnetic flux. As a consequence, the magnetic field is defined in the following way:

$$\mathbf{B} = \frac{\nabla\Psi \times \mathbf{e}_{\hat{\varphi}}}{2\pi\varpi} - \frac{2I}{\alpha\varpi}\mathbf{e}_{\hat{\varphi}}, \tag{5}$$

where $I(r,\theta)$ is the total electric current inside the region $\Psi < \Psi(r,\theta)$.

As usual, we assume that the magnetosphere contains sufficient amount of plasma to satisfy the freezing-in condition which, using the $3+1$ split language, preserves the form $\mathbf{E} + \mathbf{v} \times \mathbf{B} = 0$. On the other hand, the stationarity (as well as the condition for zero longitudinal electric field) implies that the field $\mathbf{E}$ can be written as

$$\mathbf{E} = -\frac{\Omega_{\mathrm{F}} - \omega}{2\pi\alpha}\nabla\Psi. \tag{6}$$

Substituting relation (6) into the Maxwell equations, it is easy to verify that the condition $\mathbf{B} \cdot \nabla\Omega_{\mathrm{F}} = 0$ is satisfied, i.e., Ω_{F} must be constant at the magnetic surfaces (Ferraro's isorotation law):

$$\Omega_{\mathrm{F}} = \Omega_{\mathrm{F}}(\Psi). \tag{7}$$

Next, the Maxwell equation $\nabla \cdot \mathbf{B} = 0$, the continuity equation, and the freezing-in condition allow us to write the four-velocity of matter $\mathbf{u}$ in the form

$$\mathbf{u} = \frac{\eta}{\alpha n}\mathbf{B} + \gamma(\Omega_F - \omega)\frac{\varpi}{\alpha}\mathbf{e}_{\hat{\varphi}}, \tag{8}$$

where $\gamma = 1/\sqrt{1 - v^2}$ is the Lorentz factor of matter (measured by ZAMOs), and the quantity η is the particle flux to magnetic flux ratio. Due to the relationship $\nabla \cdot (\eta\mathbf{B}_p) = 0$, η must be constant at the magnetic surfaces $\Psi(r,\theta) = $ const as well, i.e.,

$$\eta = \eta(\Psi). \tag{9}$$

The next two integrals of motion result from our assumption that the flow is axisymmetric and stationary. It yields the conservation law of energy E and the z-component of angular momentum L_z:

$$E = E(\Psi) = \frac{\Omega_{\mathrm{F}} I}{2\pi} + \mu\eta(\alpha\gamma + \omega u_\varphi); \tag{10}$$

$$L = L(\Psi) = \frac{I}{2\pi} + \mu\eta\varpi u_{\hat{\varphi}}, \tag{11}$$

where $\mu = (\rho_m + P)/n$ is the relativistic enthalpy (ρ_m is the internal energy density, and P is the pressure). Finally, in the axially symmetric case the isentropy condition yields

$$s = s(\Psi), \tag{12}$$

so that the entropy per particle, $s(\Psi)$, is the fifth integral of motion.

The five integrals of motion $\Omega_{\mathrm{F}}(\Psi)$, $\eta(\Psi)$, $s(\Psi)$, $E(\Psi)$, and $L(\Psi)$, as well as the poloidal magnetic field $\mathbf{B}_{\mathrm{p}}$, allow us to find the toroidal magnetic field $B_{\hat{\varphi}}$ and all other plasma parameters:

$$\frac{I}{2\pi} = \frac{\alpha^2 L - (\Omega_{\mathrm{F}} - \omega)\varpi^2(E - \omega L)}{\alpha^2 - (\Omega_{\mathrm{F}} - \omega)^2\varpi^2 - \mathcal{M}^2}; \tag{13}$$

$$\gamma = \frac{1}{\alpha\mu\eta} \frac{\alpha^2(E - \Omega_{\mathrm{F}} L) - \mathcal{M}^2(E - \omega L)}{\alpha^2 - (\Omega_{\mathrm{F}} - \omega)^2\varpi^2 - \mathcal{M}^2}; \tag{14}$$

$$u_{\hat{\varphi}} = \frac{1}{\varpi\mu\eta} \frac{(E - \Omega_{\mathrm{F}} L)(\Omega_{\mathrm{F}} - \omega)\varpi^2 - L\mathcal{M}^2}{\alpha^2 - (\Omega_{\mathrm{F}} - \omega)^2\varpi^2 - \mathcal{M}^2}, \tag{15}$$

where

$$\mathcal{M}^2 = \frac{4\pi\eta^2\mu}{n}. \tag{16}$$

It is easy to see that $\mathcal{M}^2$ is proportional (with the factor of α^2) to the Mach number squared of the poloidal velocity u_{p} with respect to the Alfvén velocity $u_{\mathrm{A}} = B_{\mathrm{p}}/\sqrt{4\pi n\mu}$, i.e., $\mathcal{M}^2 = \alpha^2 u_{\mathrm{p}}^2/u_{\mathrm{A}}^2$.

Since $\mu = \mu(n, s)$, definition (16) allows us to express the concentration n (and hence the specific enthalpy μ) as a function of η, s, and $\mathcal{M}^2$. This means that along with the five integrals of motion, the expressions for I, γ, and $u_{\hat{\varphi}}$ depend only on one additional quantity, namely the Mach number $\mathcal{M}$. To determine the Mach number $\mathcal{M}$, it is necessary to use the obvious relation $\gamma^2 - \mathbf{u}^2 = 1$, which, owing to equations (14) and (15), can be rewritten in the form

$$\frac{K}{\varpi^2 A^2} = \frac{1}{64\pi^4} \frac{\mathcal{M}^4(\nabla\Psi)^2}{\varpi^2} + \alpha^2\eta^2\mu^2, \tag{17}$$

where

$$A = \alpha^2 - (\Omega_{\mathrm{F}} - \omega)^2\varpi^2 - \mathcal{M}^2 \tag{18}$$

and

$$K = \alpha^2\varpi^2(E - \Omega_{\mathrm{F}}L)^2 \left[\alpha^2 - (\Omega_{\mathrm{F}} - \omega)^2\varpi^2 - 2\mathcal{M}^2\right]$$
$$+ \mathcal{M}^4 \left[\varpi^2(E - \omega L)^2 - \alpha^2 L^2\right]. \tag{19}$$

As for the Grad-Shafranov equation itself, i.e., the equilibrium equation for magnetic field lines, it can be written in the form

$$\frac{1}{\alpha}\nabla_k\left\{\frac{1}{\alpha\varpi^2}[\alpha^2 - (\Omega_{\mathrm{F}} - \omega)^2\varpi^2 - \mathcal{M}^2]\nabla^k\Psi\right\} + \frac{\Omega_{\mathrm{F}} - \omega}{\alpha^2}(\nabla\Psi)^2\frac{\mathrm{d}\Omega_{\mathrm{F}}}{\mathrm{d}\Psi}$$
$$+ \frac{64\pi^4}{\alpha^2\varpi^2}\frac{1}{2\mathcal{M}^2}\frac{\partial}{\partial\Psi}\left(\frac{G}{A}\right) - 16\pi^3\mu n\frac{1}{\eta}\frac{\mathrm{d}\eta}{\mathrm{d}\Psi} - 16\pi^3 nT\frac{\mathrm{d}s}{\mathrm{d}\Psi} = 0, \tag{20}$$

where

$$G = \alpha^2\varpi^2(E - \Omega_{\mathrm{F}}L)^2 + \alpha^2\mathcal{M}^2 L^2 - \mathcal{M}^2\varpi^2(E - \omega L)^2, \tag{21}$$

and the derivative $\partial/\partial\Psi$ acts on the integrals of motion only. Finally, expressing the term $\nabla_k\mathcal{M}^2$ in Eq. (20) according to Eq. (19), we obtain (Beskin & Pariev, 1993)

$$A\left[\frac{1}{\alpha}\nabla_k\left(\frac{1}{\alpha\varpi^2}\nabla^k\Psi\right) + \frac{1}{\alpha^2\varpi^2(\nabla\Psi)^2}\frac{\nabla^i\Psi \cdot \nabla^k\Psi \cdot \nabla_i\nabla_k\Psi}{D}\right]$$
$$+ \frac{\nabla'_k A\nabla^k\Psi}{\alpha^2\varpi^2} - \frac{A}{\alpha^2\varpi^2(\nabla\Psi)^2}\frac{1}{2D}\nabla'_k F\nabla^k\Psi + \frac{\Omega_{\mathrm{F}} - \omega}{\alpha^2}(\nabla\Psi)^2\frac{\mathrm{d}\Omega_{\mathrm{F}}}{\mathrm{d}\Psi} \tag{22}$$
$$+ \frac{64\pi^4}{\alpha^2\varpi^2}\frac{1}{2\mathcal{M}^2}\frac{\partial}{\partial\Psi}\left(\frac{G}{A}\right) - 16\pi^3\mu n\frac{1}{\eta}\frac{\mathrm{d}\eta}{\mathrm{d}\Psi} - 16\pi^3 nT\frac{\mathrm{d}s}{\mathrm{d}\Psi} = 0.$$

Here

$$D = \frac{A}{\mathcal{M}^2} + \frac{\alpha^2}{\mathcal{M}^2}\frac{B_{\hat\varphi}^2}{B_{\mathrm{p}}^2} - \frac{1}{u_{\mathrm{p}}^2}\frac{A}{\mathcal{M}^2}\frac{c_{\mathrm{s}}^2}{1 - c_{\mathrm{s}}^2}, \tag{23}$$

$$F = \frac{64\pi^4}{\mathcal{M}^4}\frac{K}{A^2} - \frac{64\pi^4}{\mathcal{M}^4}\alpha^2\varpi^2\eta^2\mu^2, \tag{24}$$

and the gradient ∇'_k denotes the action of ∇_k under the condition that $\mathcal{M}$ is fixed. Let us stress that in equation (22) the pressure P, the temperature T, the sound velocity c_{s}, and the specific enthalpy μ are to be expressed via an equation of state in terms of the entropy $s(\Psi)$ and the square of the Mach number $\mathcal{M}^2$. In turn, the quantity $\mathcal{M}^2$ is to be considered as a function of $(\nabla\Psi)^2$ and the integrals of motion

$$\mathcal{M}^2 = \mathcal{M}^2\left[(\nabla\Psi)^2, E(\Psi), L(\Psi), \eta(\Psi), \Omega_{\mathrm{F}}(\Psi), s(\Psi)\right]. \tag{25}$$

The latter relation is the implicit form of Eq. (17). The stream equation (22) coupled with definitions (5) – (11) is the desired equation for the poloidal field which contains only the magnetic flux Ψ and the five integrals of motion $\Omega_F(\Psi), \eta(\Psi), s(\Psi), E(\Psi)$, and $L(\Psi)$ depending on it.

Equation (22) is a second-order equation linear with respect to the highest derivatives. It changes its type from elliptical to hyperbolical at singular surfaces where the poloidal velocity of matter becomes equal to either fast or slow magnetosonic velocity ($D = 0$), or with the cusp velocity ($D = -1$). At the Alfvénic surface $A = 0$, the type of equation does not change. Nonetheless, the Alfvénic surface does represent a singular surface of the Grad-Shafranov equation, and the regularity condition must be satisfied there.

3. Examples

3.1 Bondi-Hoyle accretion

As a first example, we consider the hydrodynamic accretion on to a moving back hole (the Bondi-Hoyle accretion), which is one of the classical problems of modern astrophysics (Thorne et al., 1986). First of all, let us formulate the hydrodynamical limit of the Grad-Shafranov equation, where we can neglect the electromagnetic field contribution. In this case, it is convenient to introduce a new potential $\Phi(\Psi)$ satisfying the condition $\eta(\Psi) = \mathrm{d}\Phi/\mathrm{d}\Psi$. Using definition (8) we obtain

$$\alpha n \mathbf{u}_\mathrm{p} = \frac{1}{2\pi\varpi}(\nabla\Phi \times \mathbf{e}_{\hat\varphi}). \tag{26}$$

Surfaces $\Phi(r, \theta) = \text{const}$ define the streamlines of matter.

In the hydrodynamic limit, there are only three integrals of motion. These are the energy flux and the z-component of the angular momentum:

$$E(\Phi) = \mu(\alpha\gamma + \varpi\omega u_{\hat\varphi}); \tag{27}$$

$$L(\Phi) = \mu\varpi u_{\hat\varphi}, \tag{28}$$

as well as the entropy $s = s(\Phi)$. Now the algebraic Bernoulli equation (17) takes a form

$$(E - \omega L)^2 = \alpha^2\mu^2 + \frac{\alpha^2}{\varpi^2}L^2 + \frac{\hat{\mathcal{M}}^4}{64\pi^4\varpi^2}(\nabla\Phi)^2, \tag{29}$$

where the 'Mach number' squared $\hat{\mathcal{M}}^2$ is defined as $\hat{\mathcal{M}}^2 = 4\pi\mu/n$. Then the Grad-Shafranov equation (20) can be rewritten in the form (Beskin & Pariev, 1993)

$$-\frac{1}{\alpha}\nabla_k\left(\frac{\hat{\mathcal{M}}^2}{\alpha\varpi^2}\nabla^k\Phi\right) - 16\pi^3 nT\frac{\mathrm{d}s}{\mathrm{d}\Phi}$$

$$+\frac{64\pi^4}{\alpha^2\varpi^2\hat{\mathcal{M}}^2}\left[\varpi^2(E - \omega L)\left(\frac{\mathrm{d}E}{\mathrm{d}\Phi} - \omega\frac{\mathrm{d}L}{\mathrm{d}\Phi}\right) - \alpha^2 L\frac{\mathrm{d}L}{\mathrm{d}\Phi}\right] = 0, \tag{30}$$

where now

$$D = -1 + \frac{1}{u_{\mathrm{p}}^2} \frac{c_s^2}{1 - c_s^2}. \tag{31}$$

As we see, equation (30) contains only one singular surface, i.e., the sonic surface, determined from the condition $D = 0$.

To construct the solution corresponding to the Bondi-Hoyle accretion, it is possible to seek the solution of the Grad-Shafranov equation for the flux function $\Phi(r, \theta)$ in the form of the small perturbation of the spherically symmetric flow in the reference frame moving with the black hole

$$\Phi(r, \theta) = \Phi_0[1 - \cos\theta + \varepsilon_1 f(r, \theta)]. \tag{32}$$

Here we introduce a small parameter

$$\varepsilon_1 = \frac{v_\infty}{c_\infty}, \tag{33}$$

which defines the ratio of the black hole velocity to the velocity of sound at infinity. For a nonmoving gravity center we return to the spherically symmetric flow.

Since Grad-Shafranov equation (30) contains three invariants, it is necessary to specify four boundary conditions, say

1　$v_{\mathrm{p},\infty} = \mathrm{const}$,

2　$v_\varphi = 0$ (and hence $L = 0$),

3　$s_\infty = \mathrm{const}$,

4　$E_\infty = c_\infty^2/(\Gamma - 1)$.

In the last relation we neglect the terms $\sim \varepsilon_1^2$.

As a result, the Grad-Shafranov equation can be linearized:

$$-\varepsilon_1 \alpha^2 D \frac{\partial^2 f}{\partial r^2} - \frac{\varepsilon_1}{r^2}(D + 1)\sin\theta \frac{\partial}{\partial\theta}\left(\frac{1}{\sin\theta}\frac{\partial f}{\partial\theta}\right) + \varepsilon_1 \alpha^2 N_r \frac{\partial f}{\partial r} = 0, \tag{34}$$

where

$$N_r = \frac{2}{r} - \frac{\mu^2}{E^2 - \alpha^2\mu^2}\frac{M}{r^2}. \tag{35}$$

According to (26) and (31),

$$D + 1 = \frac{\alpha^2\mu^2}{E^2 - \alpha^2\mu^2} \cdot \frac{c_s^2}{1 - c_s^2}. \tag{36}$$

This means the factor α^2 is enters every term of equation (34), and equation (34) has no singularity at the horizon. In particular, this means that it is not necessary to specify any boundary conditions at $r = r_{\mathrm{g}}$. This is not surprising because the horizon corresponds to the supersonic region which cannot affect the subsonic flow.

Since all the terms in (34) contain the small factor ε_1, the functions D, c_{s}, etc. can be taken from the zeroth approximation. As for the spherically symmetric flow, the functions D, c_{s}, etc. do not depend on θ, and the solution of equation (34) can be expanded in a series of eigen functions of the operator $\sin\theta\, \partial/\partial\theta(1/\sin\theta \cdot \partial/\partial\theta)$. Thus, the solution can be presented in the form

$$f(r, \theta) = \sum_{m=0}^{\infty} g_m(r) Q_m(\theta), \tag{37}$$

the equations for the radial functions $g_m(r)$ being

$$r^2 D \frac{\mathrm{d}^2 g_m}{\mathrm{d}r^2} + r^2 N_r \frac{\mathrm{d}g_m}{\mathrm{d}r} + m(m+1)\frac{\mu^2}{E^2 - \alpha^2\mu^2}\frac{c_{\mathrm{s}}^2}{1 - c_{\mathrm{s}}^2} g_m = 0. \tag{38}$$

Here $Q_0 = 1 - \cos\theta$, $Q_1 = \sin^2\theta$, $Q_2 = \sin^2\theta\cos\theta$, $\ldots$ are the eigen functions of the angular operator.

As to the boundary conditions, they can be formulated as follows:

1 No singularity on the sonic surface (where $N_r = 0$, $D = 0$)

$$g_m(r_*) = 0. \tag{39}$$

2 The homogeneous flow $\Phi = \pi n_\infty v_\infty r^2 \sin^2\theta$ at infinity which gives

$$g_1 \to \frac{1}{2}\frac{n_\infty c_\infty}{n_* c_*}\frac{r^2}{r_*^2}, \qquad g_2, g_3, \cdots = 0. \tag{40}$$

The complete solution can be presented in the form

$$\Phi(r, \theta) = \Phi_0[1 - \cos\theta + \varepsilon_1 g_1(r)\sin^2\theta], \tag{41}$$

where the radial function $g_1(r)$ is the solution of the ordinary differential equation (38) for $m = 1$ with the boundary conditions (39) and (40).

We have constructed the analytical solution of the problem, i.e., obtained the full description of the flow structure. For example, the sonic surface now has the nonspherical form

$$r_*(\theta) = r_*\left[1 + \varepsilon_1\left(\frac{\Gamma + 1}{5 - 3\Gamma}\right)k_2\cos\theta\right], \tag{42}$$

where the numerical coefficient $k_2 = r_* g_1'(r_*)$ is expressed through the derivative of the radial function $g_1(r)$ at the sonic point. As shown in Fig. 1, the

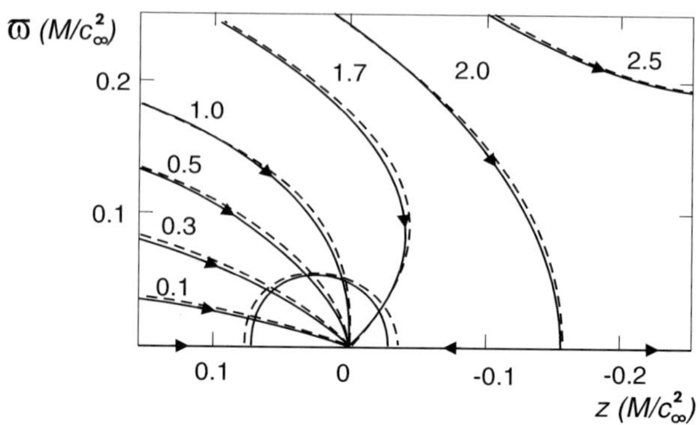

Figure 1.   Bondi-Hoyle accretion on to a moving black hole.

analytical solution fully agrees with the numerical calculations (Hunt, 1979) in spite of the parameter $\varepsilon_1 = 0.6$ here being quite large. Here $\Gamma = 4/3$, and the numbers alongside the curves denote values Φ/Φ_0; the dashed lines show the streamlines and the sonic surface form obtained numerically.

As can be easily seen, outside the capture radius $R_c \approx \varepsilon_1^{-1/2} r_*$ our main assumption, i.e., the smallness of the deviation from the spherically symmetric flow, is not valid. Nevertheless, the obtained solution remains correct. This remarkable property is due to the Grad-Shafranov equation becoming linear for constant concentration n. But as we learn from the spherically symmetric Bondi accretion, at large distances $r \gg r_*$ from the sonic surface the density of the accreting matter is virtually constant. Accordingly, the concentration is constant for the homogeneous flow as well. As a result, under the condition $R_c \gg r_*$, which holds for $\varepsilon_1 \ll 1$, near and beyond the capture radius (where the perturbation $\sim \varepsilon_1 g_1(r)$ becomes comparable to unity) equation (30) becomes linear. Therefore, the sum of the two solutions, the homogeneous and the spherically symmetric ones, is also a solution.

3.2   Thin transonic disk

As a next example, we consider the internal two-dimensional structure of a thin accretion disk. Here, for simplicity we consider the case of a non-spinning (Schwarzschild) black hole (Beskin et al., 2002). We recall that according to the standard disk model (Shakura, 1973; Shakura & Sunyaev, 1973; Novikov & Thorne, 1973) the accreting matter forms an equilibrium disk rotating around the gravitational center with the Keplerian velocity $v_K(r) = (GM/r)^{1/2}$. The disk will be thin provided that its temperature is sufficiently small ($c_s \ll v_K$) since the vertical balance of the gravity force and the pressure gradient implies that

$$H \approx r \frac{c_{\mathrm{s}}}{v_{\mathrm{K}}}. \tag{43}$$

The General Relativity effects result in two important properties: the absence of stable circular orbits for $r < r_{\mathrm{ms}}$, where for a non-spinning black hole $r_{\mathrm{ms}} = 3r_{\mathrm{g}}$, and the transonic regime of accretion. The first point means that the accreting matter passing the marginally stable orbit $r = r_{\mathrm{ms}}$ (MSO) approaches the black hole horizon sufficiently fast, namely, in the dynamical time $\tau_{\mathrm{d}} \sim [v_r(r_0)/c]^{-1/3} r_{\mathrm{g}}/c$. It is important that such a flow is realized in the absence of viscosity. The second statement results from the fact that up to the marginally stable orbit the flow is subsonic while at the horizon the flow is to be supersonic.

We consider thin disk accretion on to a black hole in the region where there are no stable circular orbits. As argued above, the contribution of viscosity should no longer be significant here (Beskin et al., 2002). Hence we may assume that the ideal hydrodynamics approach is suitable well enough for describing the flow structure in this inner area of the accretion disk.

For the sake of simplicity we adopt the polytropic equation of state $P = k(s)n^{\Gamma}$ so that temperature and sound velocity can be written as (Shapiro & Teukolsky, 1983)

$$T = k(s)n^{\Gamma-1}; \qquad c_{\mathrm{s}}^2 = \frac{\Gamma}{\mu} k(s)n^{\Gamma-1}. \tag{44}$$

Since the disk is thin, we have $c_{\mathrm{s}} \ll 1$ (see (43)). Therefore we can write $\mu = m_{\mathrm{p}} + m_{\mathrm{p}}W$, where $W = c_{\mathrm{s}}^2/(\Gamma - 1)$ is non-relativistic enthalpy and m_{p} is particle mass. For an ideal gas with $\Gamma = \mathrm{const}$ the function $k(s)$ can be shown to have a quite definite form

$$k(s) = k_0 \exp\left[(\Gamma - 1)s\right], \tag{45}$$

which for the case of the polytropic equation of state can be obtained from (44) and the thermodynamical relationship $\mathrm{d}P = n\mathrm{d}\mu - nT\mathrm{d}s$.

First, we study the **subsonic flow**. We show that the role of the dynamical terms becomes dominant with approach to the sonic surface $r = r_*(\theta)$. The problem of passing through the sonic surface (see page 69) and the supersonic flow structure (see page 71) will be consider later. For the time being, we limit our discussion to the subsonic region only, where the poloidal velocity $u_{\hat{\mathrm{p}}}$ is assumed to be much smaller than the sound velocity c_{s}. This assumption comes from the relation

$$\frac{v_r}{v_{\mathrm{K}}} \approx \alpha_{\mathrm{ss}} \frac{c_{\mathrm{s}}^2}{v_{\mathrm{K}}^2} \tag{46}$$

for the radial velocity in the accretion disk (Shapiro & Teukolsky, 1983). It is necessary to stress that in the vicinity of the MSO this estimation is apparently inapplicable. However, since the presence of a small parameter allows us to investigate the flow structure analytically, below we consider the ratio $u_{\hat{p}}/c_s$ to be small near the MSO. This makes the effect discussed more visible and at the same time allows us to obtain the results that hold for larger values of this parameter.

The presence of the small parameter allows us to simplify Grad-Shafranov hydrodynamic equation (30), in which the term $\nabla_k \hat{\mathbf{M}}^2$ has been expanded in analogy with equation (22), in the limit $u_{\hat{p}}^2 \ll c_s^2$ by neglecting the terms proportional to D^{-1} (all of them are of the order of $u_{\hat{p}}^2/c_s^2$). The resulting equation describes the subsonic flow and is elliptical. Hence, it contains no critical surfaces. Therefore, our problem requires five boundary conditions: three conditions determine the integrals of motion and the two remaining ones are the boundary conditions for the second-order Grad-Shafranov equation. Finally, to close the system, we need to supply the Grad-Shafranov equation with the relativistic Bernoulli equation which can be written as

$$u_{\hat{p}}^2 = \frac{E^2 - \alpha^2 L^2/\varpi^2 - \alpha^2 \mu^2}{\alpha^2 \mu^2}. \tag{47}$$

We assume that the α-disk theory holds outside the MSO. We adopt the flow velocity components, which this theory yields on the MSO $r = r_{\mathrm{ms}}$,[1] as the first three boundary conditions for our problem. For the sake of simplicity we consider the radial four-velocity, which is responsible for the inflow, to be constant at the surface $r = r_{\mathrm{ms}}$ and equal to u_0 and the toroidal velocity to be exactly equal to that of a free particle revolving at $r = r_{\mathrm{ms}}$:[2]

$$u_{\hat{r}}(r_{\mathrm{ms}}, \Theta) = -u_0, \tag{48}$$

$$u_{\hat{\Theta}}(r_{\mathrm{ms}}, \Theta) = \Theta u_0, \tag{49}$$

$$u_{\hat{\varphi}}(r_{\mathrm{ms}}, \Theta) = 1/\sqrt{3}, \tag{50}$$

where the new angular variable $\Theta = \pi/2 - \theta$ is counted off from the equator in the vertical direction. For the sake of convenience, we also introduce another angular variable $\Theta_{\mathrm{ms}} = \Theta_{\mathrm{ms}}(\Phi(r, \Theta))$.[3] This is a function of streamlines, it gives the Θ-coordinate of a point where a streamline pierces the surface of the MSO $r = r_{\mathrm{ms}}$. In other words, the point (r, Θ) and the point $(r_{\mathrm{ms}}, \Theta_{\mathrm{ms}})$ belong to the same streamline. Further, noting that $\Theta_{\mathrm{ms}}(r_{\mathrm{ms}}, \Theta) \equiv \Theta$ and using (26) and (48), we arrive at

$$d\Phi = 2\pi \alpha_0 r_{\mathrm{ms}}^2 n(r_{\mathrm{ms}}, \Theta_{\mathrm{ms}}) u_0 \cos \Theta_{\mathrm{ms}} d\Theta_{\mathrm{ms}}. \tag{51}$$

---

[1] A nearly parallel inflow with a small radial velocity (43).
[2] For a free particle revolving at $r = r_{\mathrm{ms}}$ around a non-spinning BH we have (Landau & Lifshits, 1987a) $u_{\hat{\varphi}}(r_{\mathrm{ms}}) = 1/\sqrt{3}, \alpha_0 = \alpha(r_{\mathrm{ms}}) = \sqrt{2/3}, \gamma_0 = \gamma(r_{\mathrm{ms}}) = \sqrt{4/3}$.
[3] A Lagrange coordinate of streamlines.

Next, we assume that the sound velocity is constant at the surface $r = r_{\rm ms}$; this yields the fourth boundary condition

$$c_{\rm s}(r_{\rm ms}, \theta) = c_0. \tag{52}$$

In the case of polytropic equation of state this means that both the temperature $T_0 = T(r_{\rm ms}, \Theta)$ and the relativistic enthalpy $\mu_0 = \mu(r_{\rm ms}, \Theta)$ are also constant at the surface $r = r_{\rm ms}$. Therefore boundary conditions (48) – (50) and (52) directly determine the invariants $E(\Phi)$ and $L(\Phi)$ (see (27), (28)),

$$E(\Phi) = \mu_0 e_0, \tag{53}$$
$$L(\Phi) = \mu_0 l_0 \cos \Theta_{\rm ms}, \tag{54}$$

where $e_0 = \alpha_0 \gamma_0 = \sqrt{8/9}$ and $l_0 = u_{\hat\varphi}(r_{\rm ms}) r_{\rm ms} = \sqrt{3} r_{\rm g}$ (see footnote 2 on page 66). Condition $E = {\rm const}$ (53) allows us to rewrite the Grad-Shafranov equation in an even simpler form,

$$\frac{\partial^2 \Phi}{\partial r^2} + \frac{\cos \Theta}{\alpha^2 r^2} \frac{\partial}{\partial \Theta} \left(\frac{1}{\cos \Theta} \frac{\partial \Phi}{\partial \Theta} \right) = -4\pi^2 n^2 \frac{L}{\mu^2} \frac{{\rm d}L}{{\rm d}\Phi} - 4\pi^2 n^2 r^2 \cos^2 \Theta \frac{T}{\mu} \frac{{\rm d}s}{{\rm d}\Phi}. \tag{55}$$

At $r = r_{\rm ms}$ the r.h.s. of equation (55) can be shown to describe the transversal balance of the pressure force and the effective potential, whereas the l.h.s. corresponds to the dynamical term $(\mathbf{v}\nabla)\mathbf{v}$ which is u_0^2/c_0^2 times smaller than each of the terms in the r.h.s. and thus may be dropped (Beskin et al., 2002). It is natural therefore to choose the entropy $s(\Phi)$ from the condition of the transversal balance of forces at the surface $r = r_{\rm ms}$,

$$r_{\rm ms}^2 \cos^2 \Theta_{\rm ms} \frac{{\rm d}s}{{\rm d}\Theta_{\rm ms}} = -\frac{\Gamma}{c_0^2} \frac{L}{\mu_0^2} \frac{{\rm d}L}{{\rm d}\Theta_{\rm ms}}, \tag{56}$$

where the value of $L(\Theta_{\rm ms})$ is determined by (54). This yields the last, fifth, boundary condition

$$s(\Theta_{\rm ms}) = s(0) - \frac{\Gamma}{3c_0^2} \ln(\cos \Theta_{\rm ms}), \tag{57}$$

whence owing to (45) we have at $r = r_{\rm ms}$ the standard shape of concentration,

$$n(r_{\rm ms}, \Theta) \approx n_0 \exp\left(-\frac{\Gamma}{6c_0^2} \Theta^2 \right) \tag{58}$$

(to be exact, $n(r_{\rm ms}, \Theta) = n_0 (\cos \Theta)^{\Gamma/3c_0^2}$).

For $c_{\rm s} \ll 1$, i.e., for non-relativistic temperature, we obtain from (47) and (44),

$$u_{\rm p}^2 = u_0^2 + w^2 + \frac{2}{\Gamma - 1} \left(c_0^2 - c_{\rm s}^2 \right) + \frac{1}{3} \left(\Theta_{\rm ms}^2 - \Theta^2 \right) + \ldots \tag{59}$$

The quantity

$$w^2(r) = \frac{e_0^2 - \alpha^2 l_0^2/r^2 - \alpha^2}{\alpha^2} \tag{60}$$

is the poloidal four-velocity of a free particle having zero poloidal velocity at the MSO. Assuming $u_{\hat{\mathrm{p}}} = c_{\mathrm{s}} = c_*$ and neglecting w^2, we find the velocity of sound c_* in the sonic point $r = r_*$, $\Theta = 0$: $c_* \approx 2c_0/\sqrt{\Gamma + 1}$. Since entropy s remains constant along the lines of flow, the gas concentration n_* at the sonic surface slightly differs[4] from the gas concentration at the MSO. In other words, the subsonic flow can be considered incompressible to a zeroth approximation. The same is true for spherically symmetric Bondi accretion (Shapiro & Teukolsky, 1983). It is important that this conclusion holds not only in the equatorial plane because the additional term $1/3(\Theta_{\mathrm{ms}}^2 - \Theta^2)$ in (59) is also of the order of c_0^2 for the range of angles corresponding to the representative disk thickness, $\Theta \lesssim \Theta_{\mathrm{disk}} \sim c_0$ (see (58)). Since the density remains almost constant and the poloidal velocity increases from u_0 to $c_* \sim c_0$, i.e., changes over several orders of magnitude for $u_0^2 \ll c_0^2$, the disk thickness H should change in the same proportion due to the continuity equation:

$$H(r_*) \approx \frac{u_0}{c_0} H(r_{\mathrm{ms}}). \tag{61}$$

Hence the purely radial motion approximation proves to be inapplicable in the vicinity of the sonic surface, and it is extremely important that both components of the dynamical force become comparable with the pressure gradient near the sonic surface,

$$\frac{u_{\hat{\Theta}}}{r} \frac{\partial u_{\hat{\Theta}}}{\partial \Theta} \approx u_{\hat{r}} \frac{\partial u_{\hat{\Theta}}}{\partial r} \approx \frac{\nabla_{\hat{\Theta}} P}{\mu} \approx \frac{c_0^2}{u_0^2} \frac{\Theta}{r}, \tag{62}$$

which follows from the analysis of the asymptotic form of (55) (Beskin et al., 2002). We can also estimate the radial logarithmic derivative (Beskin et al., 2002)

$$\eta_1 = (r/n)\,(\partial n/\partial r) \sim u_0^{-1}. \tag{63}$$

The numerical results are shown in Fig. 2. As one can see, in the subsonic region, $r_* < r \le r_{\mathrm{ms}} \equiv 3r_{\mathrm{g}}$, the disk thickness rapidly diminishes, so that it is impossible to neglect the dynamical force there. We stress that taking the dynamical force into account is indeed extremely important. This is because, unlike zero-order standard disk thickness prescription (43), the Grad-Shafranov equation has second order derivatives, i.e., it contains two additional degrees of freedom. This means that the critical condition only fixes one of these degrees

---

[4]Note that the concentration, certainly, changes from one streamline to another.

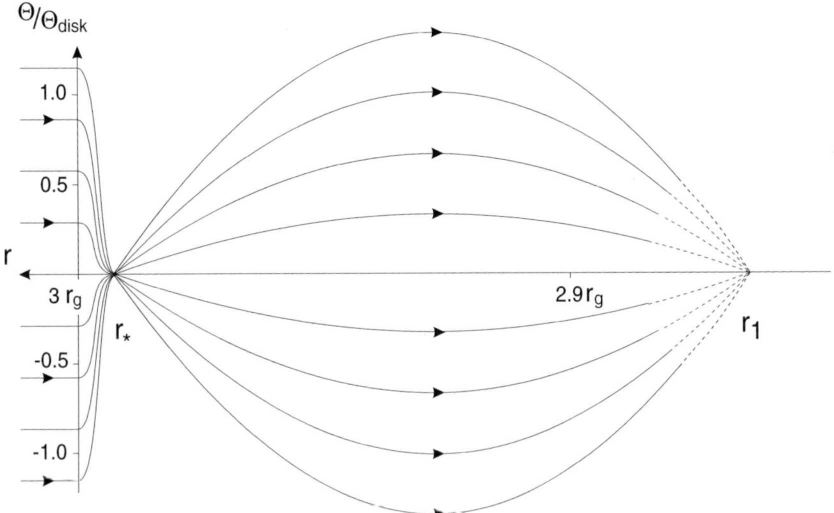

Figure 2. The structure of a thin accretion disk (actual scale) for $c_0 = 10^{-2}$, $u_0 = 10^{-5}$ after passing the MSO $r = 3r_{\mathrm{g}}$ ($a = 0$, Schwarzschild case). As sufficient dissipation can take place in the vicinity of the first node $r = r_1$, we do not prolong the flow lines to the region $r < r_1$. Here Θ_{disk} is the characteristic disk angular thickness which, according to (58), can be roughly estimated as c_0.

of freedom (e.g. imposes some limitations on the form of the flow) rather than determines the angular momentum of the accreting matter.

In order to verify our conclusions we consider the flow structure in the vicinity of the **sonic surface** in more detail. Since the smooth transonic flow is analytical at the singular point $r = r_*$, $\Theta = 0$ (Landau & Lifshits, 1987b), it is possible to write

$$n = n_* \left(1 + \eta_1 h + \frac{1}{2}\eta_3\Theta^2 + \ldots \right), \tag{64}$$

$$\Theta_{\mathrm{ms}} = a_0 \left(\Theta + a_1 h\Theta + \frac{1}{2}a_2 h^2\Theta + \frac{1}{6}b_0\Theta^3 + \ldots \right), \tag{65}$$

where $h = (r - r_*)/r_*$. Here we assume that all the three invariants E, L, and s are given by boundary conditions (53), (54), and (57) respectively. Hence, the problem needs only one more boundary condition.

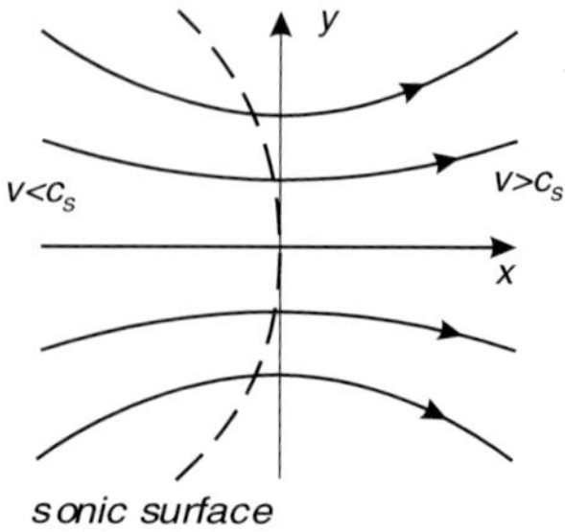

Figure 3.   Schematics of thin disk streamlines profile around the sonic point. The flow has the form of the standard nozzle. Here $x = -h$, and $y = \Theta$.

Now comparing the appropriate coefficients in Bernoulli equation (47) and full stream equation (30) we obtain neglecting the terms of the order of u_0^2/c_0^2:

$$a_0 = \left(\frac{2}{\Gamma+1}\right)^{(\Gamma+1)/2(\Gamma-1)} \frac{c_0}{u_0}, \tag{66}$$

$$a_1 = 2 + \frac{1-\alpha_*^2}{2\alpha_*^2} \approx 2.25, \tag{67}$$

$$a_2 = -(\Gamma+1)\eta_1^2, \tag{68}$$

$$b_0 = \left(\frac{\Gamma+1}{6}\right) \frac{a_0^2}{c_0^2}, \tag{69}$$

$$\eta_3 = -\frac{2}{3}(\Gamma+1)\eta_1^2 - \left(\frac{\Gamma-1}{3}\right) \frac{a_0^2}{c_0^2}, \tag{70}$$

where $\alpha_*^2 = \alpha^2(r_*) \approx 2/3$ and coefficients (66) – (70) are expressed through logarithmic derivative η_1 (63). They have clear physical meaning. So, a_0 gives the compression of streamlines, $a_0 = H(r_{\rm ms})/H(r_*)$. In agreement with (61) we have $a_0 \approx c_0/u_0$. Further, a_1 corresponds to the slope of the streamlines with respect to the equatorial plane. As $a_1 > 0$, the compression of streamlines finishes somewhere before the sonic surface, so inside the sonic radius $r < r_*$ the streamlines diverge. On the other hand, because $a_1 \ll u_0^{-1}$, for $r = r_*$ the divergency is still very weak. Hence, in the vicinity of the sonic surface the flow has a form of a standard nozzle (see Fig. 3). Finally, since $a_2 \sim \eta_3 \sim b_0 \sim u_0^{-2}$, we can conclude that the transverse scale of the transonic region $H(r_*)$ is the same as the longitudinal one. This means that the transonic region is essentially two-dimensional (a well-known fact for a nozzle, Landau & Lifshits, 1987b), and it is impossible to analyze it within the standard one-dimensional approximation.

Let us stress that it is rather difficult to connect the sonic characteristics $\eta_1 = \eta_1(r_*)$ with the physical boundary conditions on the MSO $r = r_{\mathrm{ms}}$ (for this it is necessary to know all the expansion coefficients in (64) and (65)). In particular, it is impossible to formulate the restriction on five boundary conditions (48) – (50), (52), and (57) resulting from the critical condition on the sonic surface. Nevertheless, the estimate $\eta_1 \approx u_0^{-1}$ (63) makes sure that we know the parameter η_1 to a high enough accuracy. Then, according to (66) – (70), all the other coefficients can be determined exactly.

Using expansions (64) and (65), one can obtain all physical parameters of the transonic flow. In particular, we have

$$u_{\hat{\mathrm{p}}}^2 = c_*^2 \left[1 - 2\eta_1 h + \frac{1}{6}(\Gamma - 1)\frac{a_0^2}{c_0^2}\Theta^2 + \frac{2}{3}(\Gamma + 1)\eta_1^2\Theta^2 \right], \tag{71}$$

$$c_{\mathrm{s}}^2 = c_*^2 \left[1 + (\Gamma - 1)\,\eta_1 h + \frac{1}{6}(\Gamma - 1)\frac{a_0^2}{c_0^2}\Theta^2 - \frac{1}{3}(\Gamma - 1)(\Gamma + 1)\eta_1^2\Theta^2 \right].$$

These equations yield the shape of the sonic surface, $u_{\hat{\mathrm{p}}} = c_{\mathrm{s}}$; it has the standard parabolic form

$$h = \frac{(\Gamma + 1)}{3}\eta_1\Theta^2. \tag{72}$$

Since the transonic flow in the form of a nozzle (see Fig. 3) has longitudinal and transversal scales of one order of magnitude (Landau & Lifshits, 1987b), near the sonic surface we have $\delta r_{\parallel} \approx \delta r_{\perp}$, i.e., $\delta r_{\parallel} \approx H(r_*)$. Hence for thin disks (i.e., for $c_0 \ll 1$) this longitudinal scale is always much smaller than the distance from the BH, $\delta r_{\parallel}/r_* \approx H(r_*)/r_* \ll 1$. Only by taking the transversal velocity into account do we retain the small longitudinal scale $\delta r_{\parallel} \ll r_{\mathrm{g}}$. This scale is left out during the standard one-dimensional approach.

Let us now turn to the **supersonic flow**. Since the pressure gradient becomes insignificant in the supersonic region, the matter moves here along the trajectories of free particles. Neglecting the $\nabla_\theta P$ term in the θ-component of relativistic Euler equation (Frolov & Novikov, 1998), we have (compare with Abramowicz et al., 1997).

$$\alpha u_{\hat{r}}\frac{\partial(r u_{\hat{\Theta}})}{\partial r} + \frac{(r u_{\hat{\Theta}})}{r^2}\frac{\partial(r u_{\hat{\Theta}})}{\partial\Theta} + (u_{\hat{\varphi}})^2 \tan\Theta = 0. \tag{73}$$

Using (54), toroidal four-velocity $u_{\hat{\varphi}}$ can be easily expressed in terms of radius: $u_{\hat{\varphi}} = \sqrt{3}/x$, where $x = r/r_{\mathrm{g}}$. We also introduce dimensionless functions $f(x)$ and $g(x)$:

$$\Theta f(x) = x u_{\hat{\Theta}}, \tag{74}$$

$$g(x) = -\alpha u_{\hat{r}} > 0. \tag{75}$$

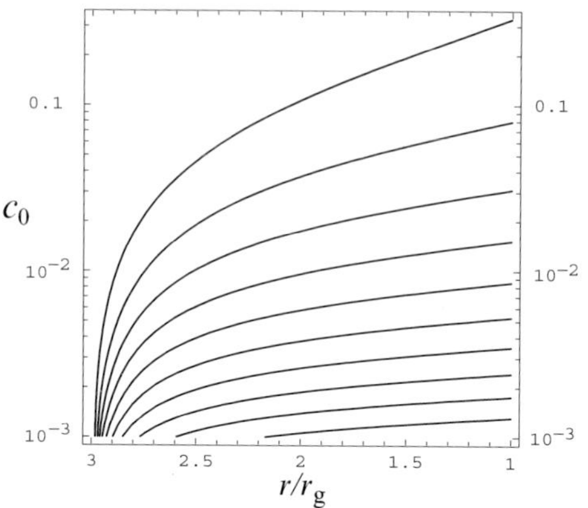

Figure 4.   The positions of nodes for a range of initial sound velocities (for $\Gamma = 4/3$). Each curve relates the radial position of a node to a value of the initial sound velocity. Intersection points of these curves with the line $c_0 = $ const give the the nodes' radial positions for this particular value of c_0.

Using now (73) and the definitions above, we obtain an ordinary differential equation for $f(x)$

$$\frac{\mathrm{d}f}{\mathrm{d}x} = \frac{f^2 + 3}{x^2 g(x)}. \tag{76}$$

Integrating (76), we obtain

$$f(x) = \sqrt{3} \tan\left[\sqrt{3} \int_{x_*}^{x} \frac{d\xi}{\xi^2 g(\xi)} + \frac{\pi}{2}\right], \tag{77}$$

where the factor $\pi/2$ has been substituted for the integration constant $\arctan[f(x_*)/\sqrt{3}]$.[5] Finally, to determine function $g(x)$, one can see that $u_{\hat{\mathrm{p}}}^2 \to w^2$ as $r \to r_{\rm g}$. On the other hand, we have $u_{\hat{\mathrm{p}}} \approx c_* \approx c_0$ for $r \lesssim r_*$. Therefore, the following approximation should be valid throughout the $r_{\rm g} < r < r_*$ region: $g(x) \approx \sqrt{(\alpha w)^2 + (\alpha c_*)^2}$, where, owing to (60), $\alpha^2 w^2(r) = (3 - x)^3/(9x^3)$.

The results of numerical calculations are presented in Fig. 2. In the supersonic region the flow performs transversal oscillations about the equatorial plane, their frequency independent of their amplitude. We see as well that the maximum thickness of the disc in the supersonic (and, hence, ballistic)

---

[5]For r just below r_*, the function f should be positive to reflect the fact that the flow diverges. Then, $f = 0$ corresponds to the point where the divergency finishes, and the flow starts to converge.

region, which is controlled by the transverse component of the gravitational force, actually coincides with the disc thickness within the stable orbits region, $r > r_{\mathrm{ms}}$, where standard estimate (43) is correct.

Once diverged, the flow converges once again at a 'nodal' point closer to the BH. The radial positions the nodes, $r_n \equiv x_n r_{\mathrm{g}}$, are given by the implicit formula $f(x_n) = \pm\infty$, i.e.,

$$\sqrt{3} \int_{x_n}^{x_*} \frac{d\xi}{\xi^2 g(\xi)} = n\pi, \tag{78}$$

where n is the node number and the node with $n = 0$ corresponds to the sonic surface. Interestingly, u_0 does not enter equation (78) in our approximation so that the positions of the nodes are entirely determined by c_0 as long as u_0 is sufficiently small, $u_0^2 \ll c_0^2$. Fig. 4 shows the positions of nodes for different values of c_0.

4.   Conclusion

In some simple cases the Grad-Shafranov equation allows us to construct the exact solutions to problems. In particular, this approach is very useful in studying the analytical properties of transonic flows and in determining the required number of boundary conditions. On the other hand, in the general case no consistent procedure exists regarding the construction of the solution within the Grad-Shafranov approach. The point is that the location of singular surfaces, at which critical conditions should be formulated, is not known beforehand and itself must be found from the solution to the problem. Moreover, it is impossible to generalize this approach to the case of nonideal, non-axially symmetric and nonsteady flows. So it is not surprising that most investigators, who are in the first place interested in astrophysical applications, have focused on a totally different class of equations, namely on time relaxation problems which can only be solved numerically (Hawley et al., 1984; Petrich et al., 1989; Ustyugova et al., 1995; Koide et al., 2000). Nevertheless, it is clear that the key physical results obtained using the Grad-Shafranov approach are independent of the computing method. For this reason one can hope that the results presented above can be useful for everyone working in this field.

Acknowledgments

This work is supported by Russian Foundation for Basic Research (grant N 02–02–16762), Dynasty fund, and ICFPM.

References

Abramowicz, M.A., A. Lanza, and M.J. Percival, 1997, *Astrophys. J.*, **479**, 179.

Beskin, V.S. and V.I. Pariev: 1993, *Physics Uspekhi*, **36**, 529.

Beskin, V.S., R.Yu. Kompaneetz, and A.D. Tchekhovskoy: 2002, *Astronomy Letters*, **28**, 543.

Frolov, V.P. and I.D. Novikov: *Black Hole Physics*, Kluwer Academic Publishers, Dordrecht, 1998.

Grad, H.: 1960, *Rev. Mod. Phys.*, **32**, 830.

Guderley, K.G.: *The Theory of Transonic Flows*, Pergamon, Oxford, 1962.

Heinemann, M. and S. Olbert: 1978, *J. Geophys. Res.*, **83**, 2457.

Hunt, R.: 1979, *Mon. Not. R. Astron. Soc.*, **198**, 83.

Hawley, J.F., L.L. Smarr, and J.R. Wilson: 1984, *Astrophys. J.*, **277**, 296.

Koide, S., D.L. Meier, K. Shibata, and T. Kudo: 2000, *Astrophys. J.*, **536**, 668.

Landau, L.D. and E.M. Lifshits: *The Classical Theory of Fields*, 4th edn. Butterworth-Heinemann, 1987.

Landau, L.D. and E.M. Lifshits: *Fluid mechanics*, 2nd edn. Butterworth-Heinemann, 1987.

Novikov, I.D. and K.S. Thorne, *Astrophysics of black holes*, in Black Holes, eds. C. DeWitt, B. DeWitt, Gordon and Breach, New York, 1973, p. 343.

Okamoto, I.: 1975, *Mon. Not. R. Astron. Soc.*, **173**, 357.

Paczyński, B. and G.S. Bisnovatyi-Kogan: 1981, *Acta Astron.*, **31**, 283.

Petrich, L.I., S. Shapiro, R.F. Stark, and S. Teukolsky: 1989, *Astrophys. J.*, **336**, 313.

Shafranov, V.D.: 1958, *Sov. Phys. JETP*, **6**, 545.

Soloviev, L.S.: *Review of Plasma Physics*, **3**, Consultants Bureau, ed. M.A. Leontovich, 1967.

Shakura, N.I.: 1973, *Sov. Astron.*, **16**, 756.

Shakura, N.I. and R.A. Sunyaev: 1973, *Astron. Astrophys.*, **24**, 337.

Shapiro, S.L. and S.A. Teukolsky: *Black Holes, White Dwarfs, and Neutron Stars*, A Wiley–Interscience Publication, New York, 1983.

Thorne, K.S., R.H. Price, and D.A. Macdonald: *Black Holes. The Membrane Paradigm*, Yale University Press, New Haven, 1986.

Ustyugova, G.V., A.V. Koldoba, M.M. Romanova, V.M. Chechetkin, and R. Lovelace: 1995, *Astrophys. J.*, **439**, L39.

von Mises, R.: *Mathematical Theory of Compressible Fluid Flow*, Academic, New York, 1958.

DISC FORMATION IN BINARY BE STARS

D. V. Bisikalo,[1] A. A. Boyarchuk,[1] P. Harmanec,[2,3] P. V. Kaigorodov,[1] and O. A. Kuznetsov[1,4]

[1]*Institute of Astronomy*
Russian Academy of Sciences
Moscow, Russia
bisikalo@inasan.rssi.ru

[2]*Astronomical Institute of the Charles University,*
Praha, Czech Republic

[3]*Astronomical Institute*
Academy of Sciences of the Czech Republic
Ondřejov, Czech Republic

[4]*Keldysh Institute of Applied Mathematics*
Moscow, Russia

Abstract   In 2002 we suggested a new hypothesis of the formation of Be envelopes in binaries, via an outflow from a rapidly rotating B star in a detached binary. Gas-dynamical simulations of flow structure in binary with parameters of typical Be star has proven the existence of such a mechanism of the formation of a disc-like envelope in binary Be stars. It is shown that an outflow of matter from the vicinity of 'asynchronous' inner Lagrangian point L_1^{rot} of a rapidly rotating B star leads to the formation of a disk-like envelope but not to any significant mass exchange between the binary components. Here, we confirm and extend our conclusions via another numerical simulation, based on a more realistic treatment of the temperature of the outflowing envelope and discuss briefly the potential of our model to explain the observed properties of Be stars.

Keywords: accretion: hydrodynamics, accretion disks

1.   Introduction

Be stars exhibit time-variable emissions in their Balmer (and some other) line profiles. Their continuum and line spectra vary on several time scales, ranging from minutes to decades or more. Actually, the longest time scale of

Alexei M. Fridman, et al. (eds.), Astrophysical Disks; Collective and Stochastic Phenomena 75–96

their variability is not known and one cannot even exclude the possibility that any B star may sometimes become a Be star. In spite of all the effort of several generations of stellar astronomers, neither the very nature of the Be phenomenon nor the physical causes of variations are well understood at present.

There is a general agreement on that the emission lines are formed in extended gaseous envelopes around these stars which have dimensions of one or even two orders of magnitude larger than the stars themselves and re-radiate the stellar radiation to all directions. Evidence has been gradually built to confirm the original Struve's (Struve, 1931) suggestion that the Be envelopes have the form of rotationally flattened disks. An ultimate proof of this statement came from spectro-interferometric observations of several Be stars which allowed a partial spatial resolution of their envelopes – see Quirrenbach et al. (1993; 1997).

Concerning the very origin of the Be envelopes, the number of the hypotheses has been growing steadily and there is no general agreement on a single one. The most promising class of models are models which assume that the Be phenomenon is somehow related to the duplicity of the Be stars.

Kříž & Harmanec (1975) and Harmanec & Kříž (1976) formulated a general hypothesis of the binary nature of Be stars, explaining the Be stars envelopes as accretion disks created from gas matter flowing to the Be stars via a Roche-lobe overflow from their unrecognized binary companions. They pointed out that in the order of an increasing orbital period one observes either a typical Algol interacting binary or a Be binary or a symbiotic star. Since a given B star occupies less and less space when placed into a binary system with longer and longer orbital period, the space available for the formation of an extended accretion disk also gets larger with the increasing orbital period. In systems with shorter orbital periods, the Roche-lobe filling and less massive secondaries are usually much less luminous than their B-type primaries. Only for very long orbital periods, the absolute dimensions of the Roche lobe around the secondary become so large that even a very cool secondary has an optical luminosity comparable to, or even larger than the B star and one observes a symbiotic binary. Kříž & Harmanec (1975) were also able to explain rapid rotation of Be stars as a consequence of tangential accretion from the disk which brings some extra angular momentum. Furthermore, they explained at least some of the observed types of variations of Be stars, for instance the phase-locked V/R, RV, line-width and luminosity changes or long-term E/C and V/R variations and offered some ideas how rapid changes could be related to corotating structures in the accretion disks, caused by resonances.

While their hypothesis certainly represents a serious attempt to address the Be phenomenon in its complexity and very probably explains indeed the nature of some of the actually observed Be binaries (for example AX Mon, RX Cas, SX Cas, KX And, V360 Lac, β Lyr etc. – see Harmanec (2001) for a catalogue of known emission-line binaries), it is now well proven that it cannot be accepted as a universal explanation of the origin of Be envelopes. Already Plavec (1976) pointed out that if all Be stars were binaries with a Roche-lobe filling secondaries, one should observe more eclipsing binaries among them than what is actually observed. While Harmanec (1987) slightly weakened this objection, there is a stronger one: Detailed studies of several known Be binaries (φ Per = HD 10516: Poeckert, 1981, Gies et al., 1998; V839 Her = 4 Her: Koubský et al., 1997, for instance) clearly demonstrated that the secondaries in those binaries are not Roche-lobe filling objects but very small stars. The same is also true about binaries composed of a Be star and a compact, X-ray companion. Harmanec (1985) came with a provocative suggestion that even in massive X-ray binaries the mass is flowing from the X-ray star towards the Be primary and presented some observational facts to support such a view. Recalling an earlier suggestion by Kříž (1982), he also argued that the contraction of the originally mass-losing star had to lead to its rotational instability near equator, leading to another phase of mass transfer from this star to its (now more massive) counterpart. He called it a case PB of the mass transfer and mentioned φ Per as a system being possibly in such a mass transfer stage. It is obvious, however, that unless somebody gets an idea how to excite X-ray emission from compact stars without allowing them to accrete mass from their optical companions, Harmanec's (1985) idea is not tenable.

Being obviously unaware of Kříž's (1982) and Harmanec's (1985) studies, Pols et al. (1991) also investigated the possibility of formation of Be stars as products of case B mass exchange in binaries. Their approach was different, however. They accepted the idea of Kříž & Harmanec (1975) that some Be stars are case B mass-exchanging binaries but argued that the majority of Be stars are remnants of case B mass exchange in intermediate-mass close binaries after the termination of mass transfer. In other words, they postulated that the Be phenomenon occurs due to some, still unknown physical mechanism which is only operational in rapidly rotating stars. The role of the mass exchange in their hypothesis is to rejuvenate and spin-up the original secondaries in binaries. They argued that Be stars in mass-exchanging binaries represent only a small fraction of all Be stars. Estimating the lifetimes of different evolutionary stages, they concluded that more than 80% of post mass-transfer Be stars should have a helium-star companion and that there should be 10 times more Be stars with a white-dwarf companion than those with a neutron-star secondary (observable than as an X-ray source). They predicted that many new helium-star and white-dwarf companions should be detectable in the XUV spectral region.

The role of duplicity was critically examined by Baade (1992). Using high-S/N IR spectra near 880 nm, he carried out a search for late-type companions of 35 southern Be stars with a completely negative result. He also expressed some doubts about the existence of many binaries with hot compact companions and his conclusion was that the cause of the Be phenomenon cannot be related to their duplicity.

For the following reasons we believe, however, that the role of duplicity was not still investigated well enough and that Baade's view cannot be accepted as the final word:

1 Perhaps most importantly, the number of known binaries among emission-line stars has been growing steadily – see, e.g., Gies (2000) or Harmanec (2001). As pointed out by Harmanec (2001), duplicity of Be stars can actually serve two different roles: (i) to explain the formation of Be envelopes via some kind of binary interaction, and (ii) to explain some of the variability patterns of Be stars.

2 Studies by Slettebak (1987) or Harmanec (2000) demostrated that Be stars are observed among stars of clearly different evolutionary ages. This may indicate that the cause of the Be phenomenon has to be sought in some external mechanism, not primarily in a physical mechanism related to the stars themselves.

3 Such a view can be also supported by another similar argument. B stars span a huge range of stellar masses and it is well known from the theory of stellar evolution that the time scales of all processes depend strongly on the stellar mass. However, as pointed by Horn et al. (1982) the time scale of the formation of a new Be envelope was found to be very similar for three Be stars of spectral classes B0, B6 and B8. This again seems to indicate an external mechanism of the formation of such envelopes.

In any case, it is obvious that a new generation of powerful optical interferometers will soon be able to resolve many close binaries, so far detectable only via spectroscopy. This should allow new stringent tests for various binary scenarios of the Be phenomenon.

General considerations and the first attempt at a 3-D gasdynamical modelling of gas outflow from hot and rapidly rotating OB stars in binaries were made in 2002 by Harmanec et al. (2002). These results were obtained for rather hot envelope with temperature of the outer layers up to 10^5 K. Observations show that temperature of the envelope does not significantly exceed 10^4 K, therefore it is necessary to re-investigate the suggested mechanism and estimate the parameters of the envelope. Here we present the results of 3D gasdynamical simulations of flow structure in binary with parameters of typical Be star where the temperature of the envelope is about 14 000 K.

2.　　Physical model

Let us consider a single, rapidly rotating star. When the velocity of rotation on equator $V_{\rm rot}$ will reach the Keplerian (also called break-up) velocity $V_{\rm br} =$

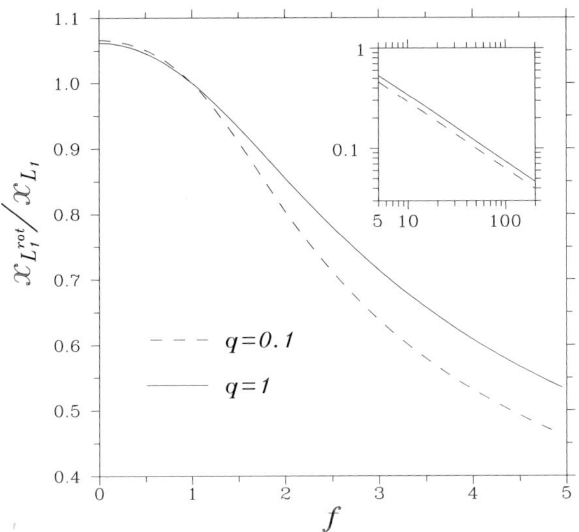

Figure 1.   Relative change in the position of the inner Lagrangian point for the systems with non-synchronous rotation of components $x_{L_1^{\mathrm{rot}}}/x_{L_1}$ as a function of the degree of non-synchronous rotation $f = \Omega_\star/\Omega$ for two values of the mass ratio: $q = 0.1$ and $q = 1$. The inserted panel shows the dependence $x_{L_1^{\mathrm{rot}}}/x_{L_1}$ vs. q for large values of f

$\sqrt{GM/R}$, the centrifugal and gravitational attractive forces will compensate each other:

$$F_{\mathrm{cfg}} \equiv \frac{V_{\mathrm{rot}}^2}{R} = \frac{GM}{R^2} \equiv F_{\mathrm{grav}}.$$

In such situations, the presence of a pressure gradient (not counterbalanced by other forces) permits the matter to outflow from the equatorial belt (called the Roche limit). A particle with specific kinetic energy $\simeq V_{\mathrm{br}}^2/2$ rotates along closed trajectories and forms an envelope. To escape to infinity the particle would need to acquire additional energy and reach the parabolic velocity $V = \sqrt{2}V_{\mathrm{br}}$.

The situation changes dramatically if the same star is a component of a binary system. Let us consider a binary with a spin-orbit synchronization (the velocities of angular rotation of both components are equal to the velocity of angular revolution of the system) and let us use a Cartesian coordinate system that rotates with an angular velocity Ω and has the origin in the centre of star 1. The X axis is directed toward star 2, Z axis is parallel to the vector of orbital revolution, and Y axis is so oriented to define a right-hand coordinate system. There are several forces acting on a test particle located between the binary components: gravitational attraction of the two binary components, the pressure gradient, and two forces related to the co-rotating frame used: centrifugal and Coriolis force. The law of motion of such a test particle was first investigated by Roche (1848; 1851) in a ballistic approach (i.e. ignoring the pressure force) as a solution of a restricted three-body problem (assuming the mass to be concentrated into two point masses); for a different formulation, see also Hill (1905). The force field (without Coriolis force and the pressure gradient) in cases of a spin-orbit synchronism can be described by the standard Roche potential Φ:

$$\Phi = -\frac{GM_1}{\sqrt{x^2 + y^2 + z^2}} - \frac{GM_2}{\sqrt{(x - A)^2 + y^2 + z^2}}$$

$$- \frac{1}{2}\Omega^2 \left(\left(x - A\frac{M_2}{M_1 + M_2}\right)^2 + y^2\right), \tag{1}$$

where M_1, M_2 are the two point masses, A is binary separation, Ω is the angular velocity of orbital motion, and x, y, z are Cartesian coordinates in the adopted frame.[1]

The presence of additional forces (absent in the case of a single star) results in violation of equilibrium in the inner Lagrangian point L_1. In particular, the pressure gradient cannot be counterbalanced there by the gradient of the Roche potential. Hence, as soon as the star expands and fills the cricital Roche lobe, matter begins to flow towards the binary companion in the vicinity of L_1 point but not from the entire equatorial zone. The position of inner Lagrangian point L_1 can be derived from equation $\nabla\Phi = 0$, which can be rewritten after Kopal (1959):

$$\left(\frac{x_{L_1}}{A}\right)^{-2} - \frac{x_{L_1}}{A} = q\left(\left(1 - \frac{x_{L_1}}{A}\right)^{-2} - \left(1 - \frac{x_{L_1}}{A}\right)\right), \tag{2}$$

where $q = M_2/M_1$ denotes the mass ratio.

In case of asynchronous rotation of star 1,[2] we should also include a centropedal acceleration and Coriolis force into equilibrium conditions. The presence of these two terms is related to the motion of the stellar matter in the adopted corotating frame. In such a case, the force field at the stellar surface is given by asynchronous Roche potential (see, e.g., Plavec, 1958; Kruszewski, 1963; or Limber, 1963):

---

[1]Note that the Roche model, based on two point masses, represents a very realistic description of equipotentials of real binaries. This is due to the fact that the mass concentration toward the centre is usually very high in real stars. Moreover, the inner equipotential surfaces below the critical Roche lobe are nearly circular and it is known that the outer potential of a mass sphere is identical to that of a mass point. The mass of the outer parts of the stars, violating the spherical symmetry, is negligible with respect to the total mass of the star in all practical applications.

[2]Hereafter, we will call the rotation of star 1 asynchronous if the absolute values of the vector of axial rotation of star 1 and vector of the orbital revolution are not equal, i.e. $|\vec{\Omega}_\star| \neq |\vec{\Omega}|$, but they are parallel to each other: $\vec{\Omega}_\star \parallel \vec{\Omega}$.

$$\Psi = \Phi - {}^1\!/_2(\Omega_\star^2 - \Omega^2)(x^2 + y^2) \tag{3}$$

where $\Omega_\star$ is the angular velocity of rotation of the star in question. Similarly as in the synchronous case it is possible to introduce a concept of the inner Lagrangian point for asynchronous rotation, L_1^{rot}, i.e. a point where the pressure gradient ceases to be counterbalanced by other forces and where the matter begins to flow towards the companion when reaching this point. The position of this point is given by the condition $\nabla\Psi = 0$ which leads to the following equation (see, e.g., Pratt & Strittmatter, 1976)

$$\left(\frac{x_{L_1^{\mathrm{rot}}}}{A}\right)^{-2} - f^2\frac{x_{L_1^{\mathrm{rot}}}}{A} = q\left(\left(1 - \frac{x_{L_1^{\mathrm{rot}}}}{A}\right)^{-2} - \left(1 - f^2\frac{x_{L_1^{\mathrm{rot}}}}{A}\right)\right), \tag{4}$$

which is similar to Eq. (2) valid for synchronous rotation. In this equation $f = \Omega_\star/\Omega$ and the sign of f does not affect the solution, i.e., the sense of the stellar rotation in the laboratory coordinate system does not affect the position of L_1^{rot}. Therefore – without loss of generality – we consider only the cases of $f \geq 0$, i.e. cases when the directions of stellar rotation and binary revolution are the same. The ratio $x_{L_1^{\mathrm{rot}}}/x_{L_1}$ as a function of q and f is shown in Fig. 1. It is obvious that for stellar rotation rates slower than the orbital revolution ($f < 1$), the 'non-synchronous' Roche lobe is larger than the standard Roche lobe, achieving maximum for $f = 0$. When the star rotates faster than the binary revolves, the 'non-synchronous' Roche lobe becomes smaller than the standard one. Note that formally $x_{L_1^{\mathrm{rot}}}/x_{L_1} \to 0$ as $f \to \infty$. In real binaries, however, the position of $x_{L_1^{\mathrm{rot}}}$ is actually limited by the break-up rotation velocity of the star in question, so that $x_{L_1^{\mathrm{rot}}}/A$ cannot be smaller than $(\Omega_{\mathrm{br}}/\Omega)^{-2/3}(q + 1)^{-1/3}$.

As discussed above, the matter can outflow from the stellar surface as it reaches L_1^{rot} point. This fact changes the limiting value of the break-up velocity when the outflow begins. In Fig. 2, the values of the linear velocity at L_1^{rot} point are plotted as a function of binary mass ratio q for different values of asynchronicity parameter f. All velocities are expressed in the units of critical velocity V_{br}, derived for a single star of the same properties. The results presented in Fig. 2 shows that in a number of cases even a small additional increase of the rotational velocity can lead to an outflow of matter in the vicinity of L_1^{rot} point. However, for values typical for Be stars, say $q \sim 0.1$ and $f \sim 100$, the outflow occurs for rotational velocities only slightly smaller than the break-up velocity of respective star. It is important to realize, however, that when V_{rot} gets close to V_{br} for Be stars, which are members of binary systems, the outflow of matter occurs only in the vicinity of L_1^{rot}, not from the the whole equatorial belt of the star (the Roche limit for single stars).

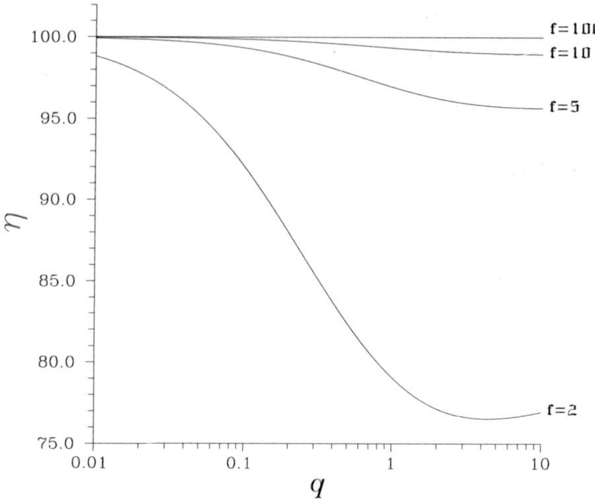

Figure 2.   The ratio of linear velocity in $L_1^{\rm rot}$ point (in observer's coordinate system) to critical velocity $\eta = \Omega_\star \cdot x_{L_1^{\rm rot}}/\sqrt{GM_1/x_{L_1^{\rm rot}}} \cdot 100\%$ vs. q for different values of asynchronicity parameter $f = \Omega_\star/\Omega$.

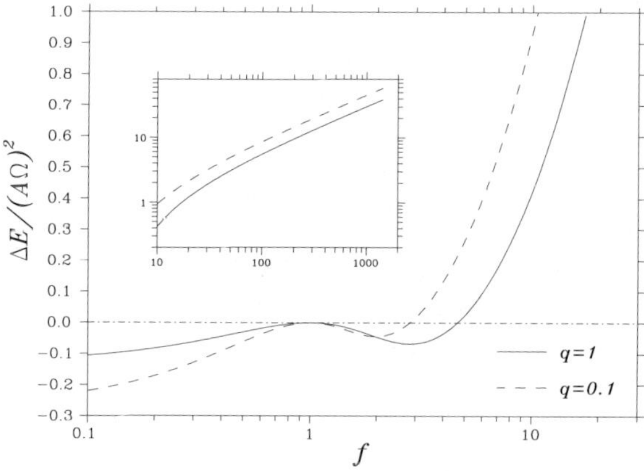

Figure 3.   Extra energy ΔE (in units of $A^2\Omega^2$) needed for a particle leaving $L_1^{\rm rot}$ to reach the L_1 point, plotted vs. asynchronicity parameter $f = \Omega_\star/\Omega$ for values of binary mass ratio of $q = 0.1$ and $q = 1$. Dash-pointed line shows value $\Delta E = 0$. The inserted panel shows the dependence of ΔE vs. f for large values of f.

The presence of a companion to the Be star results also in another notable change in the mechanism of the outflow from the surface of a rotating star. For a single star, a particle can escape to infinity if it has the parabolic velocity. For a binary star, the particle leaving via L_1^{rot} point can escape from the system if its energy is large enough to reach the L_1 point (this energy is smaller than that needed to reach the escape velocity from a single star) since after it reaches L_1, it can be captured by the gravitational field of the companion. Particles with energies insufficient to reach the L_1 point will move along closed trajectories around the star from which they escaped. The extra energy ΔE, needed to get a particle with velocity $V_{\text{rot}} = (\Omega_\star - \Omega) \cdot x_{L_1^{\text{rot}}}$ to the vicinity of L_1 point, is plotted in Figure 3 as a function of f for two mass ratios, $q = 1$ and $q = 0.1$, and is expressed in the units of the characteristic energy of the system $E_{sys} = A^2\Omega^2$. It is obvious that the sum of potential energy of a particle in L_1^{rot} point plus its kinetic energy given by the stellar rotation is much smaller than the potential energy in L_1 point

$$\Phi(L_1^{\text{rot}}) + {}^1\!/_2(\Omega_\star - \Omega)^2 \cdot x_{L_1^{\text{rot}}}^2 < \Phi(L_1).$$

The value of needed extra energy to reach L_1 is a few orders of a magnitude larger than the characteristic energy of the system E_{sys}. Considering that the effective temperatures of B stars range roughly from $10\,000$ to $30\,000$ K, it is clear that the thermal energy cannot change the overall energy balance significantly. One is, therefore, led to the conclusion that the outlow from rapidly rotating B stars in binaries via the L_1^{rot} point should lead to the formation of roughly Keplerian equatorial disks around such stars but not to a significant mass transfer towards their companions.

It is necessary to point out, however, that the above analysis of the energy balance was not exhaustive. For instance, the numerical investigations carried out by Narita, Kiguchi & Hayashi (1994) show that if the rotational velocity is close to V_{br} and viscosity is considered, a disk in a binary system may evolve from the accretion disk to the outflowing one. At the same time, their numerical simulations showed that the transfer of the angular momentum is far more pronounced than the mass loss and that the mass loss by the disk is significant only on the evolutionary time scale (i.e. $\sim 10^6$ years). Unfortunately, it is impossible to carry out 3-D gasdynamical simulations for such long time interval with the present-day computers. One only has to expect that a viscous smearing will not significantly influence the solution obtained only over a time interval comparable to the orbital period of the binary (say, less than a year).

It summary, the outlow of matter via the L_1^{rot} point seems to represent the most probable scenario of the formation of the Be envelope for a rapidly rotating B star which is a member of a binary system.

3.   Numerical model

For a numerical investigation, we have chosen a binary with parameters that are typical for binary Be stars. Since Be stars are the most abundant around the spectral class B2, we have chosen paramaters corresponding to a normal main-sequence B2 star according to the empirical calibration by Harmanec (1988) and adopted a mass ratio of 0.1 which is also typical for known Be binaries. In particular, we used the mass of Be star (star 1) $M_1 = 8.6 M_\odot$, mass of star 2 $M_2 = 0.86 M_\odot$, binary mass ratio $q = M_2/M_1 = 0.1$, binary separation $A = 121 R_\odot$, and orbital period $P = 50^{\mathrm{d}}$. Inner Lagrangian point is then located at the distance of $x_{L_1} = 87 R_\odot$ from the centre of star 1. The equatorial radius of the Be star $R_1 = 4.6 R_\odot$ was chosen. In accordance with the above considerations we assumed the Be component to be large enough to reach the 'asynchronous' Roche lobe, i.e. that $x_{L_1^{\mathrm{rot}}} = R_1$. Using the adopted values of $x_{L_1^{\mathrm{rot}}}/A$ and q, one gets the asynchronicity parameter of Be star $f = \Omega_\star/\Omega$ via Eq. (4). In this case, f is equal to 128.6 (implying the period of rotation of the Be star to be $0^{\mathrm{d}}39$). The value of linear velocity at L_1^{rot} point in the adopted coordinate system is equal to $V_{\mathrm{rot}} = (\Omega_\star - \Omega) \cdot x_{L_1^{\mathrm{rot}}} = 594$ km s^{-1}. Note that the value of critical velocity of a single star with the same mass and radius is equal to $V_{\mathrm{br}} = \sqrt{GM_1/R_1} = 599$ km s^{-1}. It means that the value of velocity in L_1^{rot} point equals to 99.2% of the critical velocity.

As pointed out above, the matter located in L_1^{rot} can reach L_1 point only when it gets some extra energy ΔE. Figure 3 shows that for the adopted values of q and f it would be necessary to add energy $\Delta E = 10.5 \cdot A^2\Omega^2$ to reach L_1. For the binary in question, the thermal energy needed to provide such an energy excess would require temperatures of at least 1.86 millions of K. In the numerical model, we actually adopt the value of effective temperature corresponding to a normal star of a similar mass, 22900 K. One then gets $\Delta E = 0.013 \cdot A^2\Omega^2$, and the matter outflowing from the L_1^{rot} point cannot get to distances larger than $R = 9.1 R_\odot \approx 2R_1$. Hence, one has to expect formation of an envelope extending to a few stellar radii.

To describe the gas flow, we have used 3-D gasdynamical equations in cylindrical coordinates. We have modified the original conservative form of equations in cylindrical coordinates to obtain a system similar to the system of gasdynamical equations in Cartesian coordinates. This approach permits us to treat the flow near the axis more accurately (see, e.g., Pogorelov, Ohsugi & Matsuda, 2000). Gasdynamical equations are written in the adopted corotating frame (i.e. in the frame where the centres of stars are in rest), so the Coriolis force is included into momentum equations. As usual, the system of gasdynamical equations is made closed using the equation of state. The equation of state of a perfect gas was adopted. To obtain solution for a cool envelope we included terms of radiative cooling and heating of the gas (Bisikalo et al., 2003) into the energy equation.

The system of gasdynamical equations was solved using a monotonic Roe's scheme (Roe, 1986) of first-order approximation with Osher's flux limiters (Chakravarty & Osher, 1985) that increases the order of approximation and retains the scheme monotonous.

Gas flow was simulated in a cylinder $r \leq 1.1 \cdot x_{L_1} = 95 R_\odot, 0 \leq z \leq 50 R_\odot$ (thanks to symmetry with respect to the equatorial plane, calculations could only be conducted in the upper half-space). Non-uniform finite-difference grids (denser near the Be star and the equatorial plane) containing $50 \times 25 \times 30$ gridpoints on r, z, and φ, respectively, were used.

As for the initial condition, we adopted a rarefied gas with $\rho_0 = 10^{-6}$, $P_0 = 10^{-4}$, and $\mathbf{u}_0 = 0$.

The boundary conditions were defined as follows: In the gridpoints that correspond to $L_1^{\rm rot}$ point we adopted the condition of injection of matter: $\rho(L_1^{\rm rot}) = 1$, $T(L_1^{\rm rot}) = 22900$ K, which corresponds to the sound velocity $c(L_1^{\rm rot}) = 14$ km s^{-1}, $v_r(L_1^{\rm rot}) = c(L_1^{\rm rot})$, $v_\varphi(L_1^{\rm rot}) = 594$ km s^{-1}, $v_z(L_1^{\rm rot}) = c(L_1^{\rm rot})$. Note that an arbitrary value of the boundary density ρ_0 can be chosen since the system of equations can be scaled with respect to ρ and P. To derive the true values of density in a specific system with known mass loss rate, the calculated densities must simply be changed in accordance with the scale determined from the ratio of the true and model mass-loss rate. The boundary conditions were derived by solving the Riemann between the gas parameters $(\rho_0, \vec{v}_0, P_0)$ in $L_1^{\rm rot}$ point and the parameters in the computation gridpoint closest to it (see, e.g., Sawada, Matsuda & Hachisu, 1986; Sawada & Matsuda, 1992; or Bisikalo et al., 1998). A full absorption of matter was assumed for the rest of the Be-star surface and for the outer boundary of the computational domain. We have verified that the outer boundary has virtually no effect on the results of computations. The boundary condition at the stellar surface is more important and less clear. However – considering the strong gravitational pull of the massive Be star and centrifugal acceleration, insufficient to cause a mass outflow – we believe that the assumption of the full absorption of matter is legitimate and a physically correct one.

4. Results and discussion

Let us first mention the results of numerical simulation obtained by Harmanec et al. (2002) for the case of the hot (up to 10^5 K) envelope. Two panels of Fig. 4 show the bird-eye view of density isosurfaces in the vicinity of the Be star at the level $\rho = 5 \times 10^{-5} \cdot \rho(L_1^{\rm rot})$. These isosurfaces are shown for two moments of time: $t = 0.514 P_{orb}$, and $0.608 P_{orb}$, respectively. Projections of the envelope into XY plane (density distribution and velocity vectors in

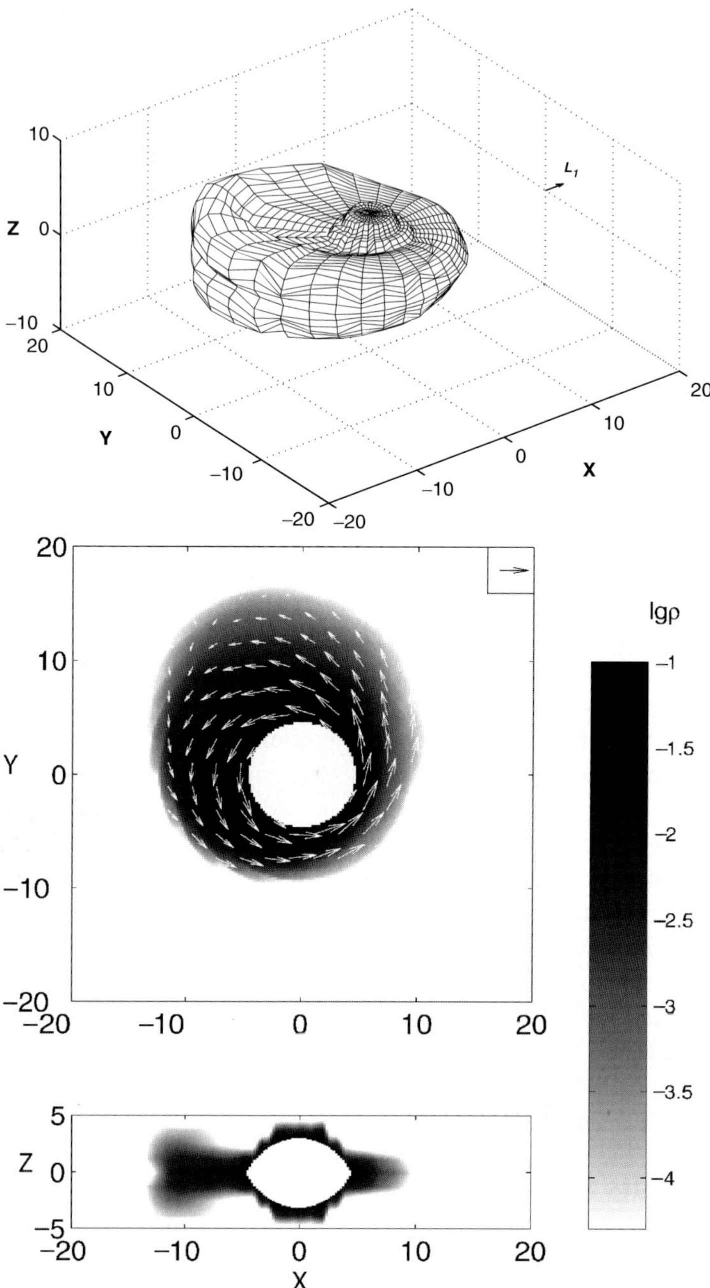

Figure 4a.   (Top panel) Bird-eye view of density isosurfaces on level $\rho = 5 \times 10^{-5} \cdot \rho(L_1^{\mathrm{rot}})$ for the moment of time $t = 0.514 P_{orb}$. (Middle panel) The slice of formed envelope by XY plane (density distribution and velocity vectors in equatorial plane). Vector in top right corner corresponds to velocity 500 km s^{-1}. (Bottom panel) The slice of the envelope by XZ (density distribution in frontal plane). Coordinates in all three panels are expressed in $R_\odot$.

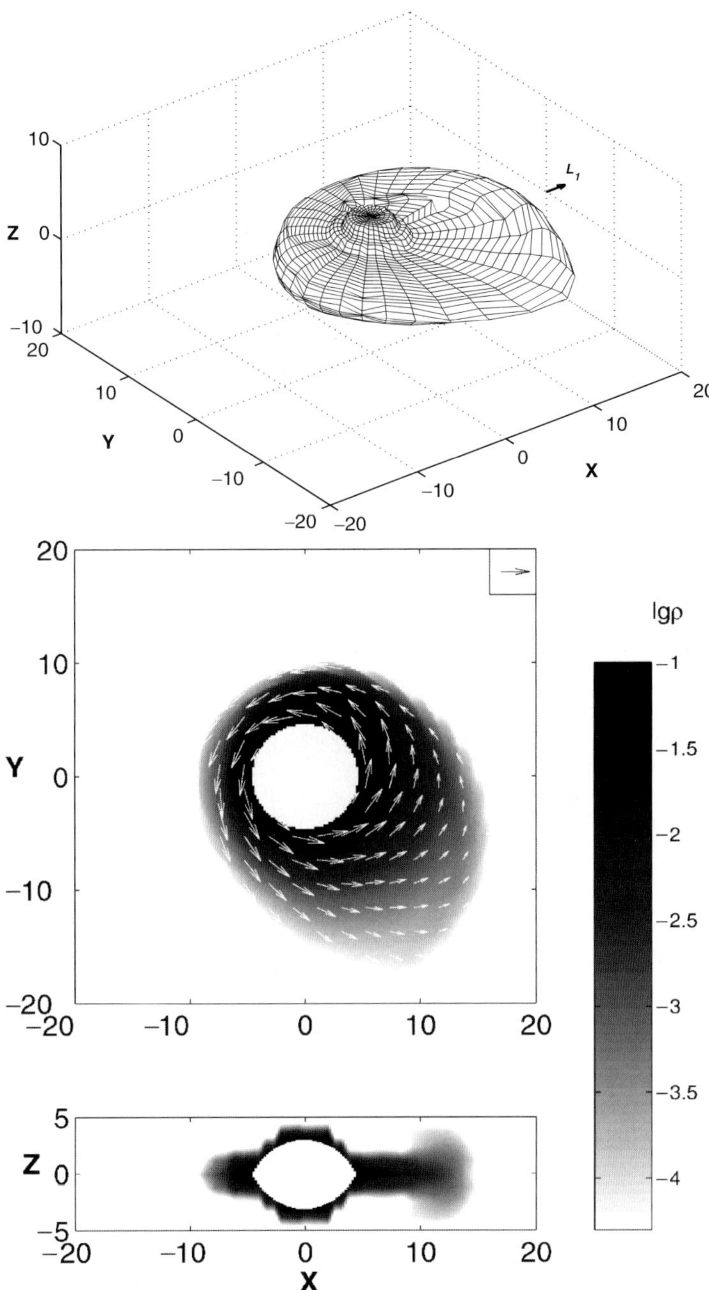

Figure 4b.   The same as in Fig. 4a but for the time $t = 0.608 P_{orb}$.

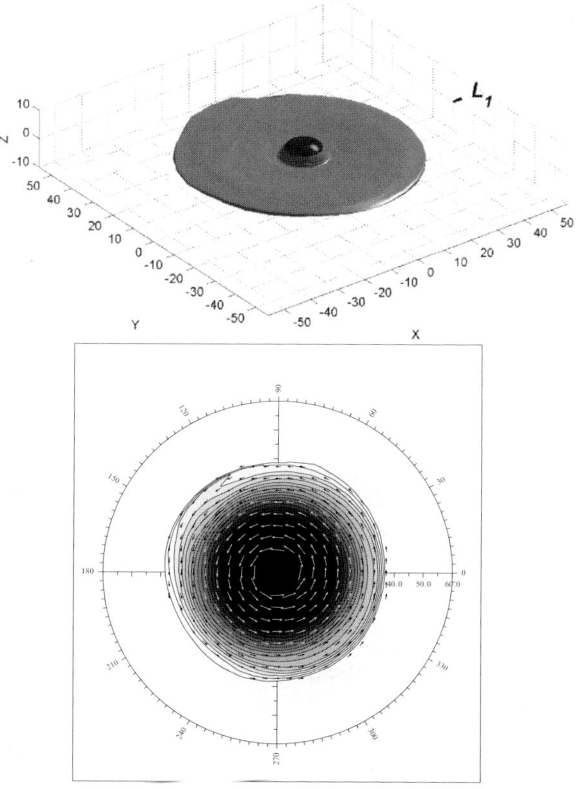

Figure 5a. (Top panel) Bird-eye view of density isosurfaces on level $\rho = 1 \times 10^{-3} \cdot \rho(L_1^{\mathrm{rot}})$ for the moment of time $t = 38.22 P_{orb}$. (Bottom panel) The slice of formed envelope by XY plane (density distribution and velocity vectors in equatorial plane). Coordinates in all panels are expressed in $R_\odot$.

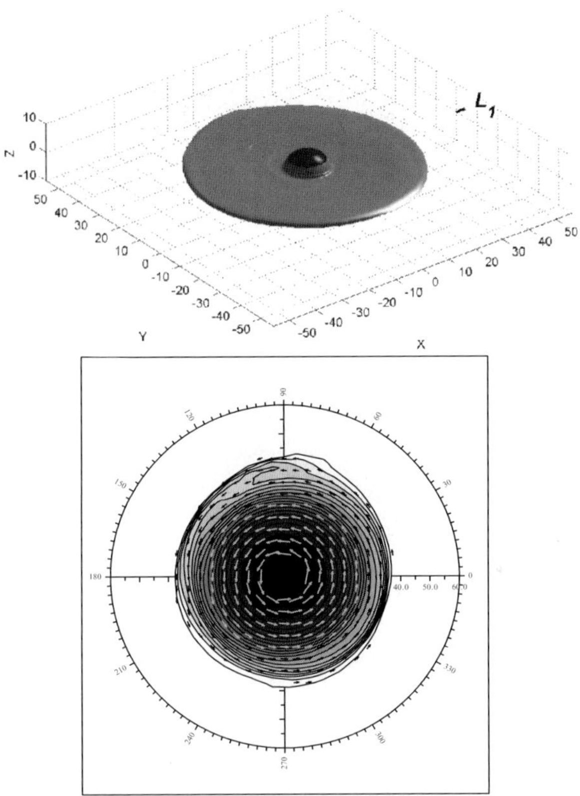

Figure 5b.   The same as in Fig. 5a but for the time $t = 38.87 P_{orb}$.

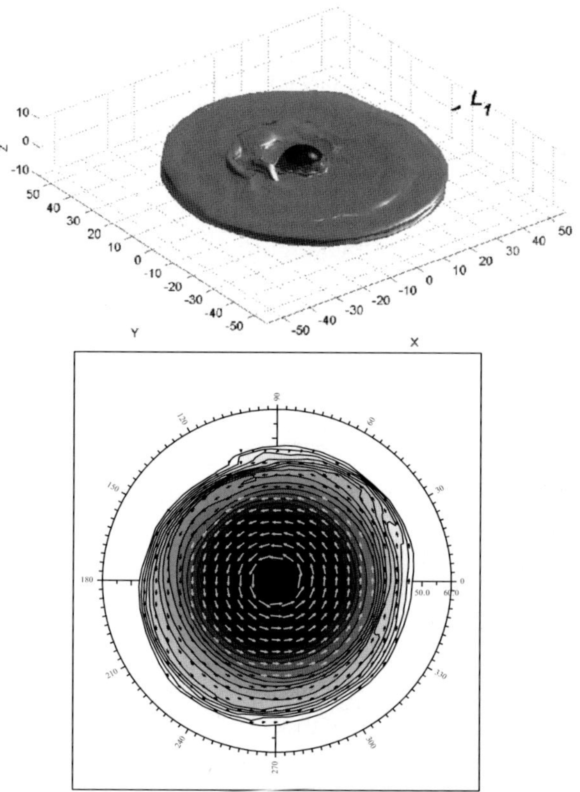

Figure 6a. (Top panel) Bird-eye view of density isosurfaces on level $\rho = 5 \times 10^{-5} \cdot \rho(L_1^{\rm rot})$ for the moment of time $t = 38.22 P_{orb}$. (Bottom panel) The slice of formed envelope by XY plane (density distribution and velocity vectors in equatorial plane). Coordinates in all panels are expressed in $R_\odot$.

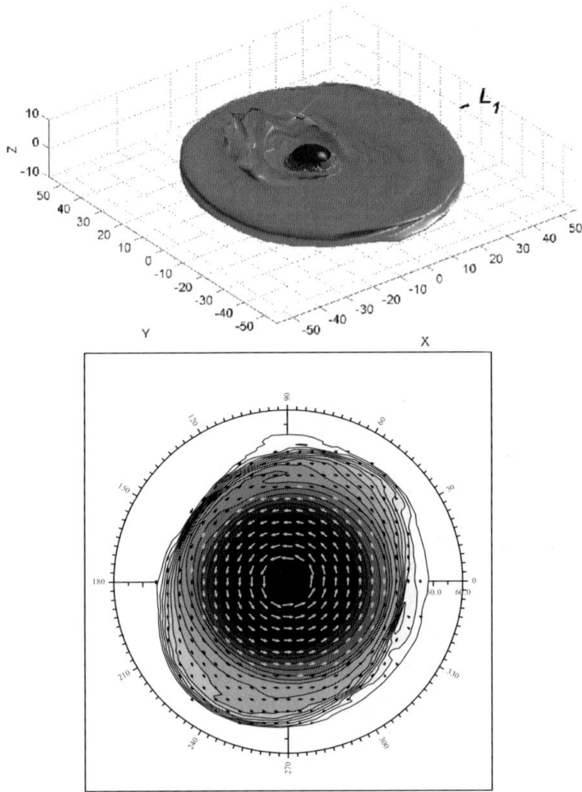

Figure 6b. The same as in Fig. 6a but for the time $t = 38.87 P_{orb}$.

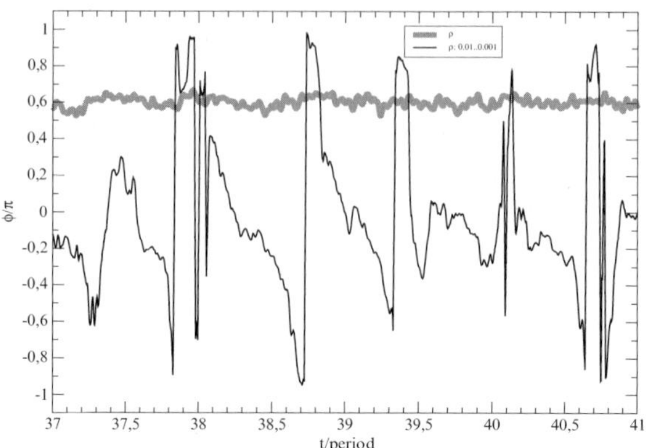

Figure 7.   A time evolution of the angle between X axis and the direction to the centre of mass of the envelope (bold grey line), and the angle between X axis and the direction to the centre of mass of the outer parts of the envelope (with values of density $\rho \in [10^{-3}, 10^{-2}]$ – solid black line).

equatorial plane) and XZ plane (density distribution in the frontal plane) are also depicted in those six frames.

An analysis of the results shows that the outflow of matter from the Be star in the vicinity of L_1^{rot} point results in formation of an envelope with a fast retrograde apsidal motion. The mean angular velocity of apsidal motion Ω_{aps} is equal to -6Ω, and $P_{\mathrm{aps}} = \frac{1}{6}P_{\mathrm{orb}}$. Consideration of the time evolution of two angles, an angle α_1 between X axis and a direction to the centre of mass of the envelope,[3] and an angle α_2 between X axis and a direction to the centre of mass of the outer part of the envelope (values of density $\rho \in [10^{-3}, 10^{-2}]$) shows that the centre of mass of the whole envelope oscillates within the interval $\alpha \in [0.35\pi, 0.65\pi]$ while the centre of mass of outer layers makes a full revolution within the interval $\alpha \in [-\pi, \pi]$. This finding seems to indicate the presence of a strong differential rotation of the envelope.

Figure 4 also shows that for a part of time (Fig. 4a) the envelope has a torus-like shape, the thickness of the envelope being $h \sim R_1$, i.e. exceeding the polar radius which equals to $\sim 0.65R_1$. In the rest of time (Fig. 4b), the elongation of the envelope and its shape becomes nearly disk-like one, with a characteristic thickness $h \sim \frac{1}{2}R_1$. It is obvious that the changes in the envelope shape result from both, the presence of the binary companion, and the interaction of the gas in the envelope (during its apsidal motion) with the stream of gas leaving L_1^{rot}. The characteristic linear sizes of the envelope in

---

[3]Angle $\alpha \in [-\pi, \pi]$ is reckoned counter-clockwise from X axis.

equatorial plane are the following: for time when it has torus-like shape it is $\sim 3R_1$ (on the level $\rho = 5 \cdot 10^{-5} \cdot \rho(L_1^{\mathrm{rot}})$), while for the time when it has disk-like shape, its size increases up to $\sim 4.5R_1$.

Let us now consider the results of numerical modeling of the outflowing envelope in a Be binary for a more realistic temperature of the envelope of some 14000 K. The two panels of Figs. 5 and 6 show the bird-eye view of the density isosurfaces at the level $\rho = 1 \times 10^{-3} \cdot \rho(L_1^{\mathrm{rot}})$ (in the vicinity of the Be star) and at $\rho = 5 \times 10^{-5} \cdot \rho(L_1^{\mathrm{rot}})$ (outer parts of the envelope), respectively. The isosurfaces in Figs. 5a, 6a are shown for the time instant $t = 38.22P_{\mathrm{orb}}$, those in Figs. 5b, 6b for $t = 38.87P_{\mathrm{orb}}$. Projections of the envelope into XY plane (density distribution and velocity vectors in the equatorial plane) are also depicted in those figures. An analysis of the results shows that for the cool case the envelope becomes larger and deviates less from the circular shape than for the hot case studied earlier. The characteristic linear size of the envelope in equatorial plane is $\sim 35R_{\odot}$ on the level $\rho = 1 \cdot 10^{-3} \cdot \rho(L_1^{\mathrm{rot}})$ and $\sim 45R_{\odot}$ on the level $\rho = 5 \cdot 10^{-5} \cdot \rho(L_1^{\mathrm{rot}})$. It is also seen that at the direction of the center of mass the envelope has a hump on the surface. A characteristic semithickness of the envelope $h \sim \frac{1}{10}R_1$ on the level $\rho = 1 \cdot 10^{-3} \cdot \rho(L_1^{\mathrm{rot}})$ and $\sim \frac{1}{3}R_1$ on the level $\rho = 5 \cdot 10^{-5} \cdot \rho(L_1^{\mathrm{rot}})$. The semithickness of the hump is about R_1 on the level $\rho = 5 \cdot 10^{-5} \cdot \rho(L_1^{\mathrm{rot}})$. The hump does not rotate and its position remains stable. The hump seems to be caused by the stream of the matter outflowing from the vicinity of L_1^{rot} point.

In Fig. 7 the time evolution of two angles, an angle between X axis and a direction to the centre of mass of the envelope (bold grey line), and an angle between X axis and a direction to the centre of mass of the outer part of the envelope (solid line; values of density $\rho \in [10^{-3}, 10^{-2}]$) are shown. It follows from the data of Fig. 7 that the center mass of the envelope oscillates within the interval $\alpha \in [0.55\pi, 0.65\pi]$, while the centre of mass of outer layers makes a full revolution within the interval $\alpha \in [-\pi, \pi]$. The behavior of the cool envelope is more stable in comparison with the hot case: almost all layers of the envelope oscillate within the same interval of α as the center of mass. The only outer layers of the envelope ($\rho \leq 5 \cdot 10^{-3} \cdot \rho(L_1^{\mathrm{rot}})$) have a retrograde apsidal motion with period $P_{\mathrm{aps}} = \frac{1}{2}P_{\mathrm{orb}}$. The mass of these layers is rather small (less than 5%) in comparison with mass of the whole envelope. Clearly, the more realistic treatment of the envelope temperature leads to the formation of a disk which is more stable that that obtained for the hot disk. For careful readers, we would like to mention that the evolution of the cool outflowing disk was followed to a stabilized situation after 38 orbital periods while the initial model for the hot disk was only followed for a fraction of one orbital period. We also continued the calculations for the hot disk to find out that its rapid retrograde motion persists even after many orbital periods. Therefore, the difference between the two cases is a real one and is indeed related to the improved treatment of the envelope temperature discussed here.

How to interpret our findings? It is clear that a real comparison of our results with the observed Be binaries will require calculation of the emission-line profiles originating from the outflowing envelopes of our gasdynamical models. The new results indicate that our model could explain some key properties of the real observed systems.

First, the fact that the orientation of the bulk of the envelope remains phase-locked within the binary system promises to explain the observed phase-locked variations of the V/R ratio of the violet and red peaks of the Balmer emission lines observed for known Be binaries (cf., e.g., 4 Her: Koubský et al., 1997).

The potential of the new binary model to solve the long-standing problem of the formation of Be envelopes has been questioned in a review paper by Porter & Rivinius (2003). They pointed out that for typical Be binaries the model requires equatorial rotation rates which are very close to the Keplerian one for the outlow via L_1^{rot} point to occur. From this they conclude that our model basically does not differ from the very first hypothesis explaining the formation of Be envelope via rotational instability at the stellar equator (Struve, 1931). Our reply is that all the time since the rotational hypothesis was published, various attempts to explain the formation of Be envelopes were mainly concentrated on the problem where to find a relatively small additional agent to facilitate outflow of matter from the stars rotating very near to the Keplerian velocity at their equators. Besides – the original Struve's model offers no obvious explanation for the secular variations of the emission strength and for their multiple appearance and disappearance on the time scale of years and decades. In contrast to it, the present binary model has the potential to explain the long-term variations of the envelopes. Our current simulations do not include the small changes of the orbital period of the binary which could be caused by the slight variations in the rotational rate, related to angular momentum changes of the rotating Be star. It is conceivable, however, that such variations could be responsible for cyclic termination of the outflow and its later reapperance. If so, the fact that the rotational speed is close to the Keplerian one becomes an advantage for such a mechanism to operate. In a sense, one can see some analogy with the much more spectacular cyclic outbursts of recurrent novae. Also for Be stars, a longer period of outflow and formation of a more extended envelope could result in a consecutive longer period of inactivity. At present, these statements are only speculative ones. They will require tests via real modelling. However, they show the great potential of the binary model to explain some of the principal problems of the Be star research.

Acknowledgments

The work was partially supported by Russian Foundation for Basic Research (projects NN 05-02-16123, 05-02-17070, 05-02-17874, 06-02-16097), by Science Schools Support Program (project N 162.2003.2), by Federal Programme "Astronomy", by Presidium RAS Programs "Mathematical modelling and intellectual systems", "Nonstationary phenomena in astronomy", and by INTAS (grant N 00-491). O.A.K. thanks Russian Science Support Foundation for the financial support. P. H. was supported from the research plans J13/98: 113200004 of Ministry of Education, Youth and Sports and AV 0Z1 003909, from project K2043105 of the Academy of Sciences of the Czech Republic and also from the grant GA ČR 205/2002/0788 of the Granting Agency of the Czech Republic.

References

Baade, D.: 1992, in *Evolutionary Processes in Interacting Binary Stars*, Proc. IAU Symp. 151, ed. by Y. Kondo, R. Sistero and R.S. Polidan, Kluwer, Dordrecht, p. 147

Bisikalo, D.V., Boyarchuk, A.A., Chechetkin, V.M., Kuznetsov, O.A. and Molteni, D.: 1998, *MNRAS*, **300**, 39

Bisikalo, D.V., Boyarchuk, A.A., Kaigorodov P.V. and Kuznetsov, O.A.: 2003, *Astronomy. Reports* **47**, 809

Chakravarthy, S.R. and Osher, S.: 1985, *AIAA Pap.* N 85-0363

Gies, D.R., Bagnuolo Jr., W.G., Ferrara, E.C., Kaye, A.B., Thaller, M.L., Penny, L.R. and Peters, G.J.: 1998, *ApJ*, **493**, 440

Gies, D.R.: 2000, in *The Be Phenomenon in Early-Type Stars*, Proc. IAU Col. 175, ed. by M.A. Smith, H.F. Henrichs and J. Fabregat, ASP Conf. Ser., p. 668

Harmanec, P.: 1985, *Bull. Astron. Inst. Czechosl.*, **36**, 327

Harmanec, P.: 1987, in *Physics of Be Stars*, IAU Coll. 92, ed. by A. Slettebak and T.P. Snow, Cambridge Univ. Press, Cambridge, p. 339

Harmanec, P.: 1988, *Bull. Astron. Inst. Czechosl.*, **39**, 329

Harmanec, P.: 2000, in *The Be Phenomenon in Early-Type Stars*, Proc. IAU Col. 175, ed. by M.A. Smith, H.F. Henrichs and J. Fabregat, ASP Conf. Ser., p. 13

Harmanec, P.: 2001, in *Interacting astronomers: A symposium on Mirek Plavec's favorite stars*, ed. by P. Harmanec, P. Hadrava and I. Hubeny, Publ. Astron. Inst. Acad. Sci. Czech Rep., No. 89, p. 9

Harmanec, P.: 2002, in *Exotic Stars as Challenges to Evolution*, Proc. IAU Col. 187, ed. by R.E. Wilson and W. Van Hamme, ASP Conf. Ser.

Harmanec, P., Bisikalo, D.V., Boyarchuk, A.A. and Kuznetsov, O.A.: 2002, *A& A*, **396**, 937

Harmanec, P. and Kříž, S.: 1976, in *Be and Shell Stars*, IAU Symp. 70, ed. by A. Slettebak, Reidel, Dordrecht, p. 385

Hill, G.W.: 1905, *Collected Works*, vol. 1, p. 290

Horn, J., Harmanec, P., Koubský, P., Žďárský, F., Božić, H. and Pavlovski, K.: 1982, *Bull. Astron. Inst. Czechosl.*, **33**, 308

Kopal, Z. 1959, *Close Binary Systems*, Chapman & Hall, London

Koubský, P., Harmanec, P., Kubát, J., Hubert, A.-M., Božić, H., Floquet, M., Hadrava, P., Hill, G. and Percy, J.R.: 1997, *A& A,* **328**, 551

Kříž, S.: 1982, *Bull. Astron. Inst. Czechosl.,* **33**, 302

Kříž, S. and Harmanec, P.: 1975, *Bull. Astron. Inst. Czechosl.,* **26**, 65

Kruszewski, A.: 1963, *Acta Astron.,* **13**, 106

Limber, D.N.: 1963, *ApJ,* **138**, 1112

Plavec, M.: 1958, *Mém. Soc. Roy. Sci. Liège* **20**, 411

Plavec, M.J.: 1976, in *Be and Shell Stars*, Proc. IAU Symp. 70, ed. by A. Slettebak, Reidel, Dordrecht, 439

Poeckert, R.: 1981, *PASP,* **93**, 297

Pogorelov, N.V., Ohsugi, Y. and Matsuda, T.: 2000, *MNRAS,* **313**, 198

Pols, O., Coté, J., Waters, L.B.F.M. and Heise, J.: 1991, *A& A,* bf 241, 419

Porter, J.M. & Rivinius, T.: 2003, *PASP,* **115**, 1153

Pratt, J.P. and Strittmatter, P.A.: 1976, *ApJ,* **204**, L29

Quirrenbach, A., Hummel, C.A., Buscher, D.F., Armstrong, J.T., Mozurkewich, D. and Elias, II N.M.: 1993, *ApJ,* , **416**, L25

Quirrenbach, A., Bjorkman, J.E., Hummel, C.A., Buscher, D.F., Armstrong, J.T., Mozurkewich, D., Elias, N.M., II. and Babler, B.L.: 1997, *ApJ,* **479**, 477

Roche, E.A.: 1848, *Mem. de l'Acad. des Sci. de Montpellier*, **1**, 243 & 333

Roche, E.A.: 1851, *Mem. de l'Acad. des Sci. de Montpellier*, **2**, 21

Roe, P.L.: 1986, *Ann. Rev. Fluid. Mech.*, **18**, 337

Sawada, K. and Matsuda, T.: 1992, *MNRAS,* **255**, 17P

Sawada, K., Matsuda, T. and Hachisu, I.: 1986, *MNRAS,* **219**, 75

Slettebak, A.: 1985 *ApJS,* **59**, 769

Struve, O.: 1931, *ApJ,* **73**, 94

ACCRETION DISKS AROUND BLACK HOLES WITH ACCOUNT OF MAGNETIC FIELDS*

Gennady Bisnovatyi-Kogan
Space Research Institute RAN, Moscow, Russia
and
Joint Institute of Nuclear Researches, Dubna, Russia

Abstract   Accretion disks are observed in young stars, cataclysmic variables, binary X-ray sources et al. Accretion disk theory was first developed as a theory with the local heat balance, where the whole energy produced by a viscous heating was emitted to the sides of the disk. Important part of this theory was the phenomenological treatment of the turbulent viscosity, known the "alpha" prescription, where the $(r\phi)$ component of the stress tensor was connected with the pressure as αP. Sources of turbulence in the accretion disk are discussed, including hydrodynamic turbulence, convection and magnetic field role. Optically thin solution and advective disks are considered. Related problems of mass ejection from magnetized accretion disks and jet formation are discussed.

Keywords: accretion disk, X-ray source, jet

1.   Introduction

Accretion is the main source of energy in many astrophysical objects, including different types of binary stars, binary X-ray sources, most probably quasars and active galactic nuclei (AGN). Accretion onto stars, including neutron stars, terminates at an inner boundary. This may be the stellar surface, or the outer boundary of a magnetosphere for strongly magnetized stars. We may be sure in this case, that all gravitational energy of the falling matter will be transformed into heat and radiated outward.

The situation is quite different for sources containing black holes, which are discovered in some binary X-ray sources in the galaxy, as well as in many AGN. Here matter is falling to the horizon, from where no radiation arrives, so

*Partial funding provided by RFBR grant 02-02-16900, INTAS grant 00491, and Astronomy Programm "Nonstationary phenomena in astrophysics"

Alexei M. Fridman, et al. (eds.), Astrophysical Disks; Collective and Stochastic Phenomena 97–130

all luminosity is formed on the way to it. A high efficiency of accretion into a black hole takes place only when matter is magnetized (Schwartzman, 1971), or has a large angular momentum, when accretion disk is formed.

Intensive development of the accretion disk theory began after birth of the X-ray astronomy, when luminous X-ray sources in binary systems have been discovered, in which accretion was the only possible way of the energy production. The X-ray astronomy was born after the rocket launch in 1961 in USA by the group of physicists headed by R. Giacconi. The first and the brightest X-ray source outside the solar system, Sco X-1, was discovered during this flight. In subsequent time the main discoveries in X-ray astronomy have been done from satellites. The main discovery of the first X-ray satellite UHURU, launched in 1970 had been X-ray pulsars - neutron stars in binary systems. The fundamental importance has also a discovery of the first real black hole candidate in the Cyg X-1 binary source. The next X-ray satellite EINSTEIN, launched in 1978 had a good angular resolution and sensitivity, and more than 50 000 new sources, mainly extragalactic, had been discovered there. These two satellites had been also constructed in the team headed by R. Giacconi.

In subsequent years more than 20 X-ray satellites had been launched, and the most advanced ones CHANDRA (USA) and NEWTON (ESA) are operating now.

2.   Standard accretion disk model

The small thickness of the disk in comparison with its radius $h \ll r$ indicate to small importance of the pressure gradient ∇P in comparison with gravity and inertia forces. That leads to a simple radial equilibrium equation denoting

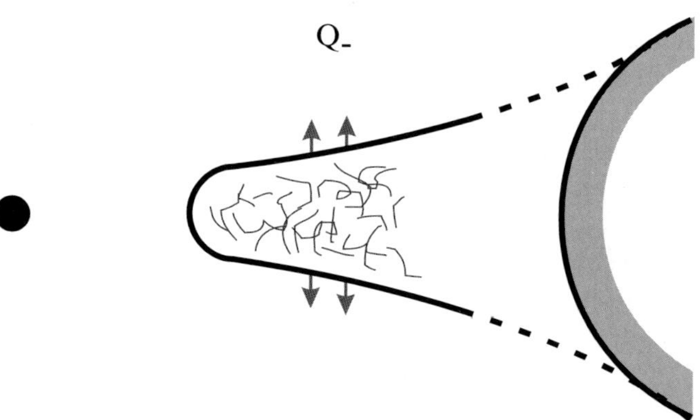

Figure 1.   Schematic picture of a disc accretion into a black hole in the standard model.

the balance between the last two forces occuring when the angular velocity of the disk Ω is equal to the Keplerian one Ω_K,

$$\Omega = \Omega_K = \left(\frac{GM}{r^3}\right)^{1/2}. \tag{1}$$

In the "standard" accretion disk model the relation (1) is suggested to be fulfilled all over the disk, with an inner boundary at the last stable orbit, $r_{in} = 3r_g$, $r_g = 2GM/c^2$ is the gravitational Schwarzschild radius of a black hole.

The equilibrium equation in the vertical z-direction is determined by a balance between the gravitational force and pressure gradient, which for a thin disk leads to algebraic one, determining the half-thickness of the disk as

$$h \approx \frac{1}{\Omega_K}\left(2\frac{P}{\rho}\right)^{1/2}. \tag{2}$$

The ϕ component of the Navier-Stokes equation has an integral in a stationary case which represents the conservation of angular momentum

$$\dot{M}(j - j_{in}) = -2\pi r^2\, 2h t_{r\phi}, \quad t_{r\phi} = \eta r\frac{d\Omega}{dr}. \tag{3}$$

Here $j = v_\phi r = \Omega r^2$ is the specific angular momentum, $t_{r\phi}$ is the component of the viscous stress tensor, $\dot{M} > 0$ is a mass accretion rate, j_{in} is an integration constant. Multiplication of j_{in} by $\dot{M}$, gives a difference between viscous and advective flux of the angular momentum in the disk. For the accretion into a black hole it is usually assumed, that on the last stable orbit the gradient of the angular velocity is zero, corresponding to zero viscous momentum flux. In that case

$$j_{in} = \Omega_K r_{in}^2, \tag{4}$$

which is the Keplerian angular momentum of the matter on the last stable orbit.

The choice of the viscosity coefficient is the most difficult and speculative aspect of the accterion disk theory. Observations of X-ray binaries had shown, that there should be high viscosity in the accretion disks. In the paper of Shakura (1972) it was suggested, that matter in the disk is turbulent, and described by a viscous stress tensor, parametrized as

$$t_{r\phi} = -\alpha\rho v_s^2 = -\alpha P, \tag{5}$$

where α is a dimensionless constant and v_s is the sound speed. This simple parametrization corresponds to a turbulent viscosity coefficient $\eta_t \approx \rho v_t l$ with an average turbulent velocity v_t and mean free path of the turbulent element l. It follows from the definition of $t_{r\phi}$ in (3), when we take $l \approx h$ from (2)

$$t_{r\phi} = \rho v_t h r \frac{d\Omega}{dr} \approx \rho v_t v_s = -\alpha \rho v_s^2, \qquad (6)$$

where a coefficient $\alpha < 1$ relates the turbulent and sound speeds $v_t = \alpha v_s$. The presentations of $t_{r\phi}$ in (5) and (6) are equivalent. Only when the angular velocity differs considerably from the Keplerian is the first relation, on the right-hand side of (6), preferable. That does not appear (by definition) in the standard theory, but may happen when advective terms are included.

Development of a turbulence in the accretion disk cannot be justified simply, because a Keplerian disk is stable in linear approximation to the development of axially symmetric perturbations, conserving the angular momentum. It was suggested by Ya.B. Zeldovich, that in presence of very large Reynolds number $Re = \frac{\rho v l}{\eta}$ the amplitude of perturbations at which nonlinear effects become important is very low, so in this situation turbulence may develop due to nonlinear instability even when the disk is stable in linear approximation. Another source of viscous stresses may arise from a magnetic field, but as was indicated by Shakura (1972), that magnetic stresses cannot exceed the turbulent ones.

It was shown by Bisnovatyi-Kogan and Blinnikov (1977), that inner regions of a highly luminous accretion discs where pressure is dominated by radiation, are unstable to vertical convection. Development of this convection produce a turbulence, needed for a high viscosity (and also leads to formation of a hot corona above this region of the disk). In the colder regions with incomplete ionization a behaviour of the accretion disk becomes more complicated, with a non-unique solutions, and convective instability (Cannizzo, Ghosh & Wheeler, 1982).

For Keplerian angular velocity the angular momentum per unit mass $j = \omega r^2 \sim r^{1/2}$ is growing outside. In this respect it is similar to the viscid flow between two rotating cylinders (Taylor experiment), when the inner cylinder is at rest. Phenomenological analysis of the Taylor experiment, and the onset of turbulence in the "stable" case of the inner cylinder at rest had been done by Zeldovich (1981).

There are arguments, both experimental and theoretical, supporting the hydrodynamic origin of the accretion disk turbulence. Non-radial perturbations ($\sim e^{im\phi}$) with a high azimuthal number $m > R/h$ are only slightly influenced by the rotation, so development of shear instability is possible for large m. The effective length in this case is $l_{eff} \sim h^2/R$, so the critical $Re_* = \rho v l_{eff}/\eta \approx 10^3$ corresponds to the actual $Re = \rho v h/\eta \approx m Re_*$. In the Taylor experiment the development of turbulence started at $Re = 10^5$. Analytically a local shear instability in the stratified accretion disk was found by Richards et al. (2001), similar results have been obtained earlier (see Glatzel (1991) and references therein).

Magnetorotational instability of Velikhov (1959) & Chandrasekhar (1960) was advocated by Balbus & Hawley (1998). Their numerical experiments which failed to find a development of hydrodynamic shear instability could not reach the required $Re \sim 10^5$ in their simulations due to high numerical viscosity. In real astrophysical objects Re could reach 10^{10}, and become even higher. Results of recent numerical simulations (Kuznetsov, 2005) confirm the idea of non-linear hydrodynamic instability.

With alpha- prescription of viscosity the equation of angular momentum conservation is written in the plane of the disk as

$$\dot{M}(j - j_{in}) = 4\pi r^2 \alpha P_0 h. \tag{7}$$

When angular velocity is far from Keplerian the relation (3) is valid with a coefficient of a turbulent viscosity

$$\eta = \frac{2}{3}\alpha \rho_0 v_{s0} h, \tag{8}$$

where values with the index "0" denotes the plane of the disk.

In the standard theory a heat balance is local, what means that all heat produced by viscosity in the ring between r and $r + dr$ is radiated through the sides of disk at the same r, see Fig. 1. The heat production rate Q_+ related to the surface unit of the disk is written as

$$Q_+ = h\, t_{r\phi} r \frac{d\Omega}{dr} = \frac{3}{8\pi}\dot{M}\frac{GM}{r^3}\left(1 - \frac{j_{in}}{j}\right). \tag{9}$$

Heat losses by a disk depend on its optical depth. The first standard disk model of Shakura (1972) considered a geometrically thin disk as an optically thick in a vertical direction. That implies energy losses Q_- from the disk due to a radiative conductivity, after a substitution of the differential eqiation of a heat transfer by an algebraic relation

$$Q_- \approx \frac{4}{3}\frac{acT^4}{\kappa\Sigma}. \tag{10}$$

Here a is the usual radiation energy-density constant, c is a speed of light, T is a temperature in the disk plane, κ is a matter opacity, and a surface density

$$\Sigma = 2\rho h, \tag{11}$$

here and below ρ, T, P without the index "0" are related to the disk plane. The heat balance equation is represented by a relation

$$Q_+ = Q_-, \tag{12}$$

A continuity equation in the standard model of the stationary accretion flow is used for finding of a radial velocity v_r

$$v_r = \frac{\dot{M}}{4\pi r h \rho} = \frac{\dot{M}}{2\pi r \Sigma}. \tag{13}$$

Equations (1), (2), (7), (11), (12), completed by an equation of state $P(\rho, T)$ and relation for the opacity $\kappa = \kappa(\rho, T)$ represent a full set of equations for a standard disk model. For power low equations of state of an ideal gas $P = P_g = \rho \mathcal{R} T$ ($\mathcal{R}$ is a gas constant), or radiation pressure $P = P_r = \frac{aT^4}{3}$, and opacity in the form of electron scattering κ_e, or Karammer's formula κ_k, the solution of a standard disk accretion theory is obtained analytically (Shakura, 1972; Novikov, Thorne, 1973; Shakura, Sunyaev, 1973). The structure of the accretion disk around the black hole in the standard model is represented in Fig. 2.

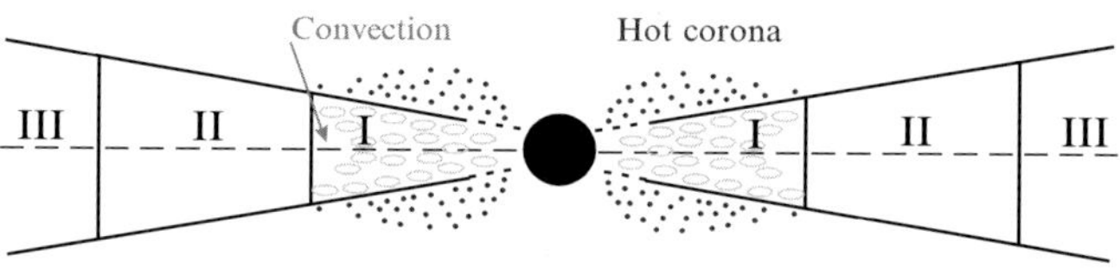

Figure 2.   Sketch of picture of a disk accretion on to a black hole at sub-critical luminosity, from Bisnovatyi-Kogan (1985).

3.   Observational evidences of the existence of Black Holes

3.1   Observational identification of black holes

Black holes (BH) of stellar masses have been observed in the galactic binary X-ray sources, and supermassive BH (SBH) had been found in the nuclei of active galaxies (AGN), having masses $10^7 - 10^9$ solar masses. In the galactic binary X-ray sources masses have been measured on the base of Kepler law: the compact stellar object is qualified as BH if its mass exceeds the mass of a stable neutron star, about 2.5 solar masses. X-ray binaries in our Galaxy containing black hole with low mass companion show global accretion disk instabilities, observed as soft X-ray transients - X-ray novae. The light curves of two X-ray novae in optical and X-ray bands are shown in Figs. 3,4, see also review of Cherepashchuk (2000). The mass functions of the compact object in X-ray novae, containing black holes are $3 - 6$ Solar masses.

SBH in AGN are found by optical, X-ray and radio observations. Optical observations show strong concentration of light to the center, and existence of the accretion disk by the distribution of rotational velocity around the center. The example of such curve for the galaxy M87 with the mass of SBH about three billion solar masses is given in Fig. 5 from Macchetto et al. (1997), see also Ho (1999). X-ray observations have revealed an existence of very broad

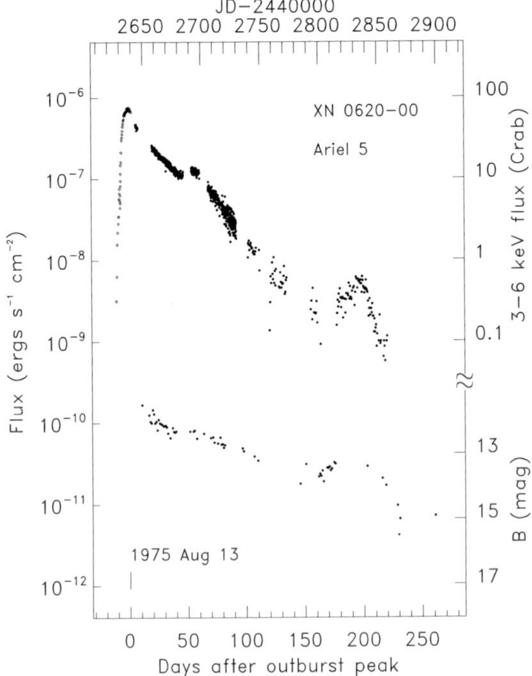

Figure 3.   X-ray and optical light curves of the black-hole LMXBT A0620-00 (from Chen et al., 1997).

emission Fe Kα lines in the X-ray spectra of AGN. The width corresponding to about one third of the speed of light may originate only near the relativistic object. The shape of the line is fitted well by the radiation from the accretion disk around SBH, which may be described by Schwarzschild or Kerr metric. This spectrum given in Fig. 6 is representing the composite spectrum of Seyfert 1 galactic nuclei, obtained by Nandra et al. (1997). The most precise measurements have been done by radio VLBI observations of the water maser line from the nucleus of the Seyfert galaxy NGC 4258. The Keplerian rotational curve is obtained with very high accuracy ($<1\%$), implying the mass of 36 millions of solar masses within 0.13 pc. This rotational curve is represented in Fig. 7 from Ho (1999).

3.2   Jets from accretion disks

Powerful, highly-collimated, oppositely directed jets are observed in AGN and quasars, and in the "microquasars" – old compact stars in X-ray binaries (Mirabel and Rodriguez, 1994). Highly collimated emission line jets are seen

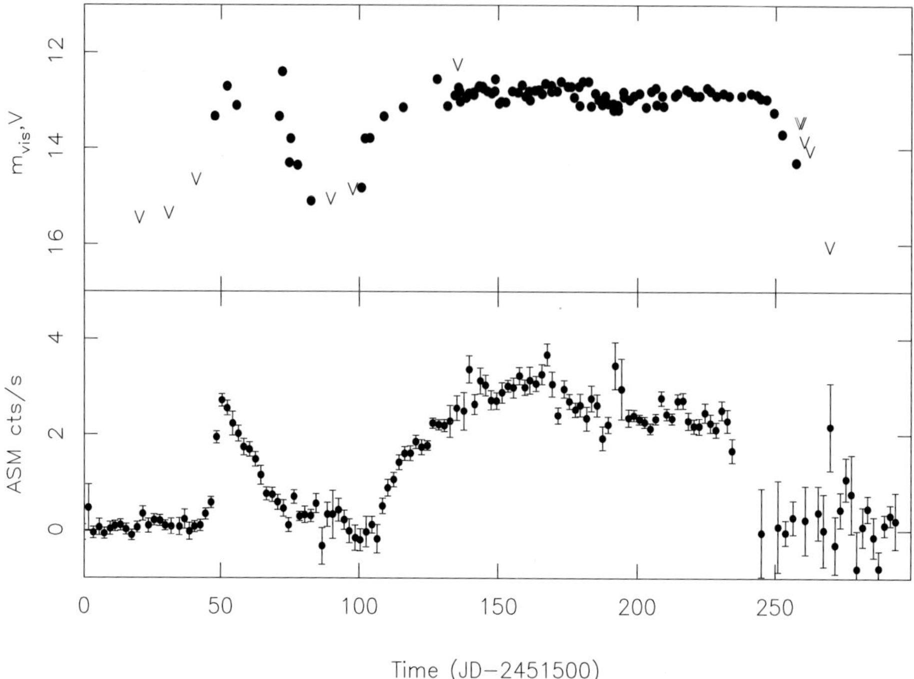

Figure 4. X-ray and optical light curves of the black-hole LMXBT XTE J1118+48 (adapted from Lasota, 2001). The 'black-hole' classification here requires confirmation.

in young stellar objects. Different ideas and models have been put forward to explain astrophysical jets (see reviews by Begelman, Blandford and Rees, 1984; Bisnovatyi-Kogan 1993; Lovelace et al., 1999). Recent observational and theoretical work favors models where twisting of an ordered magnetic field threading an accretion disk around a black hole acts to magnetically accelerate the jets as proposed by Lovelace (1976). The nature of the ordered magnetic field threading an accretion disk envisioned by Bisnovatyi-Kogan and Ruzmaikin (1976) is shown in Fig. 8, Fig. 9 shows the outflows from a disk with such an ordered field from Lovelace (1976). Jet formation in the microquasar GRS 1915 + 105 may be seen in Fig. 10 from radio observations of Fender (1999). Picture of the jet from microquasar XTE J1550-564 obtained from X-ray satellite Chanrda are given in the paper of Kaaret et al. (2003). Formation of a jet outflowing from the accretion disk in the X-ray binary is shown schematically in Fig. 11 from Fender (2001).

Observations pictures of extragalactic jets, one and two sided, are very numerous, and are obtained in several spectral bands from radio until X-rays.

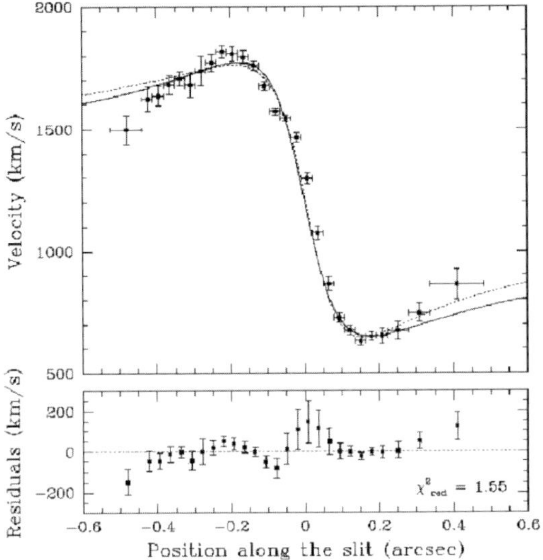

Figure 5. Optical emission-line rotation curve for the nuclear disk in M87. The two curves in the upper panel correspond to Keplerian thin disk models, and the bottom panel shows the residuals for one of the models, from Macchetto et al. (1999).

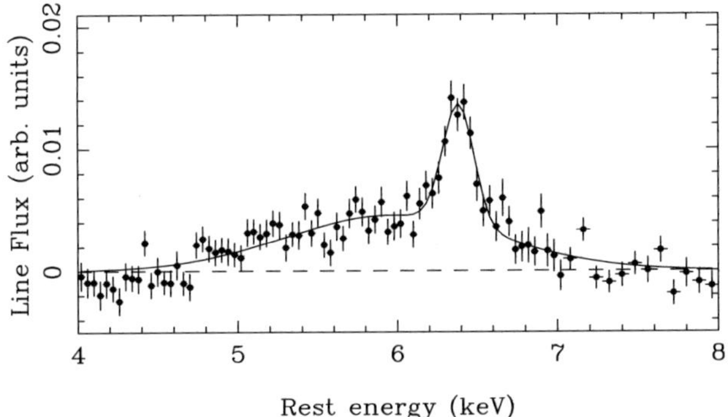

Figure 6. The Fe K line in the composite spectrum of Seyfert 1 nuclei. The solid line is a fit to the line profile using two Gaussians, a narrow component centered at 6.4 keV and a much broader, redshifted component, from Nandra et al. (1997).

Two most famous jets from AGN are shown in Fig. 12 (M 87, all bands), and Fig. 13 (quasar 3C 273, radio). Impressive pictures in X rays obtained from Chandra are represented by Marshall et al. (2001) and Harris et al. (2003) for the jet in M 87; and by Marshall et al. (2000) and Sambruna et al. (2001) for

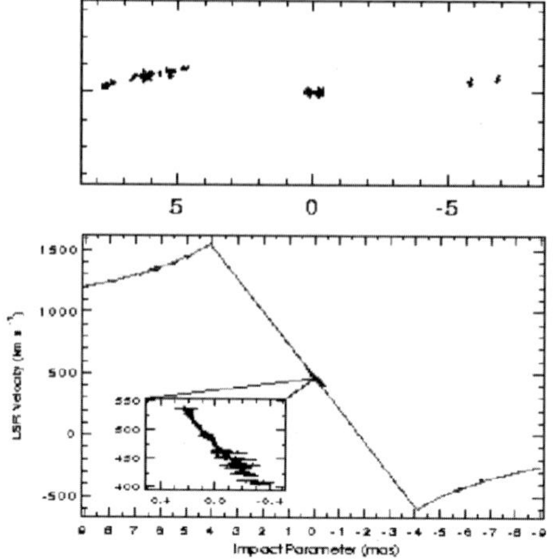

Figure 7.   Water maser emission in NGC 4258 (Miyoshi et al. 1995). Top: spatial distribution of the maser features; bottom: rotation curve. Adapted from Ho et al. (1999).

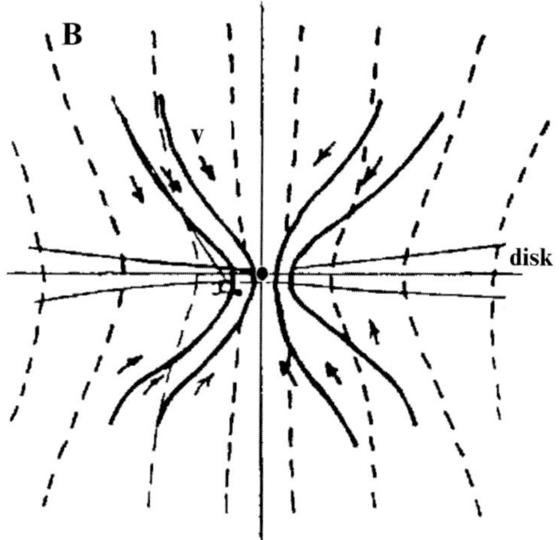

Figure 8.   Sketch of the magnetic field threading an accretion disk shown increase of the field owing to flux freezing in the accreting disk matter, from Bisnovatyi-Kogan and Ruzmaikin (1976).

the jet in the quasar 3C 273. The image of very extended radio jet from the nucleus of the galaxy IC 4296, with the length about 400 kpc is represented by Killen et al. (1986).

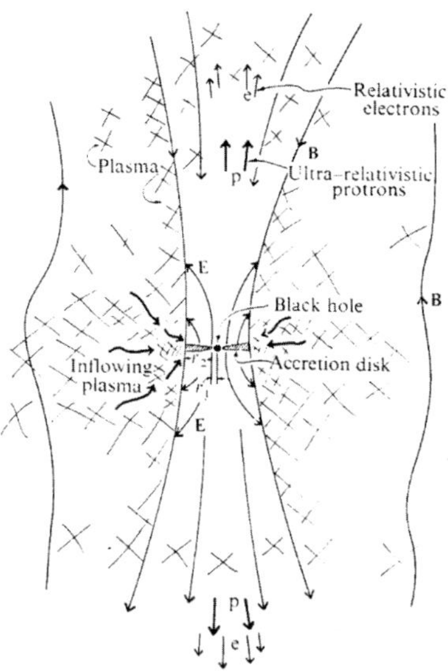

Figure 9. Sketch of the electromagnetic outflows from the two sides of the disk owing to the Faraday unipolar dynamo action of a rotating magnetized disk from Lovelace (1976).

4. Farther development of accretion theory

Few years after appearence of the standard model it was found that in addition to the opically thick disk solution there is another branch of the solution for the disk structure with the same input parameters M, $\dot{M}$, α which is also self-consistent and has a small optical thickness (Shapiro, Lightman, Eardley, 1976). Suggestion of the small optical thickness implies another equation of energy losses, determined by a volume emission

$$Q_- \approx q \rho h, \qquad (14)$$

where due to the Kirchoff law the emissivity of the unit of a volume q is connected with a Plankian averaged opacity κ_p by an approximate relation $q \approx acT_0^4\kappa_p$. In the optically thin limit the pressure is determined by a gas $P = P_g$. Analytical solutions are obtained here as well, from the same equations, with volume losses and gas pressure. In the optically thin solution the thickness of the disk is larger then in the optically thick one, and density is lower.

In order to find equations of the disc structure valid in both limiting cases of optically thick and optically thin disc, and smoothly describing transition between them, Eddington approximation had been used for obtaining formulae for a heat flux and for a radiation pressure (Artemoma et al., 1996). The following expressions had been obtained for the vertical energy flux from the disc F_0, and the radiation pressure in the symmetry plane

$$F_0 = \frac{2acT_0^4}{3\tau_0}\left(1 + \frac{4}{3\tau_0} + \frac{2}{3\tau_*^2}\right)^{-1}, \quad P_{rad,0} = \frac{aT_0^4}{3}\frac{1 + \frac{4}{3\tau_0}}{1 + \frac{4}{3\tau_0} + \frac{2}{3\tau_*^2}}, \quad (15)$$

where $\tau_0 = \kappa_e\rho h$, $\tau_* = (\tau_0\tau_{\alpha 0})^{1/2}$, $\tau_{\alpha 0} \approx \kappa_p\rho h$. At $\tau_0 \gg \tau_* \gg 1$ we have (10) from (15). In the optically thin limit $\tau_* \ll \tau_0 \ll 1$ we get

$$F_0 = acT_0^4\tau_{\alpha 0}, \quad P_{rad,0} = \frac{2}{3}acT_0^4\tau_{\alpha 0}. \quad (16)$$

Using F_0 instead of Q_- and equation of state $P = \rho\mathcal{R}T + P_{rad,0}$, the equations of accretion disc structure together with equation $Q_+ = F_0$, with Q_+ from (9), have been solved numerically by Artemova et al. (1996). It occurs that two solutions, optically thick and optically thin, exist separately when luminosity is not very large. Two solutions intersect at $\dot{m} = \dot{m}_b$ and there is no global solution for accretion disc at $\dot{m} > \dot{m}_b$ (see Fig. 14). It was concluded by Artemova et al. (1996), that in order to obtain a global physically meaningful solution at $\dot{m} > \dot{m}_b$, account of advection is needed.

It is clear from physical ground, that when a local heat production due to viscosity goes to zero near the inner edge of the disk, the heat brought by radial motion of matter along the accretion disc becomes more important. In presence of this advective heating (or cooling term, depending on the radial entropy S gradient) written as

$$Q_{adv} = \frac{\dot{M}}{2\pi r}T\frac{dS}{dr}, \quad (17)$$

the equation of a heat balance is modified to $Q_+ + Q_{adv} = Q_-$. In order to describe self-consistently the structure of the accertion disc we should also modify the radial disc equilibrium, including pressure and inertia terms. Appearance of inertia term leads to transonic radial flow with a singular point. Conditions of a continuous passing of the solution through a critical point choose a unique value of the integration constant j_{in}. In the standard local theory $j_{in} = j_{in}^{(0)}$ corresponds to the keplerian angular momentum on the last stable orbit at $r = 3r_g$, $r_g = \frac{2GM}{c^2}$. To model the effects of general relativity the gravitational potential of Paczynski & Wiita (1980) $\Phi = \frac{GM}{r - r_g}$ had been used in calculations of Artemova et al. (2001). For this potential

$$j_{in}^{(0)} \equiv l_{in}^{(0)} = \frac{3}{2}\sqrt{\frac{3}{2}}r_g c, \quad (18)$$

First approximate solution for the advective disc structure have been obtained by Paczynski and Bisnovatyi-Kogan (1981). Accretion disk models with advection in the optically thick limit have been constructed numerically by Abramovicz et al. (1988), and improved by Artemova et al. (2001). Add dynamical and radial pressure gradient term to the equation of radial equilibrium. Instead of Keplerian angular velocity we obtain radial hydrodynamic equation. Radial accretion flux becomes supersonic in the vicinity of the inner last stable orbit. The position of the critical radius is a proper value of the problem, differs noticeable from the radius of the last stable orbit at luminosity approaching the critical Eddington one $L_{Edd} = \frac{4\pi cGM}{\kappa}$ Relaxation method corrected for the existence of critical points had been used in calculations of Artemova et al. (2001), permitting to find solutions at large luminosity, formally exceeding the critical Eddington one. The results of these calculations are represented in Figs. 15-18.

4.1   Influence of the small-scale magnetic field on the accretion

While heating by viscosity is determined mainly by ions, and cooling is determined by electrons, the rate of the energy exchange between them is important for a thermal structure of the disk. The energy balance equations are written separately for ions and electrons. For small accretion rates and lower matter density the rate of energy exchange due to binary collisions is so slow, that in the thermal balance the ions are much hotter then the electrons. That also implies a high disk thickness.

It was noticed by Narayan and Yu (1995), that advection in this case is becoming extremely important. It may carry the main energy flux into a black hole, leaving rather low efficiency of the accretion up to $10^{-4} - 10^{-5}$ (advective dominated accretion flows - ADAF). This conclusion is valid only when the effects, connected with magnetic field annihilation and heating of matter due to it are neglected. To support the condition of equipartition during accretion a continuous magnetic field reconnection is necessary, leading to annihilation of the magnetic flux and heating of matter due to Ohmic heating Bisnovatyi-Kogan and Ruzmaikin (1974). The heating of electrons during reconnection is equal or larger than ion heating. While all electron energy is emitted by magneto- bremstrahlung radiation, the efficiency of accretion cannot become less than 0.25 of its standard value (0.06 for Schwarzschild metrics) (see Bisnovatyi-Kogan and Lovelace, 2001). In addition, in the highly turbulent plasma the energy exchange between ions and electrons may be strongly enhanced due to presence of fluctuating electrical fields, where electrons and ions gain the same energy. In such conditions difference of temperatures between ions and electrons may be negligible.

5.   Summary

1. Accretion into black holes (neutron stars) is a main source of energy in galactic X-ray sources and AGNs.

2. Jets, observed in AGNs and microquasars my be explained by interaction of the large-scale magnetic field with the accretion disk in self-consistent picture.

3. Accretion model deviate substantially from the standard (local) model at high accretion rates (luminosities), where advection effects are important.

4. Magnetic field action prevent formation of low-efficient flow (ADAF).

5. Turbulence in accretion disks, and turbulent viscosity is created mainly due to development of non-linear hydrodynamic instabilities at large Reynolds number.

Acknowledgments

Author is grateful to the conference organizers for support and hospitality.

References

Abramovicz, M. A., B. Czerny, J. P. Lasota, and E. Szuszkiewicz. Slim Accretion Disks *Astrophys. J.*, 332:646–658, 1988.

Artemova, I. V., G. S. Bisnovatyi-Kogan, G. Björnsson and I. D. Novikov. Structure of Accretion Disks with Optically Thick–Optically Thin Transitions. *Astrophys. J.*, 456:119–123, 1996.

Artemova, I. V., G. S. Bisnovatyi-Kogan, I. V. Igumenshchev and I. D. Novikov. On the Structure of Advective Accretion Disks at High Luminosity. *Astrophys. J.*, 549:1050–1061, 2001.

Balbus S. A. and J. F. Hawley. Instability, Turbulence, and Enhanced Transport in Accretion Disks. *Rev. Mod. Phys.*, 70:1–53, 1998.

Begelman M. C., Blandford R. D. and M. J. Rees. Theory of Extragalactic Radio Sources. *Rev. Mod. Phys.*, 56:255–351, 1984.

Bisnovatyi-Kogan, G. S. X ray Sources in Close Binary Systems: Theoretical Aspects. *Bulletin Abastumani Astrophys. Obs.*, No. 58:175–210, 1985.

Bisnovatyi-Kogan, G. S. Mechanisms of Jet Formation. In L. Errico and A. A. Vittone, editors, *Proceedings of Int. Conf. Stellar Jets and Bipolar Outflows*, 369–381, 1993. Kluwer, Dordrecht.

Bisnovatyi-Kogan G. S. and S. I. Blinnikov. Disk Accretion onto a Black Hole at Subcritical Luminosity. *Astron. Ap.*, 59:111–125, 1977.

Bisnovatyi-Kogan G. S. and R. V. L. Lovelace. Advective Accretion Disks and Related Problems Including Magnetic Fields. *New Astronomy Reviews*, 45:663–742, 2001.

Bisnovatyi-Kogan, G. S. and A. A. Ruzmaikin. The Accretion of Matter by a Collapsing Star in the Presence of a Magnetic Field. *Astrophys. and Space Sci.*, 28:45–59, 1974.

Bisnovatyi-Kogan, G. S. and A. A. Ruzmaikin. The Accretion of Matter by a Collapsing Star in the Presence of a Magnetic Field. II - Selfconsistent Stationary Picture. *Astrophys. and Space Sci.*, 42:401–424, 1976.

Cannizzo, J., P. Ghosh and J. C. Wheeler Convective Accretion Disks and the Onset of Dwarf Nova Outbursts. *Astrophys. J. Lett.*, 260:L83–L86, 1982.

Chandrasekhar, S. *Hydrodynamic and Hydromagnetic Stability*. International Series of Monographs on Physics, Oxford: Clarendon, 1961.

Chen, W., C. R. Shrader, and M. Livio. The Properties of X-Ray and Optical Light Curves of X-Ray Novae. *Astrophys. J.*, 491:312–338, 1997.

Cherepashchuk, A. M. X-ray Nova Binary Systems. *Space Sci. Rev.*, 93:473–580, 2000.

Conway, R. G., S. T. Garrington, R. A. Perley, and J. A. Biretta. Synchrotron Radiation from the Jet of 3C 273. II - The Radio Structure and Polarization. *Astron. Ap.*, 267:347–362, 1993.

Fender, R. Relativistic Jets from X-ray Binaries. *Astro-ph/9907050*, 1999.

Fender, R. Relativistic Outflows from X-ray Binaries. *Astro-ph/0109502*, 2001.

Glatzel W. Instabilities in Astrophysical Shear Flows. *Rev. in Modern Astron.*, 4:104–116, 1991.

Harris, D. E., J. A. Biretta, W. Junor, E. S. Perlman, W. B. Sparks and A. S. Wilson. Flaring X-ray Emission from HST-1, a Knot in the M87 Jet. *Astrophys. J.*, 586:L41-L44, 2003.

Ho, L. Supermassive Black Holes in Galactic Nuclei: Observational Evidence and Astrophysical Consequences. *Observational Evidence for the Black Holes in the Universe, Proc. Conference held in Calcutta, January 11-17th, 1998.*, 157–186, Kluwer, 1999.

Kaaret, P., S. Corbel, J. A. Tomsick, R. P. Fender, J. M. Miller, J .A. Orosz, A. K. Tzioumis and R. Wijnands. X-Ray Emission from the Jets of XTE J1550-564. *Astrophys. J.*, 582:945–953, 2003.

Killeen, N. E. B., G. V. Bicknell and R. D.Ekers. The radio galaxy IC 4296 (PKS 1333 - 33). I - Multifrequency Very Large Array observations. *Astrophys. J.*, 302:306–336, 1986.

Kuznetsov, O. A. On the excitation of hydrodynamical turbulence in accretion discs. INTERACTING BINARIES: Accretion, Evolution, and Outcomes. AIP Conference Proceedings, Volume 797, pp. 271–276, 2005.

Lasota, J.-P. The disc instability model of dwarf novae and low-mass X-ray binary transients. *New Astronomy Reviews*, 45:449–508, 2001.

Lovelace, R. V. E. Dynamo model of double radio sources. *Nature*, 262:649–652, 1976.

Lovelace, R. V. E., G. S. Ustyugova and A. V. Koldoba. Magnetohydrodynamic Origin of Jets from Accretion Disks. *Active Galactic Nuclei and Related Phenomena, IAU Symposium 194, eds. Y. Terzian, D. Weedman and E. Khachikian (Astron. Soc. of Pacific: San Francisco)*, 208–217, 1999.

Macchetto, F., A. Marconi, D. J. Axon, A. Capetti, W. Sparks and P. Crane. The Supermassive Black Hole of M87 and the Kinematics of Its Associated Gaseous Disk. *Astrophys. J.*, 489:579–600, 1997.

Marshall, H. L., D. E. Harris, J. P. Grimes, J. J. Drake A. Fruscione, M. Juda, R. P. Kraft, S. Mathur, S. S. Murray, P. M. Ogle, D. O. Pease, D. A. Schwartz, A. L. Siemiginowska, S. D. Vrtilek and B. J. Wargelin. Structure of the X-Ray Emission from the Jet of 3C 273. *Astrophys. J.*, 549:L167–L171, 2001.

Marshall, H. L., B. P. Miller, D. S. Davis, E. S. Perlman, M. Wise, C. R. Canizares and D. E. Harris. A High-Resolution X-Ray Image of the Jet in M87. *Astrophys. J.*, 564:683–687, 2002.

Mirabel, I. F. and L. F. Rodriguez. A Superluminal Source in the Galaxy. *Nature*, 371:46–48, 1994.

Miyoshi, M., J. Moran, J. Herrnstein, L. Greenhill, N. Nakai, P. Diamond, and M. Inoue. Evidence for a Black Hole from High Rotation Velocities in a Sub-parsec Region of NGC4258. *Nature*, 373:127–129, 1995.

Nandra, K., I. M. George, R. F. Mushotzky, T. J. Turner, T. Yaqoob. ASCA Observations of Seyfert 1 Galaxies. II. Relativistic Iron K alpha Emission. *Astrophys. J.*, 477:602–622, 1997.

Narayan, R. and I. Yi. Advection-dominated Accretion: Underfed Black Holes and Neutron Stars. *Astrophys. J.*, 452:710–735, 1995.

Novikov, I. D. and K. S. Thorne. Astrophysics of Black Holes. *Black Holes eds. C.DeWitt & B.DeWitt (New York: Gordon & Breach)*, 345–405, 1973.

Paczyński, B. and G. S. Bisnovatyi-Kogan. A Model of a Thin Accretion Disk around a Black Hole. *Acta Astron.*, 31:283–291, 1981.

Richard, D., F. Hersant, O. Dauchot, F. Daviaud, B. Dubrulle and J-P. Zahn. A powerful local shear instability in stratified disks. *Astro-ph/0110056*, 2001.

Sambruna, R. M., C. M. Urry, F. Tavecchio, L. Maraschi, R. Scarpa, G. Chartas and T. Muxlow. Chandra Observations of the X-Ray Jet of 3C 273. *Astrophys. J.*, 549:L161-L165, 2001.

Schwartsman, V. F. Halos around "Black Holes". *Soviet Astron.*, 15:377–388, 1971.

Shakura, N. I. Disk Model of Gas Accretion on a Relativistic Star in a Close Binary System. *Astron. Zh.*, 49:921–930, 1972 (1973, Sov. Astron. 16, 756).

Shakura, N. I. and R. A. Sunyaev. Black holes in binary systems. Observational appearance. *Astron. Ap.*, 24:337–355, 1973.

Shapiro, S. L., A. P. Lightman and D. M. Eardley. A two-temperature accretion disk model for Cygnus X-1 - Structure and spectrum. *Astrophys. J.*, 204:187–199, 1976.

Velikhov, E.P. Stability of an ideally conducting liquid flowing between rotating cylinders in a magnetic field. *J. Exper. Theor. Phys.*, 36:1398–1404, 1959.

Zeldovich, Ya. B. On the friction of fluids between rotating cylinders. *Proc. R. Soc. Lond., A*, 374:299–312, 1981.

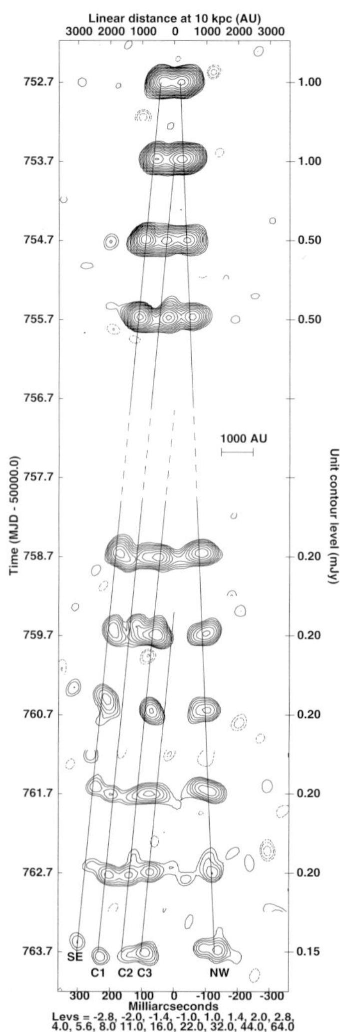

Figure 10. A sequence of ten epochs of radio imaging of relativistic ejections from the black hole candidate X-ray binary GRS 1915+105 using MERLIN at 5 GHz. The figure has been rotated by 52 degrees to form the montage. For an estimated distance to the source of 11 kpc the approaching components have an apparent transverse velocity of $1.5c$. Assuming an intrinsically symmetric ejection and the standard model for apparent superluminal motions, an intrinsic bulk velocity for the ejecta is $0.98^{+0.02}_{-0.05}c$ at an angle to the line of sight of 66 ± 2 degrees (at 11 kpc). The apparent curvature of the jet is probably real, although the cause of the bending is uncertain. (from Fender, 1999).

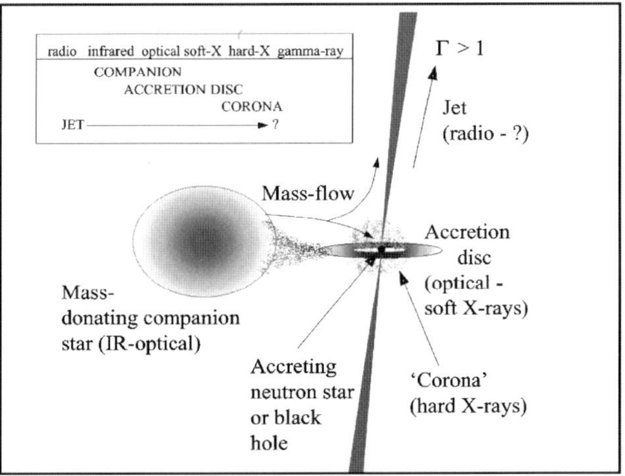

Figure 11.   Schematic picture of jet formation in binary system (from Fender, 2001).

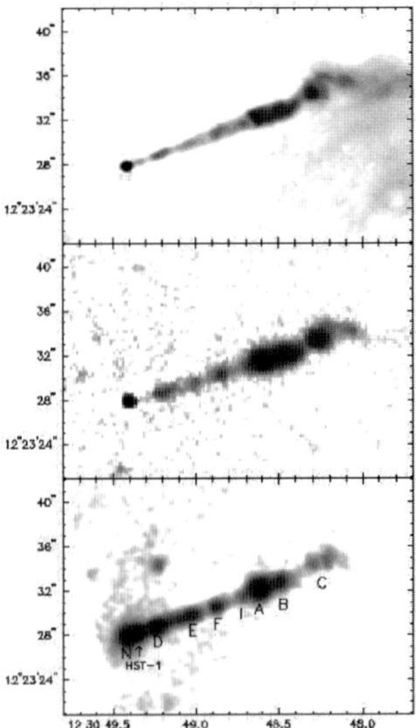

Figure 12. Gray scale representations of a 6 cm radio (top panel), an optical V band (middle panel) and the Chandra X-ray (bottom panel, 0.1 - 10 keV band) image. In the radio image, the gray scale is proportional to the square root of the brightness, in the optical image, the gray scale is also proportional to the square root of the brightness. The labels in the lower panel refer to the knots vertically above the label. N is the nucleus, from Wilson and Yang (2002).

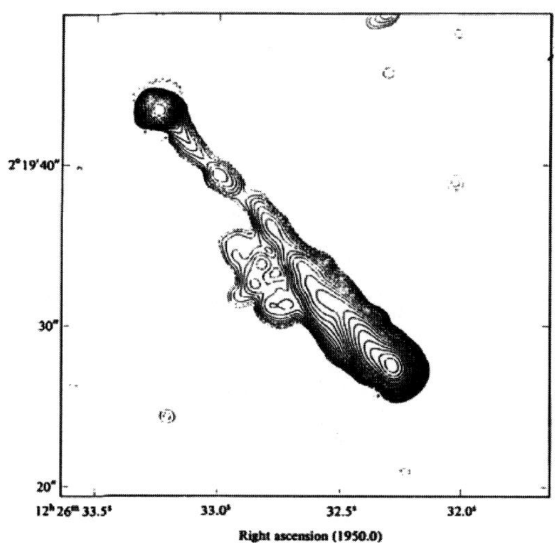

Figure 13.   MERLIN map of 3C 273 at 408 MHz. Resolution: 1.0 arcsec, from Conway et al. (1993).

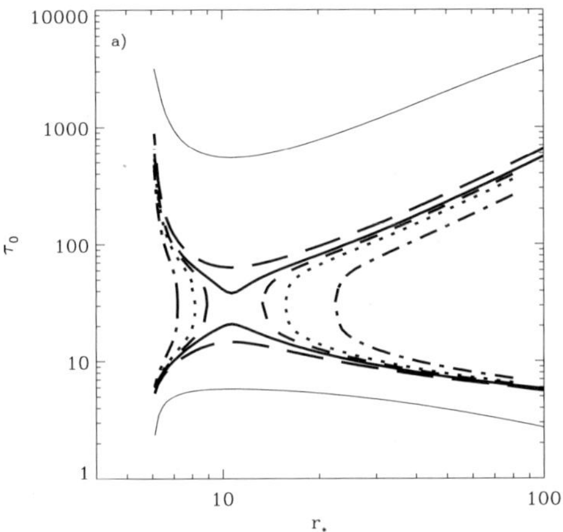

Figure 14. The dependences of the optical depth τ_0 on radius, $r_* = r/r_g$, for the case $M_{BH} = 10^8\ M_\odot$, $\alpha = 1.0$ and different values of $\dot{m}$. The thin solid, dot-triple dash, long dashed, heavy solid, short dashed, dotted and dot-dashed curves correspond to $\dot{m} = 1.0, 3.0, 8.0, 9.35, 10.0, 11.0, 15.0$, respectively. The upper curves correspond to the optically thick family, lower curves correspond to the optically thin family, from Artemova et al. (1996).

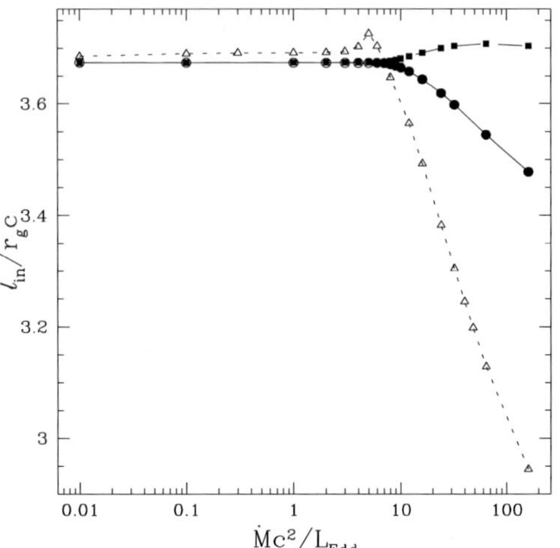

Figure 15.   The specific angular momentum $j_{in} \equiv \ell_{in}$ as a function of the mass accretion rate $\dot{M}$ for different viscosity parameters $\alpha = 0.01$ (squares), 0.1 (circles) and 0.5 (triangles), corresponding to viscosity prescription (5). The solid dots represent models with the saddle-type inner singular points, whereas the empty dots correspond to the nodal-type ones, from Artemova et al. (2001).

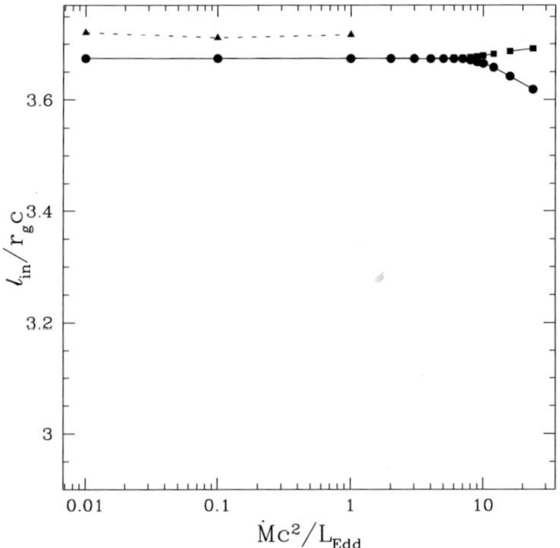

Figure 16.   The specific angular momentum $j_{in} \equiv \ell_{in}$ as a function of the mass accretion rate $\dot{M}$ for different viscosity parameters $\alpha = 0.01$ (squares), 0.1 (circles) and 0.5 (triangles), corresponding to viscosity prescription (6). The solid dots represent models with the saddle-type inner singular points, whereas the empty dots correspond to the nodal-type ones, from Artemova et al. (2001).

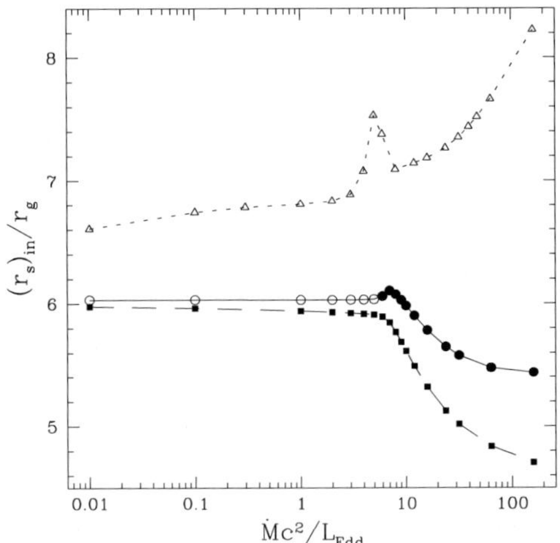

Figure 17.   The position of the inner singular points as a function of the mass accretion rate $\dot{M}$ for different viscosity parameters $\alpha = 0.01$ (squares), 0.1 (circles) and 0.5 (triangles), corresponding to viscosity prescription (5). The solid dots represent models with the saddle-type inner singular points, whereas the empty dots correspond to the nodal-type ones, from Artemova et al. (2001).

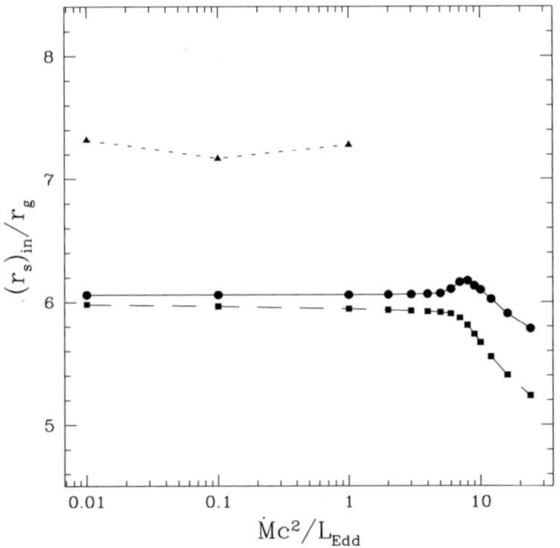

Figure 18.   The position of the inner singular points as a function of the mass accretion rate $\dot{M}$ for different viscosity parameters $\alpha = 0.01$ (squares), 0.1 (circles) and 0.5 (triangles), corresponding to viscosity prescription (6). The solid dots represent models with the saddle-type inner singular points, whereas the empty dots correspond to the nodal-type ones, from Artemova et al. (2001).

SUPERCRITICAL ACCRETION DISK IN SS433

A. M. Cherepashchuk
Sternberg Astronomical Institute, Moscow, Russia
cher@sai.msu.ru

Abstract   Characteristics of supercritical precessing accretion disk in SS433 X-ray eclipsing binary system, microquasar, that are based on optical and X-ray observations are presented. New hard X-ray observations of SS433 made from the border of the INTEGRAL observatory have shown that SS433 exhibits both precessional and eclipsing variability in the range 25–50 and 50–100 KeV. Amplitude of the hard X-ray precessing variability is about 80%. The depth of hard X-ray eclipse at the phase of maximum opening of the accretion disk is 80% which is more than that in the standard 2-10 KeV X-ray range (50%). Full width of eclipse in hard X-ray range at half intensity is 2.0–2.2 days which is close to that in 2-10 KeV range. Binary parameters of SS433 are estimated and some constrains are imposed on the structure of supercritical accretion disk.

Keywords: supercritical accretion, X-rays, massive binary systems, SS 433

1.   Introduction

SS433 object is a massive eclipsing X-ray binary system at an advanced evolutionary stage (see Margon, 1984; Cherepashchuk 1981, 1988, 2002 and references therein). An optical A-star (Gies et al., 2002) overfills its Roche Lobe. The optically bright, supercritical precessing accreting disk is formed around the relativistic object and highly collimated (opening angle $\sim 1°$) relativistic (velocity of streaming matter $v \simeq 80000$ km s^{-1}) precessing jets emanate from the central parts of the accretion disk.

Observational appearances of jets in SS433 binary system have been detected in radio, optical and X-ray spectral ranges (Kotani et al., 1996; Marshall et al., 2002; Panferov and Fabrica, 1997; Velmeulen et al., 1993). Three basic periodicities have been observed in SS433: precessional periodicity ($P_{prec} = 162^d.5$), orbital periodicity ($P_{orb} = 13^d.0821$) and nutational periodicity ($P_{nut} = 6^d.2877 \approx (2f_{orb} + f_{prec})^{-1}$). SS433 is located at the center of supernova remnant (plerion) W50 with the age of $10^4 - 10^5$ years. The distance

121

Alexei M. Fridman, et al. (eds.), Astrophysical Disks; Collective and Stochastic Phenomena 121–130
© 2006 *Springer. Printed in the Netherlands*

to SS433 is $\sim$5 Kpc, the source having high interstellar reddening ($A_v \sim 7^m.5$) and visual magnitude $V \simeq 13^m.7 - 15^m.5$ (see Catalogue by Cherepashchuk et al., 1996).

The orbital period of $13^d.0821$ exhibits no changes despite the intense mass outflow via stellar wind (v=3000 km s^{-1} , $\dot{M} \simeq 10^{-4} M_\odot yr^{-1}$) both from the supercritical accretion disk and the optical star. The light variations of SS433 are also characterized by optical flares on time scales of several days which are correlated with radio flares and pass ahead of them by one or two days.

2.   SS433 - a supercritical microquasar with hard X-ray spectrum

It is established now that SS433 is a microquasar with a black hole, a supercritical precessing accretion disk and precessing quasistationary relativistic jets. Note that the first description of supercritical accretion onto a black hole in a binary system was given by Shakura and Sunyaev (1973).

It is well known that X-ray spectra of black hole X-ray binaries in contrast to accreting neutron stars, show power-law tails extending up to many hundreds of KeV (Sunyaev et al., 1991a,b). In supercritical accretion regime the thermalization of X-ray emission within the optically thick wind outflowing from the supercritical accretion disk is expected to cut off the spectrum of an accreting black hole at the comparatively low energies, by order of several tens of KeV. At the same time, there are geometrical factors, such as the tunnel swept by the relativistic jets in the wind, that leave the hope to observe a possible hard X-ray component from SS433, at least in the precessing phases when the disk is observed mostly face-on, when the inclination angle to the line of sight $\sim 30°$.

Hard X-ray emission of SS433 was observed by RXTE satellite (Kotani et al., 2002). The IBIS detector onboard the INTEGRAL satellite offers a unique possibility of studying SS433 in hard X-rays up to hundreds of KeV (for more detail on INTEGRAL scientific payload and mission description see Wincler 1996, 1999).

Recently (Cherepashchuk et al., 2003) hard X-ray component was detected from the supercritically accreting black hole in SS433 using INTEGRAL observations carried out in March–May, 2003.

SS433 is distinctly seen in the IBIS field of view at 25–50 KeV as well as at 50–100 KeV energy bands (see Figure 1) along with the microquasar GRS 1915+105 hosting a subcritically accreting black hole. Distances for GRS1915+105 and SS433 are $\sim$12 and $\sim$5 Kpc respectively, but GRS 1915+105 at the band 50–100 KeV is seen much brighter than SS433. It is due to the fact that GRS 1915+105 containing a rapidly rotating black hole (Greiner et al., 2001) is observed at subcritical regime of accretion, in contrast with SS433. The X-ray luminosity of GRS1915+105 in the hard range (20–100

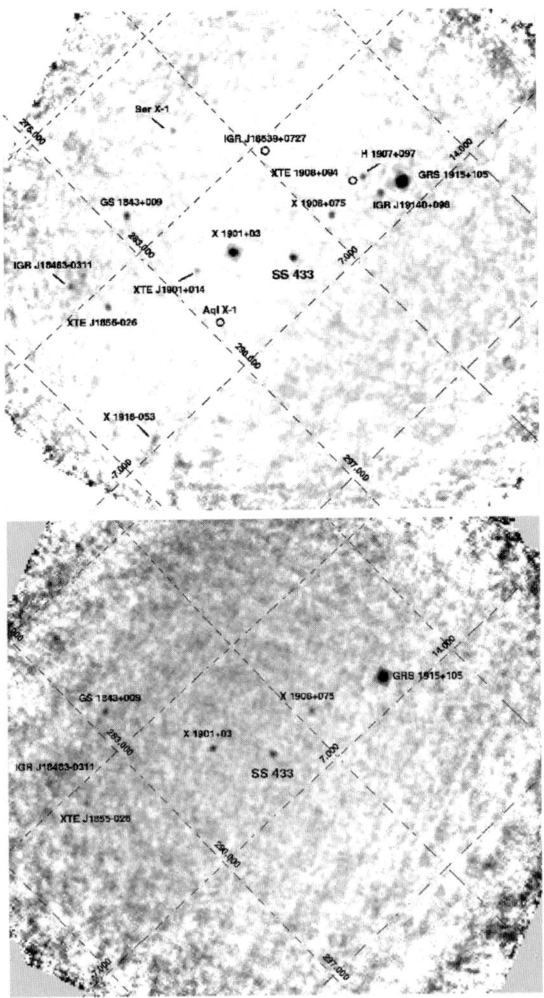

Figure 1. Map of the sky around SS433 obtained by IBIS/ISGRI telescopes of the INRTE-GRAL space X-ray observatory on May 2003 in 25-50 KeV energy band (top panel). Map of the sky around SS433 obtained by IBIS/ISGRI telescopes of the INRTEGRAL space X-ray observatory on May 2003 in 50-100 KeV energy band (bottom panel) (from Cherepashchuk et al., 2003).

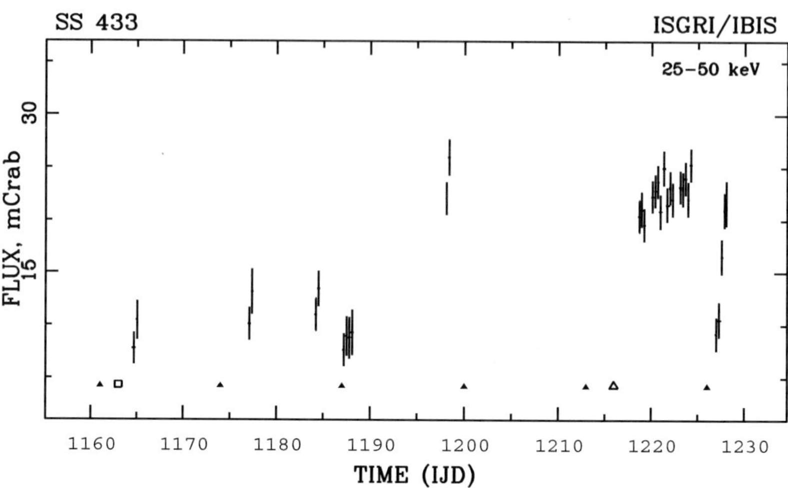

Figure 2. Light curve of SS433 (IBIS/ISGRI telescope, 25-50 KeV) in March–May 2003 (from Cherepashchuk et al., 2003). The filled triangles indicate optical eclipse minima according to Goranskii et al., (1998). The white triangle and the square indicate precessional phases of face-on and cross-over, respectively. IJD corresponds to INTEGRAL Julian Date: MJD-51544.

KeV) reaches a value of about $10^{37} - 10^{38}$erg s^{-1} (see e.g. Cherepashchuk, 2000). The X-ray luminosity of SS433 is $3 \cdot 10^{35}$erg s^{-1} in the range 25–50 KeV and $1.2 \cdot 10^{35}$erg s^{-1} in the range 50–100 KeV. Therefore hard X-ray luminosity of a subcritical accretion disk around the black hole (GRS 1915+105) is 2–3 order of magnitude higher than that of a supercritical accreting disk (SS433). The fact that SS433 is seen rather well up to 100 KeV energies evidences of the important role of the geometric factor in the structure of a supercritical accreting disk (presence of a central tunnel swept by relativistic jets) and may suggest the presence of relativistic particles in the vicinity of the supercritically accreting black hole (Cherepashchuk et al., 2003).

The hard X-ray variability at the bottom of the jets provides information on mechanisms of jet production and collimation. We did not find any evidence for periodic pulsations on time scales 400–4000 s, nor quasiperiodic oscillations (typical for subcritically accreting black holes, like GRS 1915+105) were observed.

3.   X-ray light curve and binary parameters for SS433

The light curves of SS433 obtained by the IBIS/ISGRI detector of INTEGRAL observatory exhibit precessional and orbital eclipsing variability in the 25–50 KeV band (see Figure 2).

X-ray flux from SS433 is minimum at the cross-over time and it increases gradually due to precession, beginning from the cross-over moment (the accretion disk is seen edge-on) until the T_3, corresponding to maximum separation of moving emission lines (the accretion disk is seen face-on). The X-ray flux changes with the precessional phase in the 25–50 KeV band, the highest value being over 3 as large as its lowest value. The observations of May 9 and 13, 2003 clearly show the decrease of the X-ray flux from 25 to 7 mCrab which is apparently due to an X-ray eclipse in the primary orbital minimum when the "normal" star is in front of the compact source (on May 11). Note that the X-ray flux exhibits essential decrease at all the observed moments of an expected optical eclipses (Goranskii et al., 1998). So the X-ray eclipses in the hard X-ray range are stable and regular phenomena in observational manifestations of SS433 binary system. Though the observation sequence is not continuous, an estimated full width of the X-ray eclipse at half intensity is 2.0–2.2 days. Therefore, our observations of SS433 in the hard X-ray range confirm the high durability of X-ray eclipse (Kotani et al., 1996).

Since narrow photospheric absorption lines are clearly seen in the spectrum of the optical A-star (Gies et al., 2002), the duration $\sim 2^d - 2^d.2$ is certainly due to the geometrical screening of inner parts of accretion disk and relativistic jets by the proper body of the optical star but not by stellar wind or by the interacting region of winds. Since the optical A-star fills its Roche Lobe we can estimate the components mass ratio for the SS433 binary system from the duration of an X-ray eclipse (d=$2^d - 2^d.2$) as $q = \frac{m_x}{m_v} \simeq 0.2 - 0.3$ (Antokhina et al., 1992). Using this value of q and the mass function of the optical star (Gies et al., 2002) we can estimate basic parameters of SS433 binary system. According to Gies et al. the semiamplitude of a radial velocity curve of the optical A-star in the SS433 system is $K_v = 100 \pm 15$ km s^{-1}. Corresponding values of the mass function of the optical star are $f_v(m) = \frac{m_x^3 sin^3 i}{(m_x + m_v)^2} = (0.8 \div 2)M_\odot$. A mass of the relativistic object is $m_x = f_v(m)(1 + \frac{1}{q})^2 \cdot \frac{1}{sin^3 i} = 1.06 f_v(m)(1 + \frac{1}{q})^2$, because for SS433 $i = 79°$ is determined from the analysis of moving emission lines (Margon, 1984). Taking observed value $f_v(m) = (0.8 - 2)M_\odot$ and $q = 0.2$ we get $m_x = (31 - 76)M_\odot$, $m_v = (159 - 371)M_\odot$. These values seem to be unrealistic because the temperature of the A7 star ($T_v = 8000$K) corresponds to radii range $R_v = (170 - 400)R_\odot$ for this star, which is much more than the separation between the components of the system, $a = (100 \div 120)R_\odot$. Only if we take the lowest limit for K_v=100-3σ=55 km s^{-1} we can obtain more realistic parameters for the SS433 binary system:

$$m_x = 7.6 M_\odot, \ m_v = 38 M_\odot, \ R_v \simeq 50 R_\odot \ (a \simeq 80 R_\odot)$$

An expected value of the semiamplitude of a radial velocity curve for the relativistic object should be $K_x = 275$ km s^{-1} which is too large: an observed value of $K_x = 175$km s^{-1} (obtained from stationary emission line HeII 4686).

Therefore, values of K_v and K_x should be checked by additional optical observations of SS433. A small value of the mass ratio, $q = 0.2$, for SS433 also makes some problems with interpretation of optical light curves of SS433 (Goranskii et al., 1998; Cherepashchuk, 2002). As the inclination of the orbital plane for SS433 is fixed, $i = 79°$ (Margon, 1984), total optical eclipses of the accretion disk by the optical star should be observed for $q = 0.2$ at all phases of the precessional $162^d.5$ period, that is not the case. Therefore, we can suggest that the main part of optical luminosity of the supercritically accreting disk in SS433 binary system is formed in the "thick bright optical jets" emanated from the central parts of the accretion disk. A short time lag $\sim 0^d.6$ between nutational periodicity in moving emission lines and nutational variations of optical continuum for SS433 (Cherepashchuk, 2002) support this suggestion. Time delay $0^d.6$ may be accounted for by the fact that light travels from the central part of the accretion disk and the optically emitting parts of precessing relativistic jets.

4.   Properties of the supercritical accretion disk

The spectrum of SS433 in the ranges from radio to hard X-rays is presented in Figure 3. In agreement with the theoretical prediction (Shakura and Sunyaev, 1973), maximum of the energy distribution is located near the optical range.

The mass loss rate from the accretion disk reaches very high values ($\sim 10^{-4} M_\odot yr^{-1}$) and most of the hard X-ray radiation generated in the vicinity of the accreting relativistic object is absorbed by wind and reradiated in optical and ultraviolet ranges. Optical and ultraviolet luminosity of the accretion disk is $\sim 10^{39}$ erg s^{-1} (Antokhina and Cherepashchuk, 1987). X-ray luminosity of the accretion disk and relativistic jets in the standard energy range 2-10 KeV is $\sim 10^{36}$ erg s^{-1}, which is $\sim 10^{-3}$ of its luminosity in optic and ultraviolet. Our INTEGRAL observations revealed significant X-ray luminosity of the supercritical accretion disk in the hard X-ray range: $3 \cdot 10^{35}$ erg s^{-1} for E=25-50 KeV and $1.2 \cdot 10^{35}$ erg s^{-1} for E=50-100 KeV. This new observational fact allows us to suggest that the supercritical accretion disk in the SS433 binary system has complicated structure. Its "photosphere" is not continuous but presumably has a transparent "hole" in the central part of the disk (central "cone" in which the density of wind is significantly decreased).

The depth of an X-ray eclipse in the hard X-ray range (25-50) KeV is estimated as $\sim 80\%$ which is much more than that in the standard range, 2-10 KeV ($\sim 50\%$). Therefore, the hard X-ray radiation is much more concentrated toward the center of the accretion disk than the radiation in the standard X-ray range.

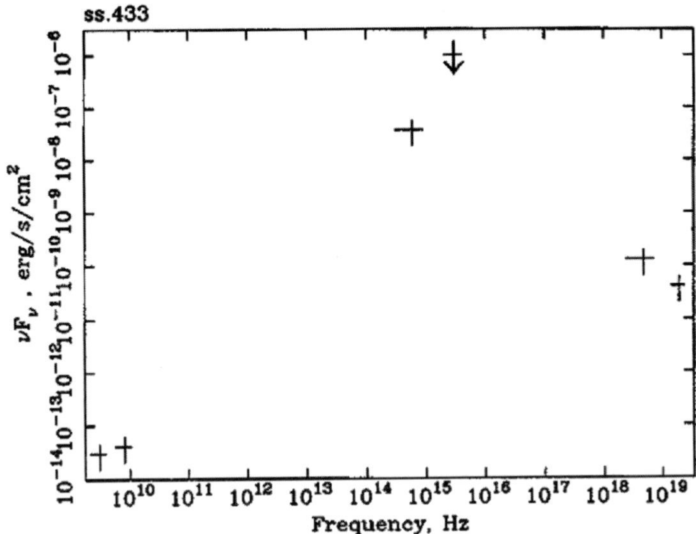

Figure 3. The total energy spectrum of SS433 in units νF_ν from radio to hard X-ray (from Cherepashchuk et al., 2003).

It should be stressed that kinetic power of the jet in SS433 is about $3 \cdot 10^{38} - 10^{39}$ erg s^{-1} which is close to the optical and ultraviolet luminosity of the accretion disk. Therefore efficiency of the energy transformation of relativistic plasma rotating around the central relativistic object into the kinetic energy of jets is very high. This fact imposes strong constrains on the mechanisms of formation and collimation of relativistic jets in a supercritical accretion disk (see e.g. Bisnovatyi-Kogan et al., 1969; Okuda, 2002; Das et al., 2003).

5. Conclusions

New INTEGRAL observations of SS433 have shown that this supercritical microquasar appears to be an eclipsing binary in the hard X-ray range with the full width of an eclipse $\sim 2^d.2$ which is close to that in the standard 2-10 KeV range. This result is quite opposite to what is found in ordinary eclipsing X-ray binaries, like Cen X-3, Vela X-1, etc in which the X-ray eclipse duration decreases with energy. This new fact may reflect a complicated structure of the innermost supercritical accretion disk in SS433. The eclipse depth in the hard X-ray range 25-50 KeV for SS433 is about 80% compared to $\sim 50\%$ in the standard range 2-10 KeV. This fact suggests smaller extension of the region where hard X-rays are formed when compared with the standard X-ray range.

Parameters of the binary system for SS433 derived from the duration of an eclipse in the hard X-ray range allow us to suggest that a value of the semiamplitude of the radial velocity curve for the optical A-star is $K_v \simeq 50 - 60$ km s^{-1}. For more precise determination of the binary parameters of SS433 further hard X-ray observations are needed.

New spectroscopic observations of the optical A-star in this microquasar are necessary too. These observations are in progress now at the INTEGRAL X-ray observatory and at the 6-meter telescope of the Special Astrophysical observatory (SAO RAS).

Acknowledgments

The author acknowledge R.A.Sunyaev, E.V.Seifina, I.E. Panchenko, S.V.Molkov and K.A.Postnov for valuable discussion.

This work is supported by the Russian Foundation for Basic Research (project 00-02-17524) and Program for State Support of Leading Scientific Schools of Russian Federation (project 388.2003.2).

References

Antokhina, E.A., Seifina, E.V., Cherepashchuk, A.M. 1992, Soviet Astron., 36, 143

Bisnovatyi-Kogan, G.S., Komberg, B.V., Freedman, A.M. 1969, Astron. Zh., 46, 465

Cherepashchuk, A.M. 1981, MNRAS, 194, 761

Cherepashchuk, A.M. 1988, Sov. Sci. Rev. Ap. Space Phys., R.A.Sunyaev (ed.), 7, 1

Cherepashchuk, A.M. 2000, Space Sci. Rev., 93, 473

Cherepashchuk, A.M. 2002, Space Sci. Rev., 102, 23

Cherepashchuk, A.M., Katysheva, N.A., Khruzina, T.S., Shugarov, S.Yu. 1996, Highly Evolved Close Binary Stars: Catalog, Gordon and Breach Publ., Amsterdam

Cherepashchuk, A.M., Sunyaev, R.A., Seifina E.V., Panchenko, I.E., Molkov S.V., Postnov, K.A. 2003, Astron. Astroph., 411, L441

Das, T.K., Rao, A.R., Vadawale, S.V. 2003, MNRAS, 343, 443

Gies, D.R., Huang, W., McSwain, M.V. 2002, Astrophys. J., 578, 67

Goranskii, V.P., Esipov, V.F., Cherepashchuk, A.M. 1998, Astron. Rep., 42, 209

Greiner, J., Cuby, J.G., McCaughrean, M.J. 2001, Nature, 414, 522

Kotani, T., Kawai, N., Matsuoka, M., Brinkman, W. 1996, Publ. Astron. Soc. Japan, 48, 619

Kotani, T., Band, D., Denissyuk, E.K., et al. In C.A. Tout and W.van Hamme, editors, Proceedings of IAU Colloqium 187, Exotic Stars as Challenges to Evolution, 2002, San Francisco, USA, A.S.P. Conf. Ser., 279, 19

Okuda, T. 2002, PASJ, 54, 253

Margon, B. 1984, Ann. Rev. Astron. Aph., 22, 507

Marshall, H.L., Canizares, C.R., Schulz, N.,S. 2002, Astrophys. J., 564, 941

Panferov, A.A., Fabrika, S.N. 1997, Astron. Zh., 74, 574

Shakura, N.I. and Sunyaev, R.A. 1973, Astron. Aph., 24, 337

Sunyaev, R.A., et al. 1991a, Astron. Aph., 247, L29

Sunyaev, R.A., et al. 1991b, Astrophys. J., 389, L49

Vermeulen, R.C., Murdin, P.G., van den Heuvel, E.P.J. et al., 1993, Astron. Aph., 270, 204
Winkler, C. 1996, Astron. Aph. Suppl., 120, 637
Winkler, C. 1999, Astrophys. Lett. and Comm., 39, 309

GALACTIC VORTICES

G. Contopoulos, and P. A. Patsis
Research Center for Astronomy, Academy of Athens
Anagnostopoulou 14, GR-10673, Athens, Greece

Abstract   Vortices are typical for the stellar and gaseous flows in the corotation region of barred and spiral galaxies. Their origin is associated with the presence of stable long period orbits around the stable Lagrangian points L_4 and L_5. However, the corotation region is also characterized by chaotic motion, mainly around the unstable Lagrangian points. Stars may follow ordered or chaotic orbits. The gas cannot follow the chaotic motion and is arranged in such a way as to follow anti-cyclonic streamlines around L_4 and L_5. In some models there are also cyclonic motions of the gas near the unstable Lagrangian points.

Keywords: galaxies: dynamics, morphology; chaos

1.   Long and Short Period Orbits

The orbits of stars near corotation have been studied extensively by Contopoulos (1973, 1978, 1981, 1983, 1988, 2002). According to these studies there are two main types of orbits. Ordered orbits, mainly around the stable Lagrangian points L_4 and L_5, and chaotic orbits, mainly around the unstable points L_1 and L_2.

The ordered orbits surround stable periodic orbits, namely the long period orbits (LPO, or banana orbits) for relatively small energies in the rotating frame (i.e. small values of the Jacobi constant), and the short period orbits (SPO, or ring orbits) for larger energies (Fig. 1).

Figure 2 shows a nonperiodic banana orbit with short period oscillations around a long period orbit. A short theoretical description of the long and short period orbits is given in Appendix A.

The long period orbits do not form a single family. The characteristic of the LPO is composed of an infinity of independent LPO families separated by gaps (Fig. 3).

These families are connected by unstable bridges with the SPO family. A bridge is generated at a bifurcation of a short period orbit described n times

Alexei M. Fridman, et al. (eds.), Astrophysical Disks; Collective and Stochastic Phenomena 131–144
© 2006 *Springer. Printed in the Netherlands*

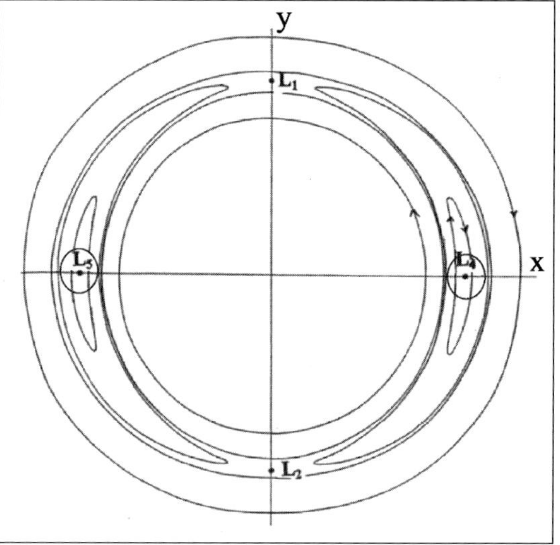

Figure 1.   Long period orbits (bananas) and short period orbits around L_4, L_5 in a barred galaxy. The bar is along the y-axis. Inside and outside corotation there are nearly circular orbits of the family x1, in direct and retrograde direction respectively.

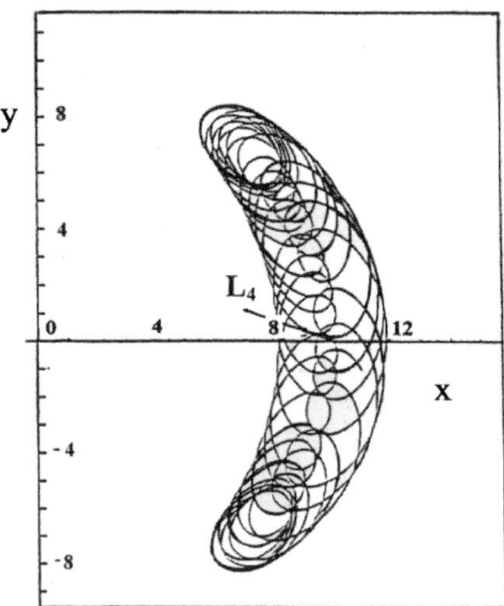

Figure 2. A nonperiodic banana orbit near L_4 consists of short period oscillations around an elongated long period orbit.

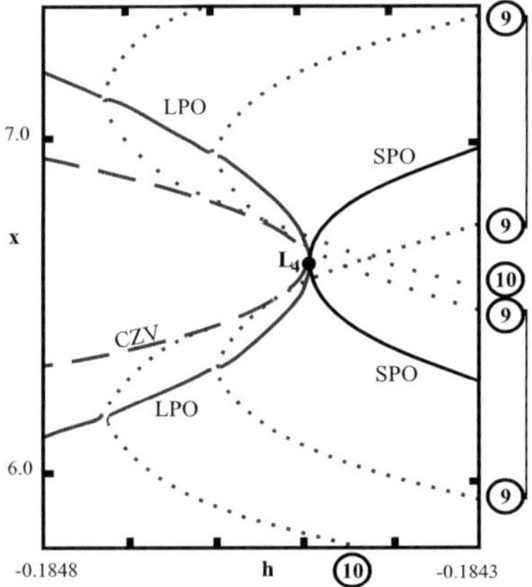

Figure 3. The characteristics of long and short period orbits near L_4 in a weak bar model. These characteristics give the distance x along an axis perpendicular to the bar as a function of the energy in the rotating frame (Jacobi constant). No orbits exist in the region between the two arcs of the curve of zero velocity (CZV). (—) stable and ($\cdots$) unstable arcs. The two lower and the two upper bridges (9) join at the same points of the SPO, outside the figure.

(e.g. n=9 in Fig. 3) at an energy h larger than the energy h_4 at L_4. It reaches a minimum energy and near this point it forms an arc consisting of stable LPO orbits. Then the characteristic continues as a bridge leading to a SPO orbit described n+1 times (n+1=10 in Fig. 3). Between the various LPO families there are gaps (Fig. 3).

When the bar is stronger the gaps between the various LPO families are larger. Such is the case of Fig. 4. Very close to L_4 there is an LPO family A.

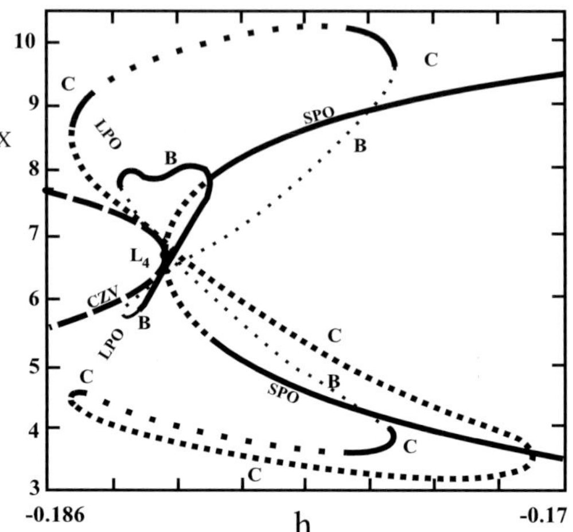

Figure 4. The LPO and SPO characteristics in a strong bar. The LPO families are: A (very close to L_4, not shown in this figure), B (generated at a 2:1 resonance from the SPO (upper branch) and terminating at a 3:1 resonance), C (generated at the same 3:1 resonance, and terminating at a 4:1 resonance) and so on.

Further away there is the LPO family B. This is generated as a bifurcation of the SPO family at a point above L_4, where the SPO family becomes unstable as h becomes smaller. Near this point the B-type orbits are close to SPO orbits described two times. This point has the maximum energy along this part of the B family. Then the B family extends to smaller energies and it is mostly stable until a minimum energy below and above L_4 and then it is continued by unstable arcs to larger energies. These unstable arcs reach the SPO characteristic at two points above, and below L_4, with energy much larger than h_4. The corresponding orbits tend to a SPO described 3 times. Beyond that point starts the C family that has a stable arc near its maximum energy. Then it reaches a minimum energy with another small stable arc and terminates at a large energy, as a SPO described 4 times.

In the same way there are further LPO families with small stable arcs for smaller energies inside and outside corotation. If we follow these arcs inside L_4 after an infinity of gaps we reach the x1 family of orbits circulating in a direct way around the center of the galaxy.

Therefore the stable banana-type periodic orbits exist only in certain intervals of distances from the Lagrangian point L_4 (or L_5). The orbits in the gaps

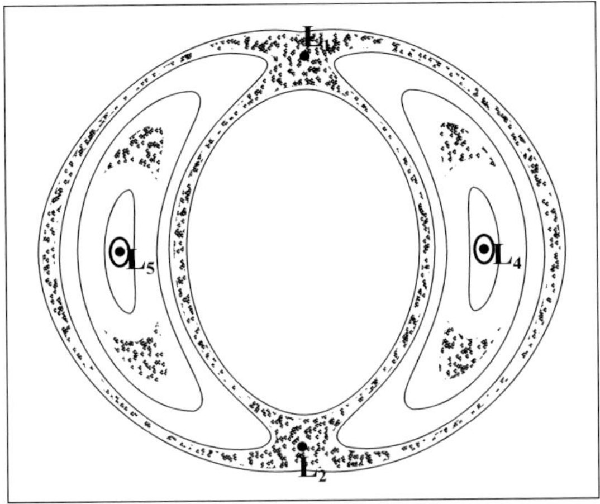

Figure 5.   Long period orbits around L_4 and L_5 and regions of chaotic orbits (schematically).

between these intervals are mostly chaotic (Fig. 5). Chaotic orbits exist mainly in a region around L_1 and L_2. The stable LPO orbits are very important, because they are followed by streamlines of gas that form the galactic vortices around L_4 and L_5. The gas cannot follow the chaotic orbits, that cross each other in an irregular way. Therefore the gas between the stable LPO orbits follows similar streamlines both close to the stable LPO periodic orbits and in the gaps between them. This explains the robustness of the gaseous vortices near L_4 and L_5, as found numerically by England (1989), Athanassoula (1992b), Piner, Stone and Teuben (1995), Fridman et al. (1997, 1999, 2001a,b) and England, Hunter and Contopoulos (2000).

2.   Vortices near L_4 and L_5

The vortices around the Lagrangian points L_4, L_5 are composed both of stars and of gas.

An example of the formation of vortices near L_4, L_5 is shown in Figs. 6, 7. These figures show the evolution of the distribution of stars in a spiral galactic model given by Contopoulos and Grosbøl (1986) after 5 and 20 revolutions respectively. The maximum spiral perturbation is 10% and it is growing over two revolutions. The initial distribution of the stars (massless test particles) is taken to be homogeneous. The stars are set initially in pure circular motion. After some time the response is strongest along the spiral arms up to the 4:1 resonance. Beyond this resonance there are weak extensions, forming double and fragmented spirals towards corotation. Near the 4:1 resonance there is a

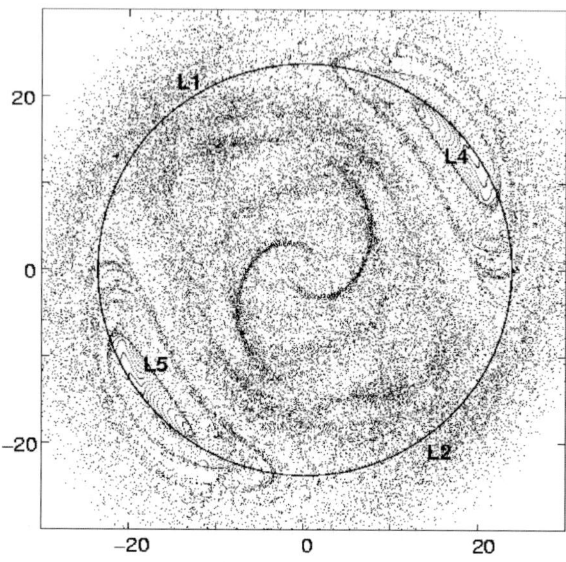

Figure 6.　　The distribution of 20000 stars in the galactic model of Contopoulos and Grosbøl (1986) consisting of an axisymmetric background and a 10% spiral perturbation, growing from zero in about 2 periods. The distribution is shown after 5 periods. In this and in all subsequent figures the patterns rotate clockwise and the circle indicates corotation.

marked concentration of orbits forming an approximate square. This configuration is consistent with the theory of the termination of strong spirals near the 4:1 resonance (Contopoulos 1985, Contopoulos and Grosbøl 1986).

Near corotation the distribution of points tends to follow the topology of the long period orbits. The lines of points in Fig. 6 are due to the fact that the evolution of the distribution around L_4 and L_5 is very slow. Due to the differential rotation of these points the maxima of density near L_4, L_5 form spirals. Such spirals have been observed also in gas responses by Athanassoula (1992b). After a longer time the distribution of points around L_4 and L_5 becomes rather smooth (Fig. 7), but still we see two overall banana-like concentrations around L_4 and L_5. In Fig. 6 we see two outermost bananas, limiting the set of banana-like orbits around L_4 and L_5. Beyond these limiting bananas towards the center there are two relative minima in the distribution. Around L_1 and L_2 there is a smooth distribution that surrounds also the corotation circle and the banana orbits. This smooth distribution is due mainly to the chaotic orbits, which are most prominent near L_1 and L_2.

In order to see better the vortices around L_4 and L_5 we give, in Fig. 8, the velocity field in this model. In this figure we see the formation of large anticyclones in an extended region around the Lagrangian points L_4, L_5. We

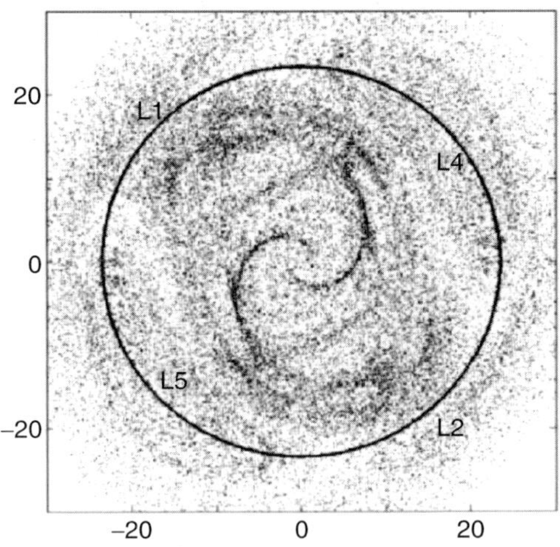

Figure 7. The distribution of the same stars after 20 periods.

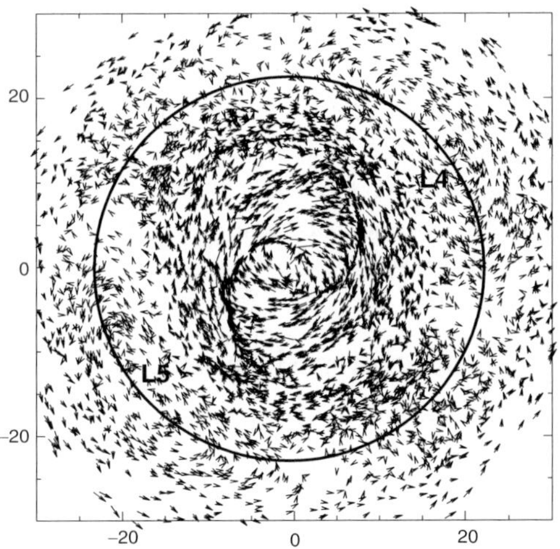

Figure 8. The velocity field of the stars in the Contopoulos-Grosbøl (1986) model in the corotating frame.

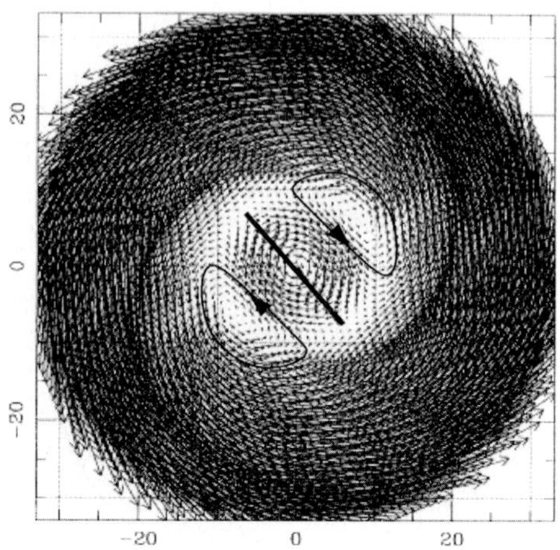

Figure 9.   Vortices around L_4, L_5 in the velocity field of gas in the Florida model (England et al. 2000) of a barred galaxy with a bar perturbation about 10% of the background. The extent of the bar is marked.

see also the abrupt change of the directions of the velocity vectors along the spirals.

Similar calculations have been made in the case of gas in a barred galaxy (England, Hunter and Contopoulos 2000). In Fig. 9 we see the velocity field in a galaxy with a relatively weak bar of amplitude 10% of the background. We see two anticyclones around L_4 and L_5 and a spiral beyond the ends of the bar.

In the case of a strong bar (35%) (Fig. 10) we see two conspicuous anticyclones around L_4 and L_5 and two cyclones along the bar, inside the Lagrangian points L_1 and L_2. The appearance of these cyclones has been explained theoretically by Fridman et al. (1999).

3.   Disk responses in barred potentials

There is a general agreement that corotation in barred galaxies, at least in early types, is located outside the bar (at 1.2 − 1.4 bar radii; Buta et al. 1996), as initially suggested by Contopoulos (1980). This gives the opportunity to investigate the location and size of vortices using response models of stellar and gaseous disks. Ferrers bars are among the most popular realistic barred potentials reproducing many dynamical and morphological properties of real bars (Athanassoula et al. 1983, Pfenniger 1984, Athanassoula 1992a,b, Skokos et al. 2002). In Fig. 11 is shown the response of a stellar disk to a two-dimensional Ferrers bar. The particular parameters used are as in model

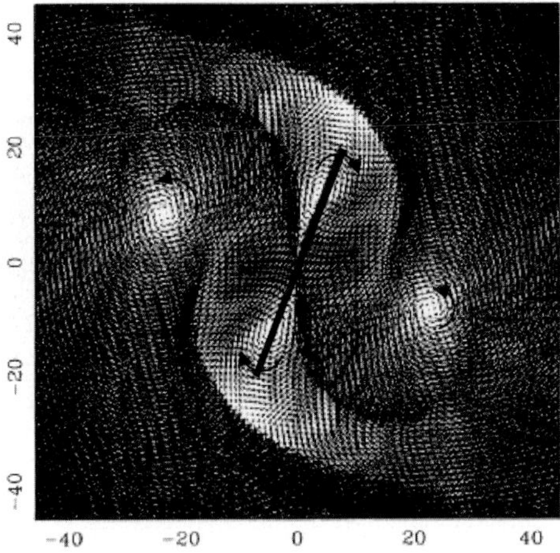

Figure 10. Vortices formed by the gas in a galaxy with a relatively strong bar, with amplitude ≈35% of the background. We see anticyclones around L_4, L_5 and cyclones around L_1 and L_2.

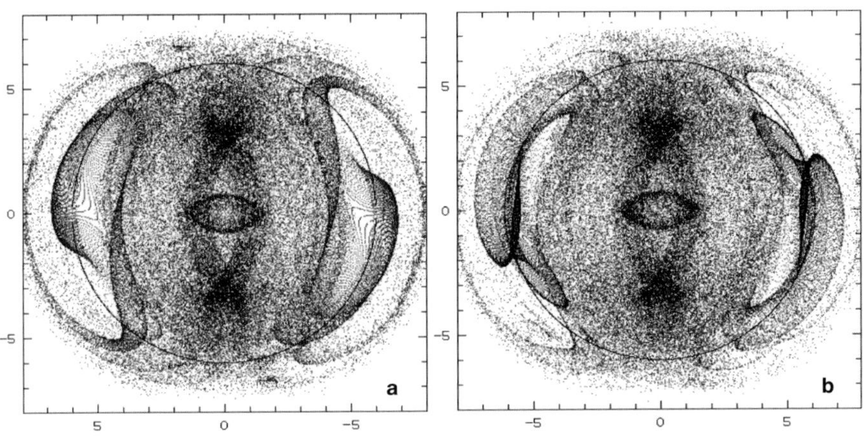

Figure 11. Response of a stellar disk to an imposed Ferrers bar potential. The disk is initially populated homogeneously. (a) after 35 and (b) after 39 pattern rotations.

No1 in the list of models given by Athanassoula (1992a). The bar grows to its maximum strength within two pattern rotations. The disk is initially populated homogeneously and the particles are initially in circular rotation. In this case

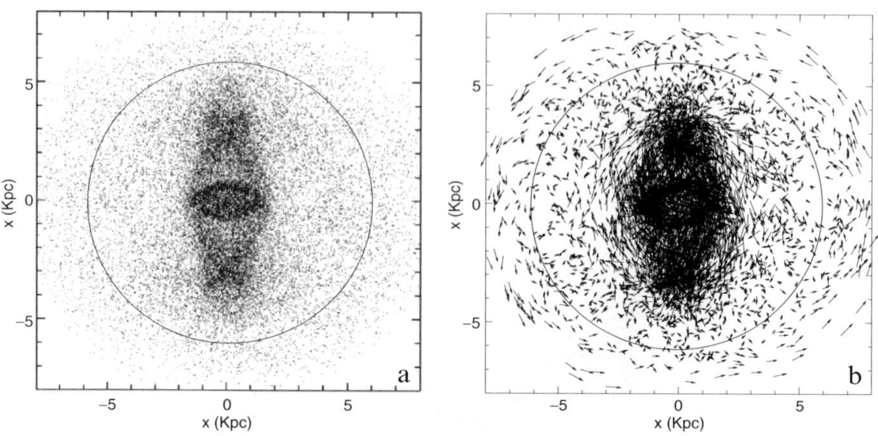

Figure 12.   (a) The response of an exponential stellar disk with radial dispersion of velocities 35 km/s to a Ferrers bar, after 30 bar rotations. (b) The corresponding velocity field. Calculations are done in the rotating frame of reference. We observe the anticyclonic motion at the sides of the bar in the corotation region.

the region around the stable L_4 and L_5 shows a changing morphology from snapshot to snapshot. Figure 11a is a snapshot after 35 bar revolutions, while Fig. 11b is a snapshot after 39 revolutions. However, the bar has the same morphology in both snapshots. The changing morphology of the vortices is partly a result of the slow dynamical evolution of the outer parts of the disk in comparison with the dynamical evolution closer to the center. After a long time the dynamical evolution leads to a smoother distribution of the particles in the region. However, the assumption of homogenously populated stellar systems is not realistic. Real stellar galactic disks show an exponential radial profile and a certain dispersion of velocities. In Fig. 12 we give the response of an exponential disk with particle velocities having a radial dispersion 35km/s. This time, after less than 10 bar revolutions one can hardly see any structure in the corotation region (Fig. 12a). Looking at the velocity field, however, (Fig. 12b) one clearly sees the anticyclonic motion at the sides of the bar in the corotation region around the stable Lagrangian points. We note that the response bar ends at about 75% of the corotation radius. Beyond the end of the bar we enter a chaotic region and no structure is supported (Fig. 12a).

In Fig. 13 we give an SPH (Smooth Particle Hydrodynamics) response of the same model. Now again we have a disk homogeneously populated at the beginning of the simulation with gaseous particles in circular motion. We assume a sound speed 25km/s and artificial viscosity parameters $(\alpha, \beta)=(1,2)$

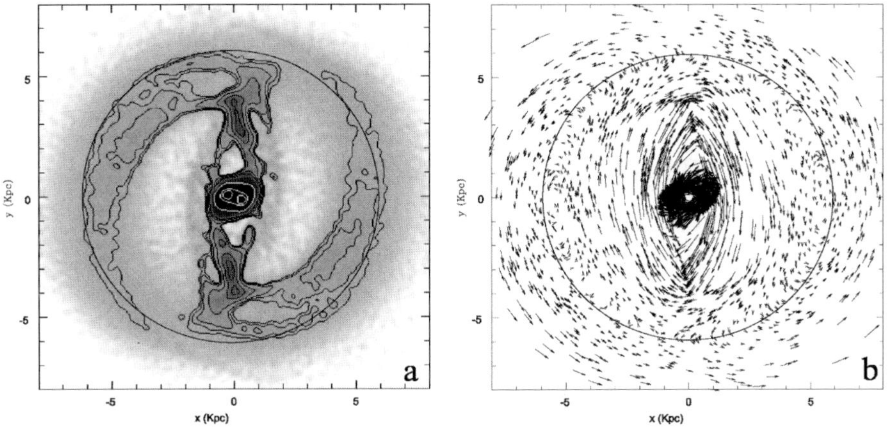

Figure 13.   Response of a gaseous disk to a Ferrers bar potential. (a) Gas density after 8 bar rotations (b) The corresponding velocity field, which shows the giant anticyclones in the corotation region.

(for definitions see e.g. Gingold and Monaghan 1977). The response is given after 8 pattern rotations and is characterized by the presence of a response spiral emerging from the ends of the bar. Details of this kind of SPH response models can be found in Patsis and Athanassoula (2000). In the velocity field (Fig. 13b) of this snapshot we see the location and the size of the anticyclones in the gaseous disk. The anticyclones have as usual their centers at the stable Lagrangian points. This configuration is typical of the cases of strong barred galaxies.

4.   Conclusions

Our main conclusions are the following:

1 In the corotation region of disk galaxies we encounter both order and chaos. Order is associated with motions around the stable Lagrangian points L_4 and L_5. Chaos is found around the unstable points L_1 and L_2, but also between the stable banana-like orbits inside L_4 and L_5.

2 The motion of stars and gas in the corotation region in spiral or barred galaxies is characterized mainly by the presence of anticyclones around the stable Lagrangian points. In gaseous models of strong bars we find also cyclonic motions near the unstable points L_1 and L_2.

3 The stable banana-like long-period orbits determine the anticyclonic flow in stellar disks. The gas follows the banana-like stable orbits and similar streamlines between them. This is verified by means of hydrodynamical simulations.

Appendix: A. Theoretical Long and Short Period Orbits

A simple model of a barred galaxy has the Hamiltonian (Contopoulos 1978)

$$H \equiv h_* + \kappa_* I_1 + \alpha_* I_1^2 + 2b_* I_1 I_2 + c_* I_2^2 + A_* \cos 2\theta_2 = h \tag{A.1}$$

where h is the energy of a star in a rotating frame (i.e. the Jacobi constant), I_1 is the radial action, $I_2 = J - J_*$ is the azimuthal action (J is the angular momentum of the star and J_* the angular momentum at corotation), and θ_2 is the angle conjugate to I_2, representing the azimuth in the rotating frame; κ_* is the epicyclic frequency , $A_* > 0$ is the amplitude of the bar and a_*, b_*, c_* are constants. The asterisks indicate quantities at corotation.

As the epicyclic tangle θ_1, which is conjugate to I_1, does not appear in this expression, the action I_1 is a second integral of motion.

The value of I_1 represents the energy of the epicyclic oscillations. If $I_1 = 0$ there are no epicyclic oscillations and we have a long period orbit.

In this case

$$I_2^2 = \frac{1}{c_*}[h - h_* - A_* \cos 2\theta_2] \tag{A.2}$$

and we have $c_* < 0$. Therefore I_2 is real only if the minimum value of h is

$$h = h_* - A_* \tag{A.3}$$

The corresponding orbits have $\theta_2 = \pi/2$, or $3\pi/2$, i.e. we have the equilibrium points L_1 , L_2, that represent unstable periodic orbits in the rotating frame. If

$$h_* - A_* < h < h_* + A_* \tag{A.4}$$

we have banana orbits of long period with their θ_2 extending from $-\theta_{max}$ to $+\theta_{max}$, where

$$\cos 2\theta_{max} = \frac{h - h_*}{A_*} \tag{A.5}$$

If

$$h = h_* + A_* \tag{A.6}$$

we have $\theta_{max} = 0$, or π, and the orbits are reduced to the Lagrangian points L_4, L_5 (stable periodic orbits in the rotating frame).

If $h > h_* + A_*$ there are no long period orbits. In such cases we find the short period orbit by taking

$$\frac{\partial H}{\partial I_2} = \frac{\partial H}{\partial \theta_2} = 0 \tag{A.7}$$

i.e.

$$b_* I_1 + c_* I_2 = 0 \tag{A.8}$$

and

$$\sin 2\theta_2 = 0 \qquad (A.9)$$

These orbits appear around L_4 (where $\theta_2 = 0$) and L_5 (where $\theta_2 = \pi$), with finite energy along the epicyclic oscillation, i.e. along a short period orbit.

In this case we have in first order approximation

$$h_* + \kappa_* I_1 + A_* = h \qquad (A.10)$$

therefore

$$I_1 = \frac{h - h_* - A_*}{\kappa_*} = -\frac{c_* I_2}{b_*} \qquad (A.11)$$

The family of short period orbits exists only for relatively large energies. Therefore these orbits are less important for the overall dynamics of galaxies.

Acknowledgments

This work has been supported by the Research Committee of the Academy of Athens.

References

Athanassoula, E.: 1992a, Mon.Not.R.Astr.Soc. 259, 328.

Athanassoula, E.: 1992b, Mon.Not.R.Astr.Soc. 259, 345.

Athanassoula, E., Bienaymé, O., Martinet, L. and Pfenniger, D.: 1983, Astron. Astrophys. 127, 349

Buta, R., Crocker, D.A. and Elmegreen, B.G. (eds): 1996, "Barred galaxies", ASP. Conf. Ser. Vol. 91

Contopoulos, G.: 1973, Astroph. J. 181, 657.

Contopoulos, G.: 1978, Astron. Astrophys. 64, 323.

Contopoulos, G.: 1980, Astron. Astrophys. 81, 198

Contopoulos, G.: 1981, Astron. Astrophys. 102, 265.

Contopoulos, G.: 1983, Cel. Mech. 31, 193.

Contopoulos, G.: 1985, Comments Astrophys. 11, 1.

Contopoulos, G.: 1988, Cel. Mech. 43, 147.

Contopoulos, G.: 2002, "Order and Chaos in Dynamical Astronomy", Springer Verlag, Heidelberg.

Contopoulos, G and Grosbøl, P.: 1986, Astron. Astrophys. 155,11.

England,M.N.: 1989, Astroph. J. 344, 669

England,M.N., Hunter, J.H.Jr. and Contopoulos, G..: 2000, Astrophys. J. 540, 154.

Fridman,A.M., Khoruzhii, O.V., Lyakhovich,V.S., Avedisova, O.K., Silchenko, O.K., Zasov, A.V., Rastorguev, A.S., Afanasiev, V.L., Dodonov, S.N. and Boulesteix, J.: 1997, Astrophys. Space Sci. 252, 11.

Fridman, A.M., Khoruzhii, O.V., Polyachenko, V.L., Zasov, A.V., Silchenko, O.K., Afanasiev, V.L., Dodonov, S.N. and Moiseev, A.V.: 1999, Phys. Lett. A. 264, 85.

Fridman, A.M., Khoruzhii, O.V., Polyachenko, V.L., Zasov, A.V., Silchenko, O.K., Moiseev, A.V., Burlak, A.N., Afanasiev, V.L., Dodonov, S.N. and Knapen, J.H. : 2001a, Mon. Not. R. Astr. Soc. 323, 651.

Fridmanm, A.M., Khoruzhii, O.V., Lyakhovich,V.S., Silchenko, O.K., Zasov, A.V., Afanasiev, V.L., Dodonov, S.N. and Boulesteix, J.: 2001b, Astron. Astrophys. 771, 538.

Gingold, R.A, Monaghan, J.J., 1977, Mon. Not. R. Astr. Soc. 181, 375

Patsis, P.A. and Athanassoula, E.: 2000, Astron. Astrophys 358, 45

Pfenniger, D: 1984, Astron. Astrophys. 134, 373

Piner, G.B., Stone, J.M. and Teuben, P.J.: 1995, Astroph. J. 449, 508

Skokos, Ch., Patsis, P.A. and Athanassoula, E.: 2004, Mon. Not. R. Astr. Soc. 333, 847

SPIRAL PERTURBATIONS IN DISK GALAXIES OBSERVED IN NIR*

P. Grosbøl

European Southern Observatory
Karl-Schwarzschild-Str. 2,
D-85748 Garching, Germany

pgrosbol@eso.org

Abstract   Near-infrared observations in the K$'$ filter of 54 spiral galaxies were used the analyze perturbations in their stellar disks. The small attenuation by dust at NIR wavelengths makes it possible to measure the distribution of luminous matter in the galaxies which otherwise is difficult to estimate for spiral galaxies with their high content of dust. Whereas the K band images mainly display the distribution of the old stellar disk population, light from young objects also contributes.

Most disk galaxies ($\sim$60%) have a symmetric, grand-design two-armed spiral pattern in their inner parts which often splits up into multiple arms in the outer regions. Upto 10% of the galaxies show complex structures in the central parts.

Non-axisymmetric perturbations in the disks were analyzed by 1D Fourier transform techniques. Relative amplitudes of bar components showed a continuous distribution down to the detection limit of $\sim$3% with only 5 of 35 SA galaxies with no bar at this level. The main two-armed spiral structures displayed a lack of tight, strong spiral. Such patterns may have so high radial forces that non-linear dynamic effects would become important and damp them.

Keywords: galaxies:spiral - galaxies:structure - infrared:galaxies

1.   Introduction

Spiral galaxies are among the most common types of galaxies in the known universe. Their two main visible components are a central, spherodial bulge containing mostly old stars and a flat disk with a mixture of old and young stars, gas and dust. Spiral arms and bars are often present in their disks.

Classical studies of galaxies were done mostly in the visual B band due to the high sensitivity of photographic plates in this band. It became possible

*Based on observations collected at the European Southern Observatory, La Silla, Chile; Programs: 63.N-0343, 65.N-2877 and 66.N-0257

Alexei M. Fridman, et al. (eds.), Astrophysical Disks; Collective and Stochastic Phenomena 145–156
© 2006 *Springer. Printed in the Netherlands*

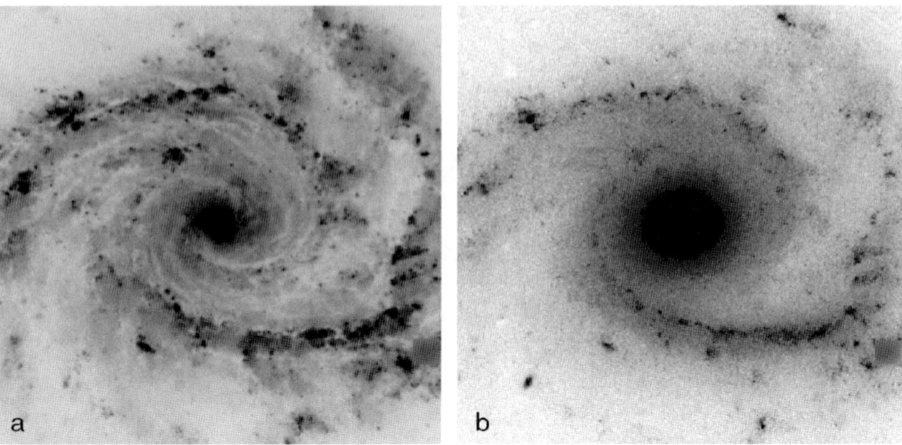

Figure 1.   Direct images of NGC 2997 with foreground stars removed in different colors: a) B image, b) K$'$ image.

to extend investigations to the full visual spectrum when CCD detectors were introduced. This gave access to many broad band colors but it was still impossible to safely separate population effects from attenuation by dust as the corrections are very similar and the detailed geometry of dust and stars in the galaxies is unknown. First with large array detectors in the near-infrared (NIR) region of 1-2.5μ, available in the early 1990's, it become feasible to observe deep surface photometry of spiral galaxies (Block and Wainscoat 1991) at wavelengths where dust attenuation starts to be insignificant. It also opened the possibility of direct studies of properties of spiral density waves (Lin and Shu 1964) in disk galaxies.

This paper presents the properties of spiral galaxies as observed in the NIR K$'$ band (at 2.1μ) including bar and spiral perturbations in their disks.

2.   Observing in Near-Infrared Bands

Observations in the NIR present several challenges compared to those in visual bands. The night sky level is significantly higher especially in K (approx. 13 mag arcsec^{-2}) where thermal emission starts to be a major component. The surface brightness of disks in spiral galaxies is much fainter and has a typical range of K = 16-19 mag arcsec^{-2}. Both absorption and emission associated to a number of atmospheric molecules (such as H_2O, CO_2 and O_2) are important and make the sky level vary at a time scale of minutes. A number of different array detectors are now available with sizes up to 2k$\times$2k pixels. Whereas the stability and cosmetic quality of these NIR detectors have improved significantly, they still do not reach the standards for CCD's used for visual bands.

Rapid varying sky and high background levels demand an elaborate observing technique including short integration times to avoid saturation of detector pixels and sequences of exposures with small offsets to ensure that the full sky area is measured even with groups of bad detector pixels. For extended objects, like galaxies, one needs to interleave target frames with exposures of blank sky fields to correct for variations in the background.

Visual images of spirals are often dominated by the light from young objects which outline the spiral arms. Especially for late-type spirals, strong absorption by dust is also seen. This is illustrated in Fig. 1 where both B- and K-band images of the spiral galaxy NGC 2997 of type Sc(r)I-II are shown. Although the galaxy in B has a patchy appearance, mainly due to dust and young stars, the K frame shows the smooth variation associated with the old stellar disk population. The dust make the bulge look smaller and the spiral arms seem to reach further in than the NIR image indicates.

For studies of the distribution of luminous matter in galaxies, the NIR bands, especially K, are much more suitable as the attenuation of dust is very small. There are still population effects as seen on Fig. 1 where several bright knots in the arm regions on the B map also appear enhanced in the K' band. The major part of light in K originates from old stellar population while a smaller fraction (up to 30%) may come from younger objects such as red super-giant (Rix and Rieke 1993).

The detailed mass distribution in disks of spiral galaxies is important for the study of their dynamic structure and possible secular evolution. For these purposes, the K' band is the better choice.

3.   Data and Basic Reductions

A total of 54 non-barred spiral galaxies was observed in the K' band with the SOFI instrument on the 3.5m NTT telescope at La Silla to study their spiral structure and estimate the fraction of bars in ordinary spirals (Grosbøl et al., 2004). The galaxies was selected to represent a wide range of morphological types with a subset of 35 SA galaxies for which no bar perturbation could be identified in visual bands. Further, they were chosen to have inclination angles $<65°$ and systemic velocities <5000 km s^{-1} to allow the detailed study of perturbations in their disks. The typical seeing was $1''$ and the limiting magnitude was around 21 mag arcsec^{-2} in K' for a signal-to-noise ratio of 3.

Before the internal structure of the galaxies could be analyzed, their sky projection parameters were determined. This was done by minimizing the constant phase component in the region of the disk occupied by the main spiral structure using 2D Fourier transforms on the polar maps of the galaxies. This algorithm is consistent with other methods (e.g. fitting of ellipses to the outer

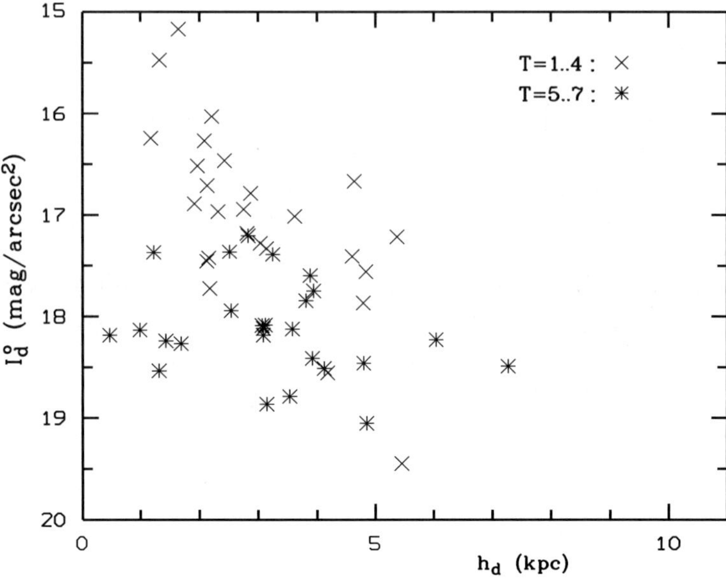

Figure 2. Distribution of central surface brightness I_d^0 and linear scale length h_d of an exponential disk fitted to the inter-arm regions of the K′ maps. Different symbols indicate early and late morphological types.

isophotes) for nice symmetric galaxies but was less effected by warps and open spiral arms which may bias the results.

4. Axisymetric Components

Fixing the projections parameters, the axisymmetric parts of the galaxies were fitted by a spherical bulge and a flat exponential disk. First the disk parameters were estimated for the region occupied by the main spiral pattern using only the inter-arm regions which better represent to old stellar disk population. Differences in the estimates of 10% were typically seen when compared them to those obtained from the full disk as the latter were biased by young object in the arms (Grosbøl and Patsis 1998).

The distribution of central surface brightness and linear scale length (assuming $H = 75$ km s^{-1} Mpc^{-1}) is shown in Fig. 2. There is a lack of centrally bright disks with long scale length which may be due to a limit in total angular momentum available during formation. Late-type spiral (i.e. $T > 4$) are, on average, fainter while the distribution of linear scale length show little correlation with type.

The central region was then fitted with a bulge component after the contribution from the main disk was subtracted. Both a modified Hubble law (Binney

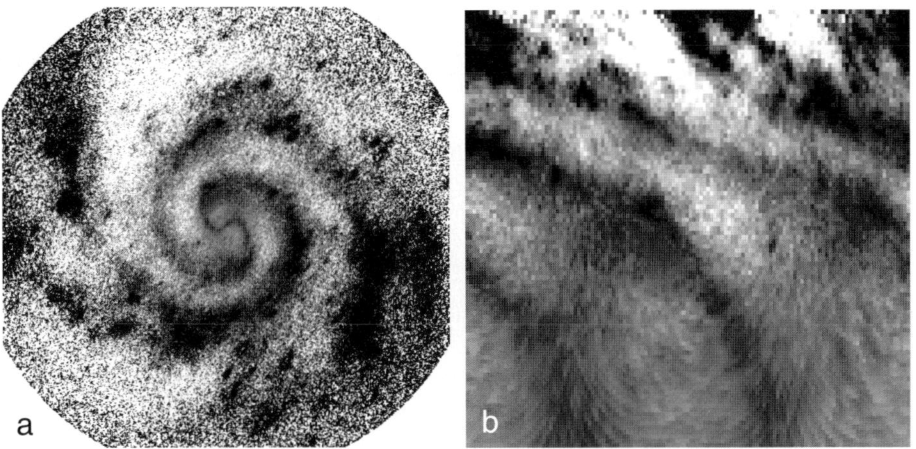

Figure 3.　　Face-on maps of NGC 6118 in K′ giving surface brightness divided by the average radial intensity profile. Two representations are shown namely: a) a direct image, and b) a polar-logarithmic θ-log(r) map. Higher intensities are darker with a range of $\pm 30\%$.

and Tremaine 1987) and a Sérsic $r^{1/n}$ power law (Sérsic 1968) were used for the spherical bulge. For most galaxies, the latter gave a better fit with the power law index n in the range of 1-2. Addition of a central point source improved the fit in several cases. Further, it was tried to include a steep central exponential disk component with projection parameters identical to those of the main disk. The fits for the majority of galaxies got better which suggests that the bulges often are oblate rather than spherical. However without detailed kinematic data, it is not possible to derive the actual 3D shape of the bulge component.

5.　　Morphology of bars and spirals

For the analysis of perturbations in the disk, it is necessary to subtract the spherical bulge component which otherwise would give raise to an artificial bar feature when the galaxy is deprojected. Deviations from the assumed bulge shape can still leave traces such as constant phase features in the central parts. Problems in defining the exact shape of the PSF and pixel interpolation errors in high gradient region in the center made the analysis of the inner $2''$ uncertain which therefore was omitted.

The general morphology of bars and spirals was checked visually on direct face-on, bulge-subtracted images in the K′ band. It is difficult to view small perturbation due to the large dynamic range of exponential disk. Thus, relative intensity maps were used in which the intensities were normalized by the average radial profile. Most spiral arms can be approximated by logarithmic

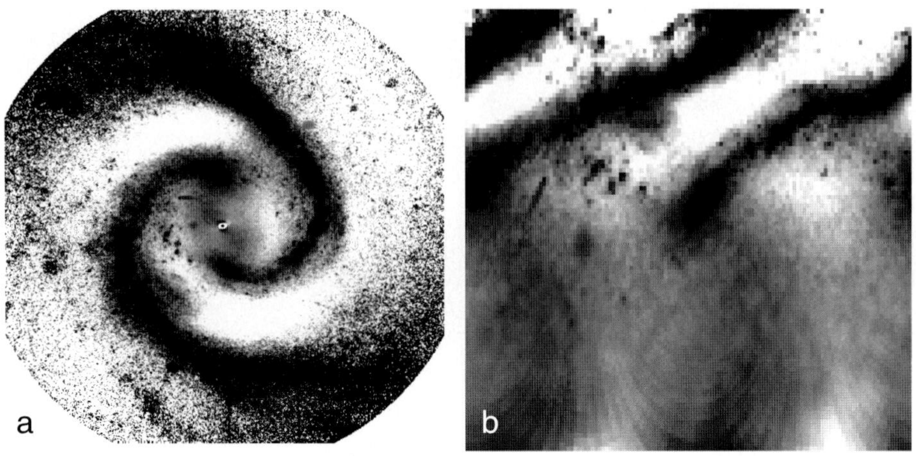

Figure 4.   Face-on maps of NGC 1566 in K′ as in Fig. 3.

spirals which appear as straight segment on logarithmic polar maps. Both of these types of presentations are shown for two galaxies in Fig. 3-4.

The major fraction of spiral galaxies (∼60%) shows a grand-design spiral structure in their inner parts while this pattern often breaks up into multiple arms in the outer regions. One such galaxy is NGC 6118 shown in Fig. 3 and classified as Sc(s)I.3. It displays a typical s-shaped spiral where the arms smooth increase pitch angle towards the center and finally almost become a bar. Although the galaxy is classified as non-barred in the visual a small bar is clearly seen on the direct map. The main two-armed spiral pattern splits up in the outer parts. Many bright knots along the arms can also be seen indicating the presence of young objects and star formation.

Another example of a grand-design spiral is NGC 1566 (see Fig. 4). It is classified as a weakly barred, s-shaped spiral galaxy with Hubble type Sc(s)I in the visual. Its K image shows a prominent bar distortion in the central parts having a phase offset of $\sim 30°$ relative to the inner ends of the spiral arms. This offset suggests that bar and spiral pattern constitute different modes with possibly separate pattern speeds (Sellwood and Sparke 1988). The spiral arms are quit symmetric and show many knots indicating strong star formation. As seen on the logarithmic polar map, one of the arms is well approximated by a logarithmic spiral while the other has a slightly more open pitch angle and displays a sudden change in its outer part.

In general, spiral structure in disk galaxies looks smoother on K images than in visual bands reflecting better the surface brightness variations in the old stellar disk population. As dust do not obscure the view in NIR, it is also

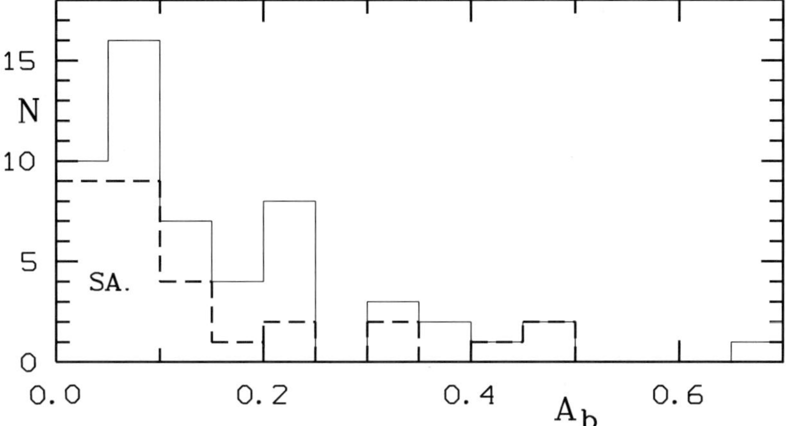

Figure 5. Histogram of relative amplitudes A_b of bars detected in the sample. The full drawn line show to full sample while the dashed line indicate the distribution of SA galaxies.

easier to estimate the density perturbation in the disk although some light from young objects still may be contribution even in the K' band.

6.   Bar Perturbations

The non-axisymmetric perturbations in the disks were studied using 1D Fourier analysis of the azimuthal K intensity variation within $1''$ wide, consecutive annuli in the plan of the disk. This provides an objective, quantitative measure which is little sensitive to noise. Although 2D Fourier transforms yield more information, 1D FFT technique was preferred as 2D transforms are more difficult to interpret if the spiral arms do not follow a simple mathematical form (e.g. a logarithmic spiral) and may depend on the choice of radial region. To reduce the effects of young objects in the arms, a median filter was applied before the 1D Fourier transform was done. Bars or oval distortions were identified by their almost constant phase of the m=2 harmonic as function of radius whereas spiral structures show a monotonous change.

The frequency of bars in disk galaxies was found to be higher in NIR than in visual bands by several studies (see e.g. Seigar and James 1998, Eskridge et al., 2000). Thus, it is of interest to quantify the distribution of weak bars and check if there exist a group of truly non-barred disk galaxies or if all have some oval distortions in their central parts. The galaxies in the current sample is a selection of SA/SX type systems (i.e. classified as either non-barred or weakly barred) and should therefore represent the lower end of the distribution of bar amplitudes.

Bars in the galaxies were identified manually by inspecting the radial phase variations of the $m = 2$ Fourier harmonic. The distribution of mean bar

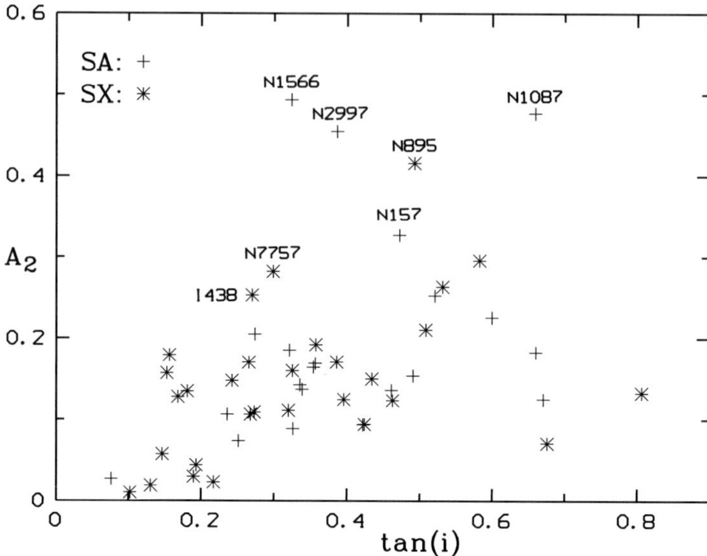

Figure 6.   Distribution of the relative amplitude A₂ of the main two-armed spiral pattern as function of the pitch angle given as $\tan(i)$. Teh symbols indicate different bar morphologies.

amplitudes is given in Fig. 5 where the inner $2''$ were omitted to reduce possible effects from an inadequate bulge subtraction. It shows a continuous distribution of amplitudes down to the detection limit of $\sim 3\%$ with only 5 galaxies below this limit. As expected, galaxies classified as non-barred have weaker bars on average than those of SX type but there is a significant overlap. This suggests that 85% of the SA galaxies have detectable bar perturbation which would indicate that the fraction of truly non-barred disk galaxies is as low as 5% of all spirals.

7.   Spiral Patterns

The extent of the main symmetric, two-armed spiral structure in the galaxies was also estimated manually from both the phase variations and the appearance on the logarithmic polar maps. The average relative amplitude of the $m = 2$ component and the pitch angle were derived (see Fig. 6).The amplitudes are somewhat overestimated compared to the true variation of the old stellar disk population due to light from young objects in the arms. The two galaxies, NGC 1566 and NGC 2997, with the highest amplitudes both show traces of significant star formation in their arms.

The pitch angles are mostly in the range of 5-30° where spirals tighter than 5° are very difficult to detect due to their small inter-arm distance. The amplitudes are generally less than 20% but there is a lack of strong, tight spiral patterns. Such spirals would have a high relative radial force perturbation ΔF_r and therefore be subject to non-linear dynamic effects which could damp such patterns. This is expected to happen for $\Delta F_r > 5\%$ (Grosbøl 1993).

8. Possible Multiple Pattern Speeds?

Direct measurements of the pattern speed of bars have been made only for a small set of early-type spiral galaxies (see e.g. Merrifield and Kuijken 1995) using the method of Tremaine and Weinberg (1983). For most other types of spiral galaxies, indirect estimates must be used such as associating morphological features (e.g. the end of the bar or the extent of the main spiral pattern) to specific stellar resonances.

It is normally assumed that bars terminate just inside their co-rotation (CO) radius (Contopoulos 1980). On the other hand, the main spiral pattern in ordinary or weakly barred galaxies is often argued to start at its Inner Lindblad Resonance (ILR) e.g. based on the location of dust lanes (Grosbøl et al., 1999). These assumptions are difficult to reconcile for normal spirals with weak bars if bar and spiral share the same pattern speed.

N-body models of Sellwood and Spake (1988) show that bars may be rotating faster than the spiral pattern and be coupled to it through resonances e.g. CO and ILR. Although one would expect to see many examples of misalignments between bar and inner spiral if they had different pattern speeds, these models indicated that such misalignments only would be observable in around 10% of the cases.

The orientation and alignment of bar and spiral can better be observed in the K band where attenuation by dust is small. In the current sample several spiral galaxies with weak bars show misalignments (e.g. NGC 1566 and NGC 6384). It is consistent with 10% expected and suggests that many weakly barred spiral galaxies with r-shaped morphology have bars rotating faster than their spiral pattern.

9. Special Cases

Around 10% of the galaxies show more complex structures in their inner parts such as multiple bars or several spiral modes. There is a significant variety of features and two examples are discussed in more detail below.

The galaxy NGC 4030 has a flocculent appearance in the K band with multiple spiral arms in its inner parts as seen in Fig. 7. On the polar-logarithmic map one can trace a weak two-armed spiral pattern in the central parts while several arm segments can be seen at larger distances. Phase shifts between these segments are also observed which could suggest that different spiral modes

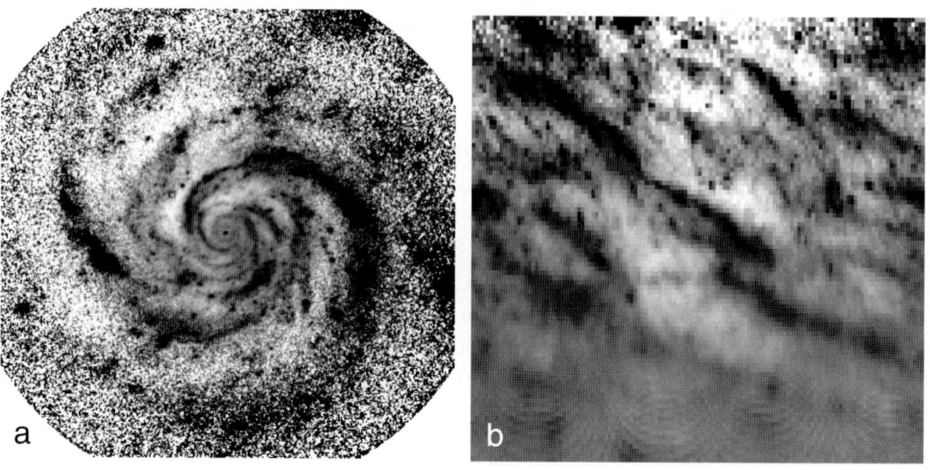

Figure 7.   Face-on maps of NGC 4030 in K' as in Fig. 3.

coexist in this galaxy. Some of the segments have a sharp appearance and could be associated to regions of star formation.

Several sets of symmetric arcs can be identified in NGC 4939 (see Fig. 8) which has the Hubble type Sbc(rs)I. A bar structure is clearly seen in the inner parts. It displays a bright knot and a phase shift near its end where arcs or tight spiral segments are seen. These features may be associated with the 'T-shaped' regions observed in n-body models (Patsis and Athanassoula 2000). Outside this inner set, at least 2 sets of symmetric arcs are visible, each with an angular offset of 90°. A symmetric two-armed spiral pattern starts from the last set of arcs. Detailed dynamic models are required to understand if these arcs are the result of stellar resonances or interaction between different spiral modes. In the very outer parts of the galaxy, a phase shift in the spiral pattern can be observed. Also in this galaxy many knots are seen along the arms.

10.   Conclusions

The surface photometry of spirals in the NIR K band gives a good estimate of the mass distribution of luminous matter in them with only minor population effects. The smooth appearance of spiral arms suggests that they are caused by density waves in the old stellar disk population.

The axisymmetric parts of the disk galaxies are well represented by a bulge and an exponential disk. The bulge component is best approximated by a Sérsic power law with a power in the range of 0.5-1. The bulge fit was improved in many cases by adding a steep exponential disk component which could indicate that many bulges are oblate rather than spherical.

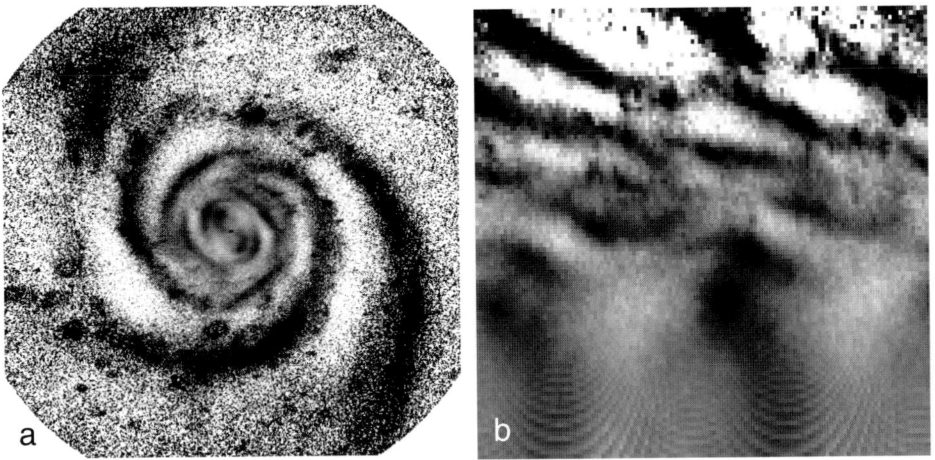

Figure 8.   Face-on maps of NGC 4939 in K′ as in Fig. 3.

Most spiral galaxies (∼60%) have a two-armed, grand-design spiral structure in their inner parts while multi-arm patterns frequently are seen in the outer
regions. A small fraction of the galaxies (∼10%) display a complex structure
in their central parts with multiple bars, arcs or spirals.

Bar amplitudes show a continuous distribution down to the detection limit of
∼3% with 5 of 35 SA galaxies for which no bar structure could be identified.
This indicates that only ∼15% of ordinary spirals classified as SA are non-
barred at this limit. It corresponds to only 5% of all spiral galaxies being
non-barred.

The distribution of amplitudes and pitch angles of the main two-armed spiral
structure shows a lack of strong, tight patterns. Such spirals would have a high
relative radial force perturbations and therefore possibly be damped by non-
linear dynamic effects.

References

Binney, J. and Tremaine, S.: 1987 *Galactic Dynamics* (Princeton, Princeton Univ. Press)

Block, D.L. and Wainscoat, R.J.: 1991 *Nature* **353** 48

Contopoulos, G.: 1980 *Astron. Astrophys.* **81** 198

Eskridge, P.B., Frogel, J.A., Pogge, R.W. et al.:2000 *Astron. J.* **119** 536

Grosbøl, P.J.: 1993 *Publ. Astron. Soc. Pacific* **105** 651

Grosbøl, P.J. and Patsis, P.A.: 1998 *Astron. Astrophys.* **336** 840

Grosbøl, P.J., Block, D.L. and Patsis, P.A.: 1999 *Astrophys. and Space Science* **269-270** 423

Grosbøl, P.J., Patsis, P.A. and Pompei, E.: 2004 *Astron. Astrophys.* **423** 849

Lin, C.C. and Shu, F.H.: 1964 *Astrophys. J.* **140** 646

Merrifield, M. R. and Kuijken, K.: 1995 *Mon. Not. R. Astron. Soc.* **274** 933

Patsis, P.A. and Athanasoula, E.: 2000 *Astron. Astrophys.* **358** 45

Rix, H.-W. and Rieke, M.J.: 1993 *Astrophys. J.* **418** 123

Seigar, M.S. and James, P.A.: 1998 *Mon. Not. R. Astron. Soc.* **299** 672

Sellwood, J.A. and Sparke, L.S.: 1988 *Mon. Not. R. Astron. Soc.* **231** 25

Sérsic, J.L.:1968 *Atlas de galaxias australes* (Cordoba: Obs. Astron. de Cordoba)

Tremaine, S. and Weinberg, M.D.: 1984 *Astrophys. J.* **282** L5

THE ROLE OF ORDERED AND CHAOTIC MOTION IN N-BODY MODELS OF ELLIPTICAL GALAXIES

N. Voglis,[1] and C. Kalapotharakos[1,2]

[1]*Academy of Athens, Research Center for Astronomy, 4 Soranou Efesiou, GR-11527, Athens, GREECE*
nvogl@cc.uoa.gr

[2]*University of Athens, Department of Physics, Section of Astrophysics*
ckalapot@cc.uoa.gr

Abstract

We investigate the fractions of mass in chaotic motion and the role this mass can play as regards the consequences of chaotic diffusion in a series of N-body models of non-rotating triaxial elliptical galaxies. Two cases of models with smooth centers are examined with maximum ellipticities E4 and E7. The mass in chaotic motion in these models depends on the ellipticity, but is relatively small, being respectively less than about 1/4 and 1/3 of the total mass. Chaotic diffusion in these models can not produced any systematic secular evolution. These models maintain a remarkably stable equilibrium configuration for times considerably longer than a Hubble time.

Then a central mass, of size $m \gtrsim 0.0005$ (in units of the mass of the galaxy), assumed to be due to a massive central black hole, is inserted in these models. The mass in chaotic motion reaches initially the level of $\simeq 50\%$ and $\simeq 80\%$ of the total mass for E4 and E7, respectively, by converting the ordered orbits passing near the center to chaotic. This extra chaotic mass is anisotropically distributed forming a bar in each system. We find that the mean rate of exponential divergence (the mean level of the Lyapunov Characteristic Numbers) of this mass has a narrow correlation with m, being roughly proportional to $m^{0.5}$.

Provided that m is large enough (e.g. $m \gtrsim 0.005$), chaotic diffusion of the orbits in the bar, can be so efficient that can alter the shapes of the self-consistent equipotential surfaces of the system in time scales less than a Hubble time, causing a secular evolution towards a new equilibrium in which the bar tends to be dissolved.

In order to measure the effectiveness of chaotic diffusion, we define a new parameter called 'effective diffusion momentum' $\mathscr{L}$ as a product of the fraction of the anisotropically distributed chaotic mass with the mean logarithmic

157

Alexei M. Fridman, et al. (eds.), Astrophysical Disks; Collective and Stochastic Phenomena 157–194
© *2006 Springer. Printed in the Netherlands*

divergence of the orbits of this mass. From our numerical results it comes out that secular evolution due to chaotic diffusion is negligible even for times larger than a Hubble time, if $\mathscr{L} < 0.0045$.

If $\mathscr{L} > 0.0045$, the models evolve following a process of self-organization in which chaotic orbits are converted to ordered orbits. The entropy reduction due to the production of order out of chaos is balanced by the increase of the mean level of logarithmic divergence of the remaining chaotic orbits. The anisotropically distributed chaotic mass decreases in time, resulting to smaller values of $\mathscr{L}$. The evolution ceases and equilibrium is established when $\mathscr{L} < 0.0045$.

Keywords: galaxy formation, chaos

1.   Introduction

It is well known that in galaxies the two body relaxation time considerably exceeds the Hubble time. Collisions or close encounters of stars are so rare that their effect in a Hubble time can be neglected. Stars move under the forces of the global gravitational field created by their current distribution in space (self-consistent field). As a consequence, galactic structures can be approximated by course-grained distribution functions that are solutions of the collisionless Boltzmann equation combined with the Poisson equation. The self-consistent potential derived by these solutions is smooth.

Such solutions can be provided by N-body codes based on a conservative technique (e.g. Allen, Palmer, and Papaloizou, 1990), or Smooth Field Codes (SFC) (e.g. Hernquist and Ostriker, 1992). Such codes can give the global potential of the system analytically as an expansion in terms of basis functions. If we start from an initial distribution of particles in phase space, the coefficients of the expansion of the potential can be evaluated from the corresponding configuration of particles.

From a potential given by analytic expressions one can write analytic expressions for the variational equations as well. In terms of these equations the stability of the orbits of individual particles in the system can be studied and libraries of particles moving in ordered or in chaotic orbits in self-consistent models of galaxies can be constructed.

Following this technique we have investigated the orbital structure of models of non-rotating elliptical galaxies.

Two types of models are examined. First, we use non-rotating triaxial equilibrium models with smooth centers. In these models the density near the center is flat. The self-consistent potential near the center is parabolic. The lowest energy orbits are combinations of three harmonic oscillations along the corresponding principal axes, but at higher energy levels the frequencies depend on the amplitude of oscillations.

This potential has no chance to be completely integrable, but it is remarkably near to an integrable potential. The fundamental frequencies of the three oscillations are in general irrational, so the orbits fill a parallelepiped with curved sides. For this reason these orbits are called Box orbits. Such orbits can pass arbitrarily close to the center. In smooth center models many resonant orbits can be also found, preferably at not very low energy levels. These orbits form tubes that in principle avoid passing very near to the center. The various families of ordered orbits are separated by chaotic layers in which particles wander describing chaotic orbits.

Smooth center models of galaxies are rather idealized models. More than forty years ago has been known that black holes possibly exits at the centers of galaxies (Salpeter, 1964; Zeldovich, 1964; Lynden-Bell, 1969). Recent investigations, via high quality observational data, show that black holes at the center of galaxies must be very common (Kormendy and Richstone, 1995; Kormendy et al., 1997; van der Marel, de Zeeuw, and Rix, 1997; van der Marel and van den Bosch, 1998; Kormendy et al., 1998; Magorrian et al., 1998; Cretton and van den Bosch, 1999; Gebhardt et al., 2000). In some cases these black holes may be quite massive, reaching the order of 1% of the mass of the hosting galaxy. Furthermore, in many elliptical galaxies the density profiles show a cuspy character near the center (Ferrarese et al., 1994; Lauer et al., 1995; Gebhardt et al., 1996; Faber et al., 1997).

Thus, smooth center models may not be good models for a wide class of galaxies. For this reason, it is interesting to construct and investigate models of elliptical galaxies, in which the central region is occupied by some extra mass, so that a central force field dominates locally. The combination of the central force field with the global self-consistent field produces very interesting dynamical complications, that under certain conditions may have serious implications on the stability of galaxies.

A main point of this dynamics regards the motion of low angular momentum stars that can pass near the center. Such stars, moving in ordered orbits in the smooth global potential, as they approach the center, can be scattered by the locally dominating central field and their orbits become chaotic (Gerhard and Binney, 1985; Merritt and Fridman, 1996; Merritt and Valluri, 1996; Fridman and Merritt, 1997; Valluri and Merritt, 1998; Merritt and Quinlan, 1998; Siopis, 1999; Siopis and Kandrup, 2000; Holley-Bockelmann et al., 2001; Holley-Bockelmann et al., 2002; Kandrup and Sideris, 2002; Poon and Merritt, 2002; Kandrup and Siopis, 2003; Poon and Merritt, 2004; Kalapotharakos, Voglis, and Contopoulos, 2004).

In this paper we examine the role of the ordered and the chaotic motion both in models with a smooth center and in models with an extra central mass for comparison. We examine also the implications of the chaotic diffusion on the stability of these two types of models.

In sections 2 the main features of the N-body models with a smooth center and the N-body models with a central mass are described, respectively. In section 3 a method for distinguishing the particles moving in ordered orbits from the particles moving in chaotic orbits in a particular snapshot of the self-consistent potential, introduced earlier (Voglis, Kalapotharakos, and Stavropoulos, 2002), is summarized and an example is given of how this method is applied in this paper. In section 4 the fractions of mass in chaotic motion found in the various models are given and the role played by this mass in every case is discussed. The conditions under which chaotic diffusion can be effective enough so that it can cause a remarkable dynamical secular evolution in the life time of galaxies is also examined. Dynamical secular evolution proceeds by a self-organization mechanism that leads the system towards an equilibrium configuration. This mechanism is explained in section 5. Our conclusions are summarized in section 6.

2.   The N-body models

2.1   N-body models with a smooth center

Two different non-rotating triaxial equilibrium models of elliptical galaxies are used as smooth center models. A detailed description of the derivation of these models from cosmological initial conditions is given in Contopoulos, Voglis, and Kalapotharakos (2002).

In brief, the first model called Q000 is produced from quiet cosmological initial conditions, i.e. a radial Hubble motion of an initially spherical distribution of particles perturbed by a radial perturbation in the density

$$\frac{\delta \rho}{\rho} \sim r^{-\frac{n+3}{2}} \tag{1}$$

The adopted value of n is $n = -2$. This value is consistent with the power spectrum

$$P(k) \sim \mid \vec{k} \mid^{n} \tag{2}$$

of the density fluctuations in the galactic mass scale in a ΛCDM (Cold Dark Matter with $\Lambda > 0$) cosmological scenario.

This model reaches a maximum expansion and then turns around to collapse. The radial motion of particles during this phase dominates. In agreement with the radial orbit instability criterion $\eta = 2T_r/T_t > 1.7$ (Polyachenko and Shukhman, 1981; Fridman and Polyachenko, 1984; Palmer and Papaloizou, 1987; Efthymiopoulos, 1999), during the collapse and relaxation this model presents a strong radial orbit instability and finally relaxes to an equilibrium configuration developing a relatively strong bar. In the equilibrium configuration η is stabilized at a value slightly above the critical value of $\eta = 1.7$.

The second model, called C000, is derived from clumpy cosmological initial conditions, i.e. the density perturbation is given by a superposition of a large number of plane waves with amplitudes correlated with their wave number $\vec{k}$ according to (2).

This density perturbation profile reduces considerably the initial value of η. The radial orbit instability is weaker. The value of η in the equilibrium configuration is somehow smaller than the critical value $\eta = 1.7$.

The simulation of these models from the stage of the cosmological initial conditions to the final equilibrium configuration is obtained in two steps. The phase of rapid evolution (collapse and relaxation) is followed by a tree-code and leads to a virialized system. The subsequent evolution of the virialized system, composed of $N \simeq 1.5 \times 10^5$ particles, is followed by a code based on the conservative technique (Allen, Palmer, and Papaloizou, 1990). In a standard version, the code gives a smooth potential containing 120 terms. Namely, in spherical coordinates (r, θ, ϕ) the expansion of the potential $V(r, \theta, \phi)$ contains 20 monopole terms (depending on r only), 20 quadrupole terms (containing r and θ) and 80 triaxial terms (containing the three coordinates). The potential can be written as

$$
V(r, \theta, \phi) = \underbrace{\sum_{l=0}^{19} B_{l00} V_{l00}(r)}_{20 \; monopole \; terms} + \underbrace{\sum_{l=0}^{19} B_{l20} V_{l20}(r, \theta)}_{20 \; quadrupole \; terms} +
$$

$$
\underbrace{\sum_{l=0}^{19} B_{l21} V_{l21}(r, \theta) \cos \phi + \sum_{l=0}^{19} C_{l21} V_{l21}(r, \theta) \sin \phi +}_{}
$$

$$
\underbrace{\sum_{l=0}^{19} B_{l22} V_{l20}(r, \theta) \cos 2\phi + \sum_{l=0}^{19} C_{l22} V_{l20}(r, \theta) \sin 2\phi}_{80 \; triaxial \; terms} \tag{3}
$$

where $B_{l00}, B_{l20}, B_{l21}, C_{l21}, B_{l22}, C_{l22}$, with $l = 0, ... l_{max} = 19$ are the corresponding coefficients.

The unit system is as follows. The unit of length is equal to the half mass radius R_h of the system Q000. The unit of energy is so defined that the deepest value of the potential in Q000 is $V_0 = -100$. The unit of time is the half mass crossing time defined as $T_{hmct} = (2R_h^3/GM_g)^{\frac{1}{2}}$ in terms of R_h and the mass M_g of the galaxy. In terms of this time unit the Hubble time is estimated to be $t_{Hub} \approx 300 T_{hmct}$.

In a self-consistent run the coefficients of the above expansion are re-evaluated at regular small time steps $\Delta t = 0.025 T_{hmct}$.

This potential is smooth everywhere. Near the center is roughly of an harmonic type, expressing the flatness of the density in this region.

The two models (Q000 and C000) are triaxial equilibrium configurations and we define the principal axes, shortest, intermediate and longest, to be along the x, y, z cartesian axes, respectively.

In Figs.1a, b, c, d the projections on the planes x-z and y-z are shown for the model Q000 (Figs.1a, b) and C000 (Figs.1c, d), respectively.

In the Q000 model the maximum ellipticity at the half mass radius is about 0.7 (E7) and decreases outwards reaching a value of 0.5 (E5) in the outer parts. In the C000 model the maximum ellipticity at the half mass radius is about 0.4 (E4) and decreases outwards reaching a value of 0.2 (E2) in the outer parts.

The triaxiality index T is defined by

$$T = \frac{c^2 - b^2}{c^2 - a^2} \tag{4}$$

where a, b, c are the lengths of the principal axes (shortest, middle, longest, respectively) of the equidensity surface with longest axis equal to the half mass radius $r = 1$. The value of T is about 0.9 in the Q000 and about 0.8 in the C000 model.

Both systems are in a well established virial equilibrium and they are so stable that only small time variation of the coefficients in (3), around their mean values, appear. These variations correspond to a noise depending on the finite number N of particles and it is of the order of $1/\sqrt{N} \lesssim 1\%$.

If the coefficients of the potential are fixed at a given snapshot we can write an autonomous 3D Hamiltonian

$$H = \frac{\dot{r}^2}{2} + \frac{L_\phi^2}{2r^2 \sin \theta^2} + \frac{L_\theta^2}{2r^2} + V(r, \theta, \phi) \tag{5}$$

By this Hamiltonian we can investigate the phase space structure of the system at the particular snapshot using test particles to study the various possible types of orbits.

Furthermore, we can find the orbits of the real particles of the system, as they move in the fixed potential of the corresponding snapshot and compare with the orbits of test particles. The stability of orbits can also be checked (see section 3) to distinguish between the real particles moving in ordered orbits and those moving in chaotic orbits at this snapshot. Thus we can construct libraries of particles in one or the other type of motion.

Of course, the orbit of a real particle in the fixed potential is not exactly the same with the orbit of the same particle in the self-consistent run. However, if the system is in equilibrium, as it is the case in the models Q000 and C000, the statistics of the various types of orbits in two different snapshots remains the same (within a small uncertainty). This has been tested by fixing the coefficients of the potential at another snapshot, hundreds of dynamical times later and repeating the investigation. We have found again almost the same statistics of ordered and chaotic orbits. Only a fraction of 1–2% of the total mass has jumped from one type of orbit to another.

A considerable amount of the mass in ordered motion is in Box orbits. The shortest, the intermediate and the longest axes of the boxes are in average oriented along the three principal axes x, y, z, respectively of the triaxial configuration. There are also other types of ordered orbits, for example, **I**nner **L**ong **A**xis **T**ube orbits (ILAT), i.e. orbits filling a tube-like region with maximum size along the longest axis. They have a hole along the same axis, due to the non-zero component of the angular momentum of these orbits along the longest axis. For a number of ILAT orbits this hole may be small, so that they can pass quite close to the center. In this case the ILAT orbits resemble the box orbits. For this reason we call them box-like orbits. If the angular momentum along the longest axis is large enough, another family of ordered LAT orbits surrounding this axis is formed, called **O**uter **L**ong **A**xis **T**ube (OLAT) orbits. Another important family of ordered orbits is the family of **S**hort **A**xis **T**ube (SAT) orbits surrounding the shortest axis. They support the flatness of the system along the shortest axis. There are also ordered tube orbits, corresponding to various higher order resonances, called **H**igher **O**rder **R**esonance **T**ube (HORT) orbits.

2.2 N-body models with a Central Mass

The construction of N-body models with a **C**entral **M**ass (CM) is based on the smooth center models, described above. We add to their self-consistent potential (3) the central potential model

$$V_{cm}(r) = \frac{GM_{cm}}{a}\left[\arctan\left(\frac{r}{a}\right) - \frac{\pi}{2}\right] \tag{6}$$

where a is a softening length given by

$$a = 0.05\frac{M_{cm}}{M_g}R_g = 0.05mR_g \tag{7}$$

where R_g is the radius of the galaxy and $m = M_{cm}/M_g$ is the ratio of the central mass M_{cm} with respect to the mass of the galaxy M_g. Notice that the force derived from this potential is

$$F_{cm}(r) = -\frac{GM_{cm}}{r^2 + a^2}, \tag{8}$$

i.e. it is of a Keplerian nature only for orbits with pericentres much larger than a. The force at the center of this model tends to a finite constant. Stars with pericentres below the softening length a are not deflected strongly by the CM. This softening length does not significantly alter the global behavior of the system, since the number of orbits with pericentres less than a is quite small.

Notice that the density profile derived from the above potential is

$$\rho_{cm}(r) = \frac{GM_{cm}a^2}{2\pi r(r^2 + a^2)^2}, \tag{9}$$

i.e. this model for $r < a$ gives an r^{-1} cuspy density profile.

At the moment when the central mass is inserted the time is reset to $t = 0$. Hereafter, the time is measured from this origin. We have examined four cases regarding the relative mass parameter m. Namely, we examine models with values $m = 0.0005,\ 0.0010,\ 0.0050,\ 0.0100$ resulting from both Q000 and C000. Thus, including the original models with the smooth centers ($m = 0$), we have a series of ten models to compare. We call these models Q000, Q005, Q010, Q050, Q100, and C000, C005, C010, C050, C100, respectively.

In general the new systems are not in equilibrium any more. The reason is that the spherical geometry of the central field does not support the geometry of all the types of ordered orbits appearing in the smooth center models, particularly the geometry of Box orbits and box-like orbits that pass near the center. As a consequence, these orbits are deflected by the central mass and are converted to chaotic, by losing their integrals of motion. The chaotic orbits diffuse and tend to fill the whole space inside the equipotential surfaces. In some cases this diffusion can alter the shape of the equidensity (and hence the equipotential surfaces) causing a secular evolution of the system towards a new equilibrium.

As an example, in Fig. 2 the time evolution of the most important coefficients, i.e. the coefficients of the monopole and the quadrupole terms of the expansion of the potential (3) in the Q000 model (left column) and in Q100 (right column) are shown.

In Q000 both the monopole and the quadrupole coefficients remain remarkably constant even for a very long run that considerably exceeds the Hubble time. On the other hand the time evolution of the monopole coefficients in the Q100 model are also constant but the quadrupole coefficients show a serious variability. These coefficients have a systematic decrease for a period of about one Hubble time, and they reach a value of about three times smaller than their initial value. Then they remain roughly constant until the end of the run.

This behavior of the quadrupole coefficients in the Q100 model corresponds to a remarkable change of the equipotential surfaces of the system in a period of about a Hubble time. In fact the values of the quadrupole coefficients measure the strength of the bar of the system. Due to the presence of the imposed central field of mass $m = 0.01$ the bar is gradually destroyed.

The evolution of the shapes of the systems can be seen if we plot the triaxiality index T as a function of time. In Fig. 3, this evolution is shown for the models of the Q-family: Q000, Q010, Q050, Q100 and the models of the C-family: C050, C100.

The models Q000 and Q010 evolve so slowly in this diagram that their evolution in a Hubble time (=300 units) is negligible. The same happens for the models C000 and C010 (not drawn in Fig. 3).

All the other models (Q050, Q100, C050 and C100) have a fast initial evolution, i.e. the triaxiality decreases considerably in the first 100 time units and then continues to decrease more slowly tending to a final value. This final value is very close to zero in the models Q050, Q100, C100. The fact that the traxiality index reaches an almost zero constant final value shows that a new almost oblate spheroidal equilibrium configuration is established in these systems.

In the model C050 the equilibrium value of the triaxiality index is about 0.3. This model achieves an equilibrium configuration that is not an oblate spheroid but it preserves some considerable triaxiality. This indicates that triaxiality can be compatible with a central field in self-consistent models under some circumstances.

Another remarkable feature in Fig. 3 is the fact that equlilibrium is achieved in different times depending on the relative size of the central mass m and on the type of the model. In Q100 model, equilibrium is reached in about one Hubble time, while in the other models it is reached in much longer times (six or seven Hubble times).

In order to understand the role of chaotic orbits in our systems and the mechanism by which secular evolution proceeds, we first identify the particles moving in chaotic orbits in these systems in the next section.

3.   Distinction between ordered and chaotic motion

As a consequence of the non-integrability of the self-consistent potential (3) a non-negligible part of orbits in the system are chaotic. In Voglis, Kalapotharakos, and Stavropoulos (2002), we have developed a method to separate particles moving in chaotic orbits from particles moving in regular orbits in the autonomous Hamiltonian corresponding to a given snapshot of the potential of a N-body system. Notice that running the orbits in such a Hamiltonian has nothing to do with the self-consistent run of the orbits and the real time evolution of the system. What we do is to check the type of motion (ordered or chaotic) that would be described by the particles, if their positions and velocities at a given snapshot were given as initial conditions in the fixed potential of this snapshot. Of course, the method can be repeated at any snapshot.

In brief, this method combines two different tools. One tool is the Finite Time Specific Lyapunov Characteristic Number (FT-SLCN), denoted by L_j. This is the mean rate of divergence, between an orbit described by a particle j and a neighboring orbit, measured in units of the inverse radial period $(1/T_{rj})$ of the orbit j. (Radial period T_{rj} of an orbit is the time needed for the particle to go from the apocenter to the pericenter and back to apocenter). L_j is calculated for a fixed number N_{rp} of radial periods for all the orbits. (If N_{rp} tends to infinity, L_j tends to the maximal Lyapunov Characteristic Number, LCN).

In the case of ordered motion L_j decreases in average as N_{rp}^{-1}. Whenever the value of L_j along an orbit takes a roughly constant value, (not decreasing any more as N_{rp}^{-1}), we infer that the orbit is chaotic, with a specific LCN approximately equal to this constant value. If L_j continues to decrease as N_{rp}^{-1} until a maximum number of radial periods $N_{rp} = N_M$ is achieved the orbit is considered (conventionally) as an ordered orbit. In fact, along ordered orbits, L_j tends to zero, as N_{rp} tends to infinity. The minimum detectable specific LCN of a chaotic orbit is of the order of N_M^{-1}.

In our applications below we use $N_M = 1200$. The corresponding minimum detectable LCN (the detection limit) of chaotic orbits is found $L_j = 10^{-2.8}$. This minimum, of course can be pushed to lower values just by increasing N_M, but the behavior of chaotic orbits having L_j smaller than this limit is in practice so close to the behavior of ordered orbits even for times of about 10^3 radial periods, that, in a good approximation, they can be considered as ordered orbits.

The advantage of using L_j is the fact that this quantity is almost independent of the characteristic time scales of individual orbits. This property is important for the distinction between regular and chaotic orbits.

However, L_j does not tell us directly the efficiency of chaotic diffusion in a given period of time e.g. in a Hubble time. A better index for this purpose can be defined as

$$L_{cuj} = L_j \frac{T_{hmr}}{T_{rj}}, \tag{10}$$

where T_{hmr} is the radial period of an orbit with energy equal to the mean value of the potential at the half mass radius. The index L_{cuj} measures the values of L_j in a common unit $1/T_{hmr}$.

Another tool is used in parallel to the above tools. It has been shown (Voglis, Contopoulos, and Efthymiopoulos, 1998; Voglis, Contopoulos, and Efthymiopoulos, 1999; Skokos, 2001) that in a 3D autonomous Hamiltonian system, two arbitrary deviation vectors $\xi_{j1}(t)$ and $\xi_{j2}(t)$ of the same orbit j, as they evolve along the orbit, behave in a different way with respect to each other if the orbit is ordered than if the orbit is chaotic.

Namely, if the orbit is chaotic, the two deviation vectors tend exponentially to acquire the same direction (parallel or antiparallel to each other), pointing along the direction of the invariant manifold with the largest real eigenvalue. If the two deviation vectors are normalized to unity, the amplitude of their difference (if they are parallel), or of their sum (if they are antiparallel) called alignment index (AI_j), i.e.

$$AI_j = |\; \xi_{j1}(t) \mp \xi_{j2}(t)\; | \tag{11}$$

tends exponentially to zero.

On the other hand, if the orbit is ordered the two deviation vectors oscillate with respect to each other. In this case the alignment index AI_j varies around a finite mean value that remains always not very far from unity (it almost never becomes less than 10^{-3}).

We present here an example of how this method is applied in the model C100. In the models with $m \neq 0$ the total potential is a superposition of the potential (3) produced by the distribution of the particles of the N-body system and the potential (6) produced by the imposed central mass. We fix the coefficients of the potential (3) at a given snapshot of this model, e.g. at $t = 150$ and we run the orbits of the system and the corresponding variational equations in the autonomous Hamiltonian

$$H = \frac{\dot{r}^2}{2} + \frac{L_\phi^2}{2r^2 \sin\theta^2} + \frac{L_\theta^2}{2r^2} + V(r, \theta, \phi) + V_{cm}(r) \tag{12}$$

The results regarding the evolution of the particles on the plane $\log AI_j - \log L_j$ are shown in Figs.4a, b, c. Three different snapshots are given, after running the orbits for $N_{rp} = 20, 100, 1200$ radial periods, respectively. In Fig. 4a all the particles are concentrated at about the same region on this plane. This is due to two reasons. First, because L_j is almost independent of the characteristic time scales of the orbits and second because the run time of $N_{rp} = 20$ radial periods is too short to separate the particles. The separation of particles on this plane starts after longer running times, e.g. $N_{rp} = 100$ radial periods (Fig. 4b). For $N_{rp} = N_M = 1200$ radial periods two different groups are clearly formed on the plane $\log AI_j - \log L_j$ (Fig. 4c).

The first group corresponds to $\log L_j < -2.8$ and $\log AI_j > -3$. This group contains particles moving either in ordered orbits or moving in very weakly chaotic orbits (with $L_j < 10^{-2.8}$, that need longer integration times to be detected). As mentioned above, such very weakly chaotic orbits behave as ordered orbits for times exceeding a Hubble time ($\simeq 300 T_{hmct}$), thus they can be considered as ordered orbits for all practical purposes.

The second group contains particles with their L_js stabilized at $\log L_j \gtrsim -2.5$, with $\log AI_j \leq -10$, i.e. moving in clearly chaotic orbits. (The sharp cut off on the right hand side of this group is due to the fact that for many particles $\log AI_j$ becomes less than -10 before N_{rp} reaches the maximum value N_M and the integration of these particles is stopped).

There is also a lane of points joining the two groups above. This lane contains particles moving in more weakly chaotic orbits or particles moving in sticky orbits, i.e. chaotic orbits trapped between various cantori before they escape to a chaotic sea. This lane is detached from the group of ordered motion at $L_j = 10^{-2.8}$. This value is adopted as the threshold of separation between ordered and chaotic orbits in what follows.

Using the method described above, we have made lists of particles in chaotic motion and of particles in ordered motion of all our models, as they result from the selected particular snapshots of the self-consistent potential of the models.

4.   The role of chaotic motion

In the Q000 and C000 models the fraction of mass in chaotic motion is about 32% and 23%, respectively and remains quite constant in time.

This remarkable stability of the above self-consistent triaxial configurations is due to the following three properties. First, the fractions of mass in ordered orbits are relatively large (68% and 77% respectively in Q000 and in C000). Second, the mass in chaotic motion is almost spherically distributed. In Figs.5a, b, for example, the particles of the Q000 model moving in ordered motion (column (a)) and the particles moving in chaotic motion (column (b)) are projected separately on the planes x-z, y-z and x-y. It is clear that the chaotic component approaches a spherical distribution in contrast to the ordered component that is strongly non-spherically distributed. The chaotic component contributes mainly to the spherical (monopole) part of the potential. Thus, diffusion of this component can not seriously affect the shapes of the equipotential surfaces. Furthermore, the part of the chaotic mass with $L_{cuj} > 10^{-2}$, that is the mass able to develop chaotic diffusion in a Hubble time, is quite small, i.e. less than 8% and 2% of the total mass in the two models Q000 and C000, respectively.

In the models with $m \neq 0$ the situation is different. The fraction of mass in chaotic motion derived from the potential of the snapshot at $t = 0$ (at the time when the central mass is inserted) is about 80% for the Q-family and about 50% for the C-family models, almost independently of the adopted values of m.

These fractions at later snapshots become different for different m. In Fig. 6 the percentage of the mass in chaotic motion (with respect to the total mass of the system) is given as a function of m, at their snapshot of $t = 150$. The solid line is for the Q-family models and the dashed line is for the C-family models. The fractions of mass in chaotic motion are always smaller in the C-family than in the Q-family models.

From this diagram it is clear that the fraction of mass in chaotic motion detected at the snapshot $t = 150$ of every model is not a monotonic function of m along the same family. The maximum values of these fractions appear in the models Q010 and C010, respectively. In the models with larger m (Q050, Q100, C050, C100) the fraction of mass in chaotic orbits decreases with m. The smaller fraction of mass in chaotic motion for larger m in these models is due to the fact that a number of chaotic orbits are converted to ordered orbits during the period from $t = 0$ to $t = 150$. These systems evolve by a process of self-organization that is discussed in section 5.

We examine next, several statistical features of the mass in chaotic motion in our models. In Figs.7a, a$'$, b, b$'$, c, c$'$, d, d$'$, e, e$'$ we see the distributions of this mass along the $\log L_j$ axis (left column, a,b,c,d,e) and the distributions along the $\log L_{cuj}$ axis (right column, a$'$, b$'$, c$'$, d$'$, e$'$) at $t = 150$ for all the models of the Q-family, as indicated in the figure. These distributions are normalized with respect to the mass in chaotic motion of the Q010 model.

The distributions along the $\log L_j$ axis for $m > 0$ (Figs.7b, c, d, e) have a characteristic maximum at different values of $\log L_j = \log L_M$, that is shifted towards larger values of $\log L_M$ as m increases. Such a maximum does not appear for $m = 0$ (Fig. 7a). The main part of the mass around this maximum corresponds to particles describing Box orbits for $m = 0$. These particles form the bar of the system.

The distributions along the $\log L_{cuj}$ axis (Figs.7a$'$, b$'$, c$'$, d$'$, e$'$) are remarkably different from the corresponding distributions along the $\log L_j$ axis, but the values of $\log L_{cuj} = \log L_{cuM}$ at which the maxima appear in b', c', d', e' are not much different than $\log L_M$.

Similar remarks can be made for the C-family models shown in Figs.8a, a$'$, b, b$'$, c, c$'$, d, d$'$, e, e$'$.

It is clear from Figs.7 and Figs.8 that the values of $\log L_M$ and $\log L_{cuM}$ (for $m \neq 0$) depend on m. The mean value of L_{cuj}, denoted by $< L_{cuj} >$, also depends on m. We find that these three quantities have a narrow correlation with m, that can be approximated by power laws as it is shown in Figs.9a, b, c. Stars refer to the Q-family and dots to the C-family models. For L_M (Fig. 9a), L_{cuM} (Fig. 9b) and $< L_{cuj} >$ (Fig. 9c) the power laws are

$$L_M \propto m^\alpha, \qquad L_{cuM} \propto m^{\alpha'}, \qquad < L_{cuj} > \propto m^\beta \qquad (13)$$

where $\alpha = 0.564 \pm 0.022$, $\alpha' = 0.57 \pm 0.07$ and $\beta = 0.45 \pm 0.03$. The quantities L_M, L_{cuM} and $< L_{cuj} >$ are different measures of the mean level of the Lyapunov numbers of the orbits that become chaotic because of the presence of the central mass. Grossly speaking, one can say that the typical Lyapunov numbers grow roughly proportional to the square root of m. This property appears here as an empirical statistical result. We do not know yet a statistical theory to explain this correlation, but it should be attributed to the strength of deflection of the orbits passing near the center and the respective variations of the actions.

In Fig. 9c we see that the models with $m < 0.0050$, i.e. the models Q005, Q010, C005, C010 have $< L_{cuj} >$ close to 10^{-2}. These models do not evolve considerably in a Hubble time. On the other hand the models with $m \geq 0.0050$, i.e. the models Q050, Q100, C050, C100 that develop a detectable secular evolution in a Hubble time have $< L_{cuj} >$ close to $10^{-1.5}$, i.e. about three times larger. In principle, this difference can be a reason to explain the faster rate of evolution in the latter case, since diffusion of a chaotic orbit is observable in a Hubble time, only when the value of its L_{cuj} is $L_{cuj} > 10^{-2}$.

A more careful examination, however, shows that there are two prerequisites for such a system to present a remarkable rate of dynamical secular evolution in a Hubble time due to the chaotic diffusion of its particles. Namely,

i) the fraction of mass in chaotic motion with $L_{cuj} \gtrsim 10^{-2}$ must not be too small and

ii) this mass must have an anisotropic (non-spherical) initial distribution in the configuration space. (If this mass has a roughly spherical distribution in the configuration space, it simply contributes to the spherical background of the potential. Diffusion of this mass can not have serious consequences on the shape of the system, as we have seen in the case of the Q000 model above, for example).

In order to collect the particles in chaotic motion that are most responsible for the secular evolution in our models we collect only the particles for which the values of $\log L_j$ belong to a window of width ± 0.3 around the value of $\log L_M$ in Figs. 7b, c, d, e and Figs. 8b, c, d, e. (The width ± 0.3 of the window is an estimated measure of the size of the maxima along the horizontal axis). The majority of the particles in this window move in Box or box-like orbits in the corresponding model with smooth center, and they still maintain much of their initial aspherical distribution.

This can be seen, for example, in Figs. 10a, b, c regarding the model Q100 at the snapshot of $t = 150$. The letters a,b,c label the columns of this figure. Column (a) gives the projections on the planes $x - z$, $y - z$, $x - y$ of the particles in ordered motion. This is a fraction of $\simeq 42\%$ of the total mass. Column (b) gives similar projections, but for the chaotic mass selected from the window around $\log L_M = -1.6$ in Fig. 7e. This mass is about 38% of the total mass and shows a non-spherical distribution not much different than the distribution of the ordered orbits in the column (a). Column (c) gives the distribution of the rest of the mass in chaotic motion (outside the selected window). The distribution of this mass, which is about 20% of the total mass forms an almost spherical background.

The secular evolution of the system Q100 is due to the mass in column (b). In the subsequent evolution the main part of this mass is organized preferably in SAT orbits. The rest part remains chaotic taking an almost spherical distribution. This is seen in Figs. 11a, b, that give separately the ordered orbits ($\simeq 78\%$, shown in column (a)) and the chaotic orbits ($\simeq 22\%$, shown in column (b)) for the snapshot of the Q100 model at $t = 300$, when an oblate spheroidal equilibrium has been established. Ordered orbits in (a) form an oblate spheroidal configuration, while chaotic orbits in (b) form an almost spherical distribution.

In Fig. 12 we compare the distributions along the $\log L_j$ axis of the detected mass in chaotic motion of the Q100 model at three different snapshots, namely

$t = 0$ (dot-dashed line), $t = 150$ (solid line) and $t = 300$ (dashed line). The three distributions are normalized with respect to the mass of the distribution at $t = 150$. The pronounced peak at about $\log L_M = -1.6$ of the first distribution is considerably smaller in the second distribution and much smaller but detectable in the last distribution, at $\log L_M = -1.5$. The decrease of this peak in time is a further verification that the part of the mass in chaotic motion that is organized and causes the secular evolution of the system comes mainly from this peak.

It is remarkable that the distribution at $t = 300$ (dashed line in Fig. 12) has an extensive right tail, exceeding the values of $\log L_j = -1$ which is roughly the maximum value for the two previous distributions. This extension is due to thermodynamical reasons and it is discussed in the next section.

In order to demonstrate that the position of the maximum of the distributions shown in Figs. 7b, c, d, e and Figs. 8b, c, d, e is important for the secular evolution to be observable in a Hubble time, we construct Figs. 13a, b, c, which are similar to the Figs. 10a, b, c, but for the model Q005 at the snapshot $t = 150$. The ordered part of the mass shown in column (a) of Fig. 13 is $\simeq 20\%$ of the total mass and has a strongly bar-like distribution, composed mainly of resonant orbits (HORT and ILAT). The chaotic mass inside the window (± 0.3) around $\log L_M = -2.3$ (see Fig. 7b) is about $\simeq 50\%$ and is shown in column (b). This mass is distributed in a quite similar way as the mass in column (a). Column (c) shows the mass in chaotic motion ($\simeq 30\%$) with values of $\log L_j$ outside the window. This mass is clearly more dispersed than the other two parts. Due to the small value of $\log L_M$ the chaotic diffusion rate in this model is so slow that the corresponding secular evolution is not detectable even for times larger than a Hubble time.

The rate of the dynamical secular evolution of our systems depends mainly on the amount of the non-spherically distributed chaotic orbits and on the mean value of their Lyapunov numbers L_{cuj}. As mentioned above, spherically distributed mass in chaotic motion cannot be efficient for secular evolution. The fraction of the non-spherically distributed chaotic mass in our models can be estimated by $\frac{\Delta N_w}{N_{total}}$ where ΔN_w is the mass in a window ± 0.3 around the maximum of the distributions in Fig. 7b, c, d, e and Fig. 8b, c, d, e. If the mean value of L_{cuj} of the mass in this window is L_w, a measure of the efficiency of chaotic diffusion can be obtained by the quantity

$$\mathcal{L} = \frac{\Delta N_w}{N_{total}} L_w \tag{14}$$

that we call "effective diffusion momentum". $\mathcal{L}$ combines the two effects, i.e. the fraction of mass that can alter the system by diffusion and the typical rate of its diffusion measured by L_w. The correlation of $\mathcal{L}$ with the evolution rate is shown in Fig. 14, where $\mathcal{L}$ is plotted vs m. Stars correspond to the Q family models and dots to the C family models. The models that show secular evolution in a Hubble time, i.e. the "evolving" models Q050, Q100, C050, C100, are plotted twice at two different snapshots, i.e. at time $t = 150$ and at their equilibrium snapshots ($t = 300$ for Q100 and $t = 2200$ for the others) as indicated in the figure. In this figure we see that for the "non-evolving" models (that do not show secular evolution in a Hubble time) Q005, Q010, C005, C010 the values of $\mathscr{L}$ are less than about 0.0045. For the evolving models $\mathscr{L}$ is initially larger than this value but falls below this limit when equilibrium is established. It seems that secular evolution due to chaotic diffusion is possible only if the effective diffusion momentum $\mathscr{L}$ is larger than 0.0045.

Notice that the non-evolving models maintain about the same fractions of mass in chaotic motion ($\simeq 80\%$ in Q005, Q010 and $\simeq 50\%$ in C005, C010) for times considerably exceeding a Hubble time. The fractions of mass in chaotic motion in the evolving models Q050, Q100, C050 and C100, at their equilibrium configurations, reach the values $\simeq 25\%$, $\simeq 22\%$, $\simeq 19\%$ and $\simeq 12\%$, respectively.

5.   Self-organization

Along their secular evolution the systems convert gradually chaotic orbits to ordered orbits. The mass in chaotic motion decreases in time. As an example, we give in Fig. 15 the fraction of mass in chaotic motion in the Q100 model as a function of time. At $t = 0$ this fraction starts from a high value of about 80% and decreases to a value of about 22% at $t = 250$. By this time the secular evolution ceases. The system has almost achieved an oblate spheroidal equilibrium configuration at which the fraction of mass in chaotic motion remains practically constant.

The majority of chaotic orbits in Q100 that are organized to ordered orbits come from Box or box-like orbits in Q000 which have lost their integrals of motion by the action of the central mass at $t = 0$ and have become chaotic.

In Figs. 16a, b, c we give an example of the phase space portrait of the projection on the plane $(|y|, \dot{y})$ of the 4D Poincaré surface of section $(x, y, \dot{x}, \dot{y})$ for $z = 0$ and energy $h = -74$.

The projections of invariant tori shown in Fig. 16a have been derived by running test particles in the potential of the Q000 model fixed at $t = 0$. These tori correspond to Box orbits of energy $h = -74$ in Q000 and they are quite constant in time, i.e. they are almost the same if they are constructed from the potential of any snapshot in this model. The big dots in the same figure, projected approximately on one of these tori, are the Poincaré consequents (successive intersections with the surface of section) of a particular Box orbit, described by a real particle in the Q000 system with energy $h \simeq -74$, during the self-consistent run.

In a portrait corresponding to the portrait in Fig. 16a but in the potential of the Q100 model fixed at $t = 0$ invariant tori are almost absent. Test particles describe chaotic motion almost everywhere. At later snapshots this portrait appears different. If the potential of the Q100 model is fixed at $t = 90$, the portrait is shown in Fig. 16b. During the period from $t = 0$ until $t = 90$ the self-consistent potential has reached a new form. Fig. 16b shows that in this potential the only type of ordered orbits are SAT orbits (represented by the island of ordered motion around the periodic orbit at $(|y|, \dot{y}) \simeq (0.4, 0.0)$. All the area on the left of this island is chaotic.

The successive Poincaré consequents of the orbit of the real particle in Fig. 16a are projected by stars in Fig. 16b in the chaotic sea. These consequents are collected during the self-consistent run of the Q100 model from $t = 0$ to $t = 180$. During this period the real particle describes a chaotic orbit.

At $t = 300$ the self-consistent potential of the Q100 model has been further evolved so that the island of SAT orbits is increased to occupy the major part of the portrait and the area of chaotic orbits is squeezed to only small values of $|y|$ (Fig. 16c). The successive Poincaré consequents of the orbit of the same real particle, collected during the self-consistent run from $t = 180$ until $t = 350$ of this model are shown by big dots in Fig. 16c. They are projected inside the island of SAT orbits, indicating that this orbit has finally been converted to an ordered orbit of SAT type.

From the above we conclude that the mechanism of the secular evolution discussed in this paper is as follows. The presence of the central mass destabilizes the box or box-like orbits passing very close to the center. These orbits become chaotic and diffuse. Diffusion can initiate a rate of secular evolution. During this evolution the self-consistent potential gradually alters to a new form, that allows a smaller number of chaotic orbits and favors the SAT type of ordered orbits. Thus, many of the diffused orbits are trapped by the new form of the potential and converted to ordered orbits. This process relaxes and equilibrium is achieved when diffusion can not be effective any more, i.e. the effective diffusion momentum becomes less than a critical limit, either because the remaining chaotic orbits are almost spherically distributed or because their Lyapunov numbers are too small.

The oblate spheroidal equilibrium configuration at which the system Q100 relaxes is mainly supported by a large number of SAT orbits trapped in this type by the new form of the self-consistent potential.

It is interesting to examine the following question. During the process of self-organization described above, order is produced out of chaos. The question is how this self-organization can be compatible with the second law of thermodynamics. In fact, the total entropy of the system increases. However, self-organization proceeds by transferring energy and entropy from one region of phase space to another. Namely, this transfer occurs from the region of SAT orbits to the region of chaotic orbits. As a thermodynamic consequence, the Lyapunov numbers in the remaining chaotic orbits spread to larger values, or chaotic orbits diffuse to larger distances in order to balance the reduction of entropy in the organized areas.

In Fig. 12 we have seen already that the distribution of the mass in chaotic motion at $t = 300$ (dashed line) has a longer tail on the right end than the distributions at $t = 0$ and $t = 150$. This difference can be better seen in Figs.17a,b where the distribution of all the particles in the Q100 model are plotted on the plane of L_j and their binding energy E at the snapshot of $t = 150$ (Fig. 17a) and at the snapshot of $t = 300$ (Fig. 17b). We see that the number of particles with $L_j > 0.1$ is clearly larger in the latter case although the total number of particles in chaotic motion is more than two times smaller. The remaining chaotic orbits although considerably fewer than initially absorb entropy by increasing their Lyapunov numbers.

6.   Conclusions

The fractions of mass in ordered and in chaotic motion and the resulting implications have been examined, in a series of N-body models simulating non-rotating elliptical galaxies. Our conclusions are as follows:

Models with smooth centers in virial equilibrium, with maximum ellipticities E4 and E7, contain a fraction of mass in chaotic motion (with specific Lyapunov Numbers L_j larger than $10^{-2.8}$) of about 23% to 32%, respectively. Chaotic diffusion in these systems is not able to cause a dynamical secular evolution for two reasons. First, the fraction of mass in chaotic motion is relatively small. Especially the part of this mass with $L_{cuj} > 10^{-2}$ that can give observable diffusion in a Hubble time is less than 2% or 8% respectively. Second, the mass in chaotic motion is roughly spherically distributed.

If a central mass, assumed to be mainly due to a massive central black hole of sizes $m \gtrsim 0.0005$ in units of the total mass of the galaxy, is inserted the mass in chaotic motion increases to the levels of 50% and 80%, respectively for the two models.

The main part of this mass comes from the Box or box-like orbits in the original models that are converted to chaotic as these orbits pass near the central mass. They are anisotropically distributed along the bar of the system. The mean level of the Lyapunov numbers of this part of mass (measured either in terms of L_j or in terms of $< L_{cuj} >$) has a narrow correlation with m, increasing roughly proportionally to $m^{0.5}$.

Dynamical secular evolution of the models with $m \neq 0$ depends on the mean level of the Lyapunov Numbers of the chaotic orbits along the bar, but also depends on the relative fraction of this mass with respect to the total mass of the galaxy. A quantity called "effective chaotic momentum" $\mathscr{L}$ has been introduced to measure the combined action of these two effects. We have estimated numerically that, if the effective chaotic momentum $\mathscr{L}$ takes values below the limit of ≈ 0.0045, secular evolution of the self-consistent equipotential surfaces is negligible, at least in a Hubble time. In these cases the systems are practically stable. For larger values of $\mathscr{L}$, however, the systems are remarkably unstable. Chaotic diffusion causes serious changes in the self-consistent equipotential surfaces. The systems evolve by a process of self-organization, converting chaotic orbits to ordered orbits of SAT type. They approach an equilibrium state of smaller triaxiality, or an oblate spheroidal equilibrium. This evolution ceases and equilibrium is established, when the effective chaotic momentum falls below the limit of 0.0045.

The mean rate of converting chaotic to ordered orbits is initially slower and becomes faster later. An example is given to explain how an orbit, of Box type in the original system that becomes chaotic in the corresponding system with central mass, diffuses. This orbit, as the potential varies in time, is finally trapped and organized to a SAT form. In this way order is produced out of chaos. In thermodynamical terms this process is feasible, although the total entropy of the system increases in time according to the second law of Thermodynamics, by transferring energy and entropy from one region of phase space to another. Namely, in the region of SAT orbits the entropy is reduced by the increase of the number of ordered orbits. This entropy reduction is partly balanced by an increase of the level of the Lyapunov numbers in the remaining chaotic orbits.

Acknowledgments

This work is partly supported by a research program of the EMPEIRIKEION Foundation. C.K. wishes to thank the Greek State Scholarship Foundation (I.K.Y) for financial support.

References

Allen, A. J., Palmer, P. L. and Papaloizou, J.: 1990, 'A conservative numerical technique for collisionless dynamical systems - Comparison of the radial and circular orbit instabilities', *MNRAS*, **242**, 576–594

Contopoulos, G., Voglis, N., Kalapotharakos, C.: 2002, 'Order and Chaos in Self-Consistent Galactic Models', *Celestial Mechanics and Dynamical Astronomy*, **83**, 191–204

Cretton, N., van den Bosch, F. C.: 1999, 'Evidence for a Massive Black Hole in the S0 Galaxy NGC 4342', *Astrophysical Journal*, **514**, 704–724

de Zeeuw, T.: 1985, 'Elliptical galaxies with separable potentials', *MNRAS*, **216**, 273–334

Efthymiopoulos, C.: 1999, PhD thesis, University of Athens, Greece

Faber, S. M., Tremaine, S., Ajhar, E. A., Byun, Y., Dressler, A., Gebhardt, K., Grillmair, C., Kormendy, J., Lauer, T. R., Richstone, D.: 1997, 'The Centers of Early-Type Galaxies with HST. IV. Central Parameter Relations', *Astronomical Journal* , **114**, 1771–1796

Ferrarese, L., van den Bosch, F. C., Ford, H. C., Jaffe, W., O'Connell, R. W.: 1994, 'Hubble Space Telescope photometry of the central regions of Virgo cluster elliptical galaxies. 3: Brightness profiles' *Astronomical Journal*, **108**, 1598–1609

Fridman, T., Merritt, D.: 1997, 'Periodic Orbits in Triaxial Galaxies with Weak Cusps' *Astronomical Journal*, **114**, 1479–1487

Gebhardt, K., Richstone, D., Ajhar, E. A., Lauer, T. R., Byun, Y., Kormendy, J., Dressler, A., Faber, S. M., Grillmair, C., Tremaine, S.: 1996, 'The Centers of Early-Type Galaxies With HST. III. Non-Parametric Recovery of Stellar Luminosity Distribution', *Astronomical Journal*, **112**, 105–113

Gebhardt, K., Richstone, D., Kormendy, J., Lauer, T. R., Ajhar, E. A., Bender, R., Dressler, A., Faber, S. M., Grillmair, C., Magorrian, J., Tremaine, S.: 2000, 'Axisymmetric, Three-Integral Models of Galaxies: A Massive Black Hole in NGC 3379', *Astronomical Journal*, **119**, 1157–1171

Gerhard, O. E., Binney, J.: 1985, 'Triaxial galaxies containing massive black holes or central density cusps', *MNRAS*, **216**, 467–502

Hernquist, L., Ostriker, J.: 1992, 'A self-consistent field method for galactic dynamics', *Astrophysical Journal*, **386**, 375–397

Holley-Bockelmann, K., Mihos, J. C., Sigurdsson, S., Hernquist, L.: 2001, 'Models of Cuspy Triaxial Galaxies', *Astrophysical Journal*, **549**, 862–870

Holley-Bockelmann, K., Mihos, J. C., Sigurdsson, S., Hernquist, L., Norman, C.: 2002, 'The Evolution of Cuspy Triaxial Galaxies Harboring Central Black Holes', *Astrophysical Journal*, **567**, 817–827

Fridman, A. M, Polyachenko, V. L.: 1984, *Physics of gravitating systems*, New York, Springer-Verlag

Kalapotharakos, C., Voglis, N., Contopoulos, G.: 2004, 'Chaos and Secular Evolution of Triaxial N-Body Galactic Models due to an Imposed Central Mass', *Astronomy & Astrophysics*, (in press)

Kandrup, H. E., Sideris, I. V.: 2002, 'Chaos in cuspy triaxial galaxies with a supermassive black hole: a simple toy model', *Celestial Mechanics and Dynamical Astronomy*, **82**, 61–81

Kandrup, H. E., Siopis, Ch.: 2003, 'Chaos and chaotic phase mixing in cuspy triaxial potentials', *MNRAS*, **345**, 727–742

Kormendy, J., Richstone, D.: 1995, 'Inward Bound—The Search For Supermassive Black Holes In Galactic Nuclei', *Annual Review of Astronomy and Astrophysics*, **33**, 581–624

Kormendy, J., Bender, R., Magorrian, J., Tremaine, S., Gebhardt, K., Richstone, D., Dressler, A., Faber, S. M., Grillmair, C., Lauer, T. R.: 1997, *Astrophysical Journal Letters*, **482**, 139–144

Kormendy, J., Bender, R., Evans, A. S., Richstone, D.: 1998, 'The Mass Distribution in the Elliptical Galaxy NGC 3377: Evidence for a $2 \times 10^8 M_\odot$ Black Hole', *Astronomical Journal*, **115**, 1823–1839

Lauer, T. R., Ajhar, E. A., Byun, Y. I., Dressler, A., Faber, S. M., Grillmair, C., Kormendy, J., Richstone, D., Tremaine, S.: 1995, 'The Centers of Early-Type Galaxies with HST.I.An Observational Survey', *Astronomical Journal*, **110**, 2622–2654

Lynden-Bell, D.: 1969, 'Galactic Nuclei as Collapsed Old Quasars', *Nature*, **223**, 690

Magorrian, J., Tremaine, S., Richstone, D., Bender, R., Bower, G., Dressler, A., Faber, S. M., Gebhardt, K., Green, R., Grillmair, C., Kormendy, J., Lauer, T.: 1998, 'The Demography of Massive Dark Objects in Galaxy Centers', *Astronomical Journal*, **115**, 2285–2305

Merritt, D., Fridman, T.: 1996, 'Triaxial Galaxies with Cusps', *Astrophysical Journal*, **460**, 136–162

Merritt, D., Quinlan D. G.: 1998, 'Dynamical Evolution of Elliptical Galaxies with Central Singularities', *Astrophysical Journal*, **498**, 625–639

Merritt, D., Valluri, M.: 1996, 'Chaos and Mixing in Triaxial Stellar Systems', *Astrophysical Journal*, **471**, 82–105

Palmer, P. L., Papaloizou, J.: 1987, 'Instability in spherical stellar systems', *MNRAS*, **224**, 1043–1053

Polyachenko, V. L., Shukhman, I. G.: 1981, 'General Models of Collisionless Spherically Symmetric Stellar Systems - a Stability Analysis', *SOVIET ASTRONOMY*, **25**, 533

Poon, M. Y., Merritt, D.: 2002, 'Triaxial Black Hole Nuclei', *Astrophysical Journal*, **568**, 89–92

Poon, M. Y., Merritt, D.: 2004, 'A Self-Consistent Study of Triaxial Black Hole Nuclei', *Astrophysical Journal*, **606**, 774–787

Salpeter, E. E.: 1964, 'Accretion of Interstellar Matter by Massive Objects', *Astrophysical Journal*, **140**, 796–800

Siopis, Ch.: 1999, PhD thesis, University of Florida

Siopis, Ch., Kandrup, H. E.: 2000, 'Phase-space transport in cuspy triaxial potentials: can they be used to construct self-consistent equilibria?', *MNRAS*, **319**, 43–62

Skokos Ch.: 2001, 'Alignment indices: a new, simple method for determining the ordered or chaotic nature of orbits', *Journal of Physics A*, **34**, 10029–10043

Statler, T. S.: 1987, 'Self-consistent models of perfect triaxial galaxies', *Astrophysical Journal*, **321**, 113–152

Valluri, M., Merritt, D.: 1998, 'Regular and Chaotic Dynamics of Triaxial Stellar Systems', *Astrophysical Journal*, **506**, 686–711

van der Marel, R. P., de Zeeuw, P. T., Rix, H. W.: 1997, 'Improved Evidence for a Black Hole in M32 from HST/FOS Spectra. I. Observations', *Astrophysical Journal*, **488**, 119–135

van der Marel, R. P., van den Bosch F. C.: 1998, 'Evidence for a $3 \times 10^8 M_\odot$ Black Hole in NGC 7052 from Hubble Space Telescope Observations of the Nuclear Gas Disk', *Astronomical Journal*, **116**, 2220-2236

Voglis, N., Contopoulos, G., Efthymiopoulos, C.: 1998, 'Method for distinguishing between ordered and chaotic orbits in four-dimensional maps', *Physical Review E*, **57**, 372–377

Voglis, N., Contopoulos, G., Efthymiopoulos, C.: 1999, 'Detection of Ordered and Chaotic Motion Using the Dynamical Spectra', *Celestial Mechanics and Dynamical Astronomy*, **73**, 211–220

Voglis, N., Kalapotharakos, C., Stavropoulos, I.: 2002, 'Mass components in ordered and in chaotic motion in galactic N-body models', *MNRAS*, **337**, 619–630

Zeldovich, Y.: 1964, 'The Fate of a Star and the Evolution of Gravitational Energy upon Accretion', *Soviet PhysicsÖDoklady* , **9**, 195

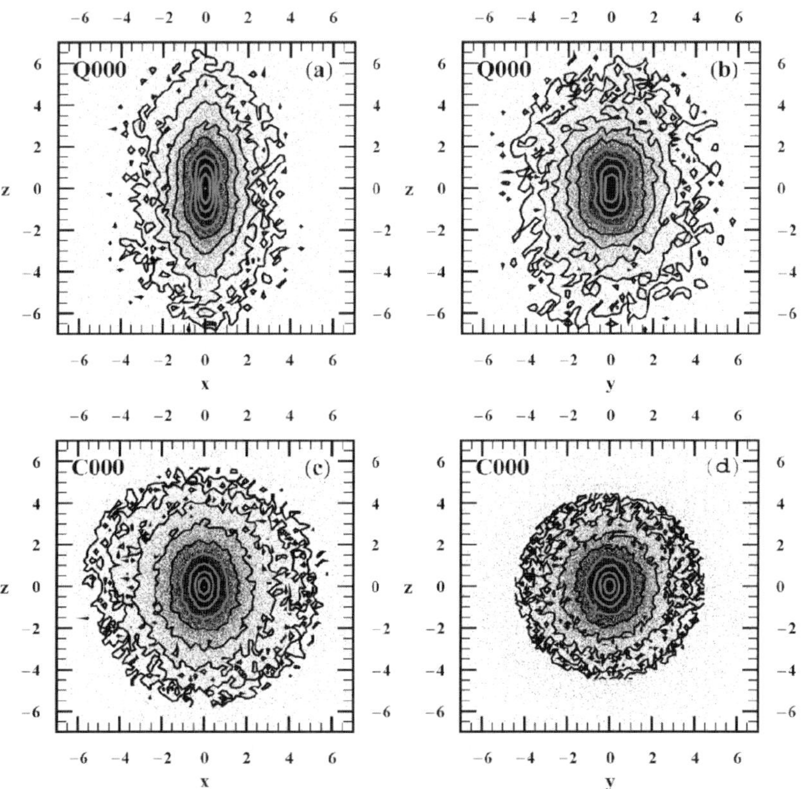

Figure 1. **(a)** and **(b)**: Projections on the planes x-z and y-z of the Q000 model. The maximum ellipticity is E7. **(c)** and **(d)**: as in (a) and (b) but for the C000 model. The maximum ellipticity is E4.

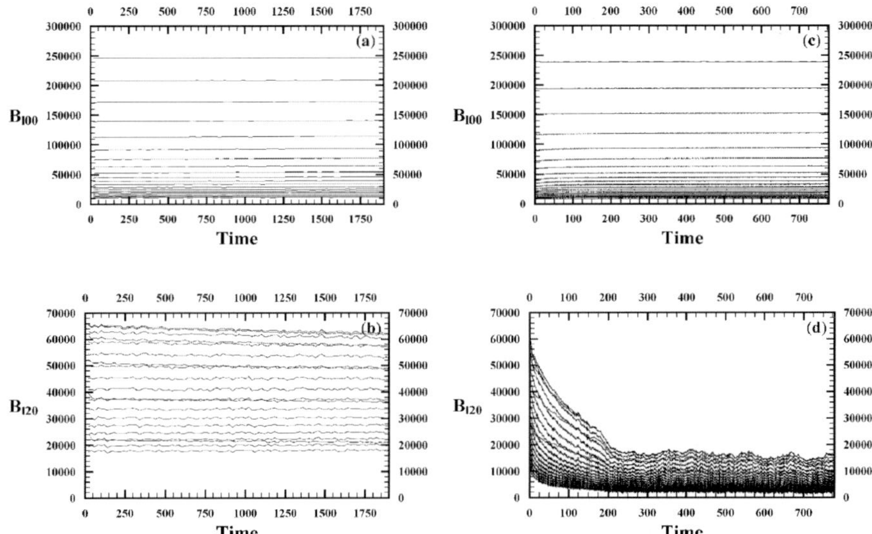

Figure 2. In (a) and (b). Time evolution of the coefficients of the most important terms, monopole in (a) and quadrupole in (b) of the expansion of the potential in the Q000 model. These coefficients remain remarkably constant even for times much longer than a Hubble time. In (c) as in (a), but for the Q100 model. The monopole coefficients in this model are also constant in time. In (d) the quadrupole coefficients of the Q100 model vary considerably in about a Hubble time due to serious changes of the shape of the self-consistent equipotential surfaces during this period. Then they relax to almost constant values, indicating that an equilibrium configuration is achieved.

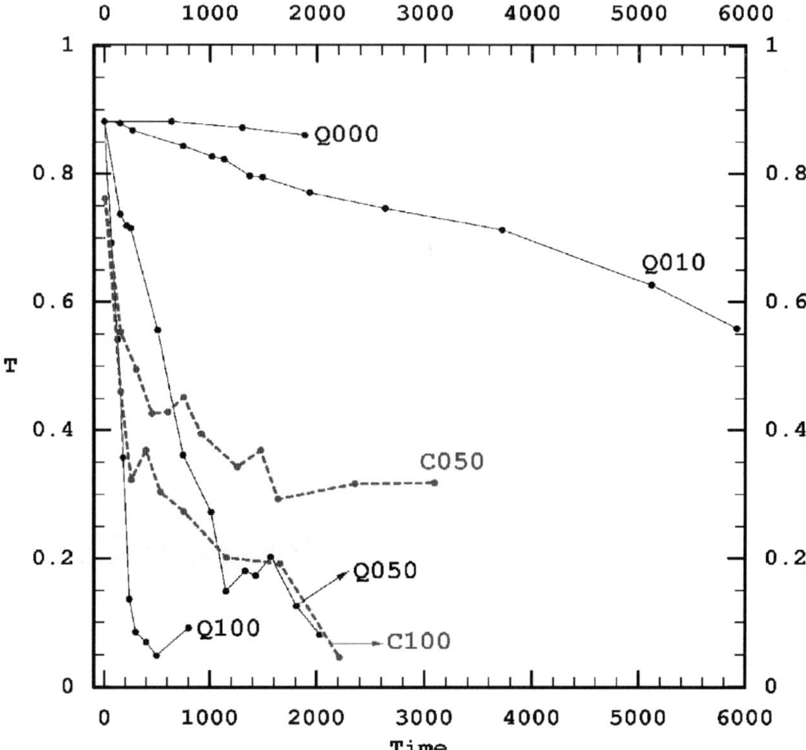

Figure 3.   Time evolution of the triaxiality index T for the models indicated in the figure. In the models Q050, Q100, and C100 the triaxiality index reaches a constant almost zero value corresponding to an oblate spheroidal equilibrium configuration. In the C050 model T is stabilized at a value of about 0.3 that corresponds to a triaxial equilibrium configuration. For all the other models the evolution is so slow, as for example in the Q000 and Q010 models shown in the figure, that is neglected in a Hubble time $\approx 300 T_{hmct}$.

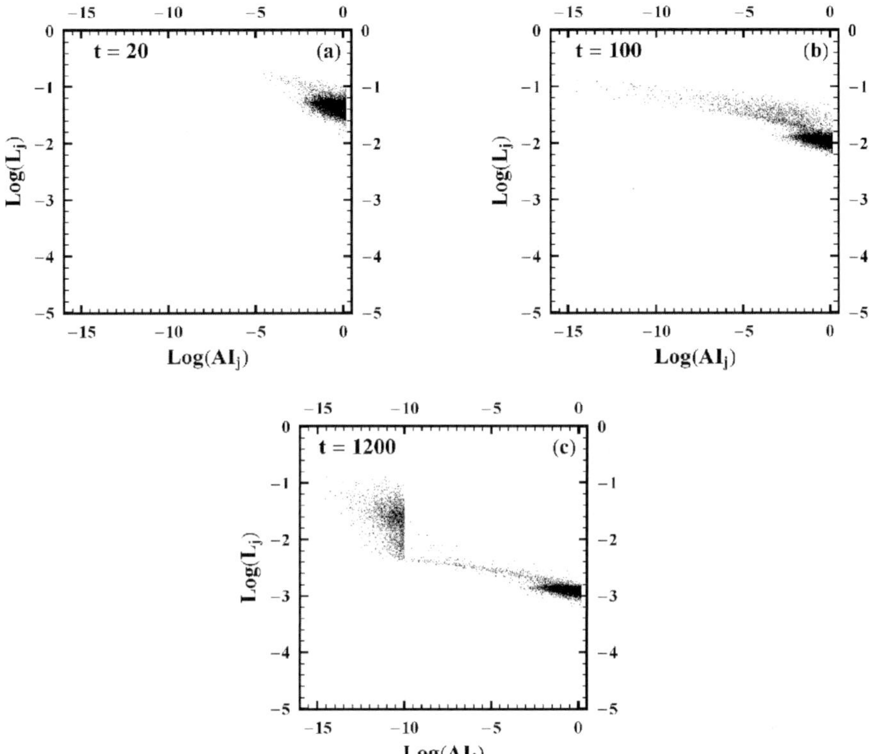

Figure 4. An example of distinction between ordered and chaotic orbits in terms of AI_j and L_j at the snapshot of the potential at $t = 150$ of the C100 model. **(a)**: A number of radial periods $N_{rp} = 20$ is too small for a distinction. All the orbits are concentrated in a small region because L_j is almost independent of the characteristic dynamical time of the orbit. **(b)**: For $N_{rp} = 100$ a number of chaotic orbits tend to stabilize their L_j and move towards very small values of AI_j. For the rest of the orbits L_j decreases as $\sim N_{rp}^{-1}$, while their AI_j is kept almost constant. **(c)**: For $N_{rp} = 1200$ one can separate orbits with $\log L_j < -2.8$ as ordered orbits from the rest of the orbits that are chaotic. (See the text for more details).

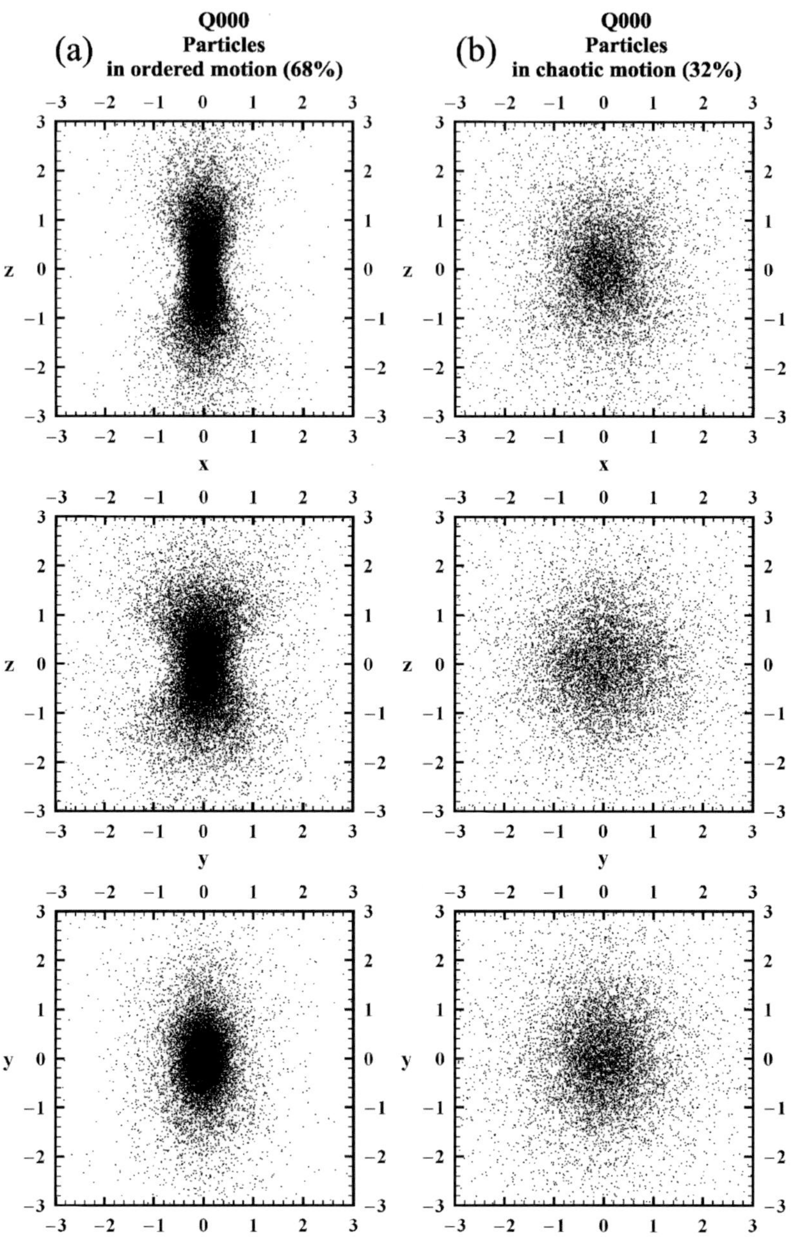

Figure 5.   **Column (a)**: Projections on the planes x-z, y-z, x-y of the mass in ordered motion in the Q000 model. This mass is $\simeq 68\%$ and forms a triaxial strong bar elongated along the z axis. **Column (b)**: Projections on the same planes of the mass in chaotic motion ($\simeq 32\%$) of this model. This mass has an almost spherical distribution.

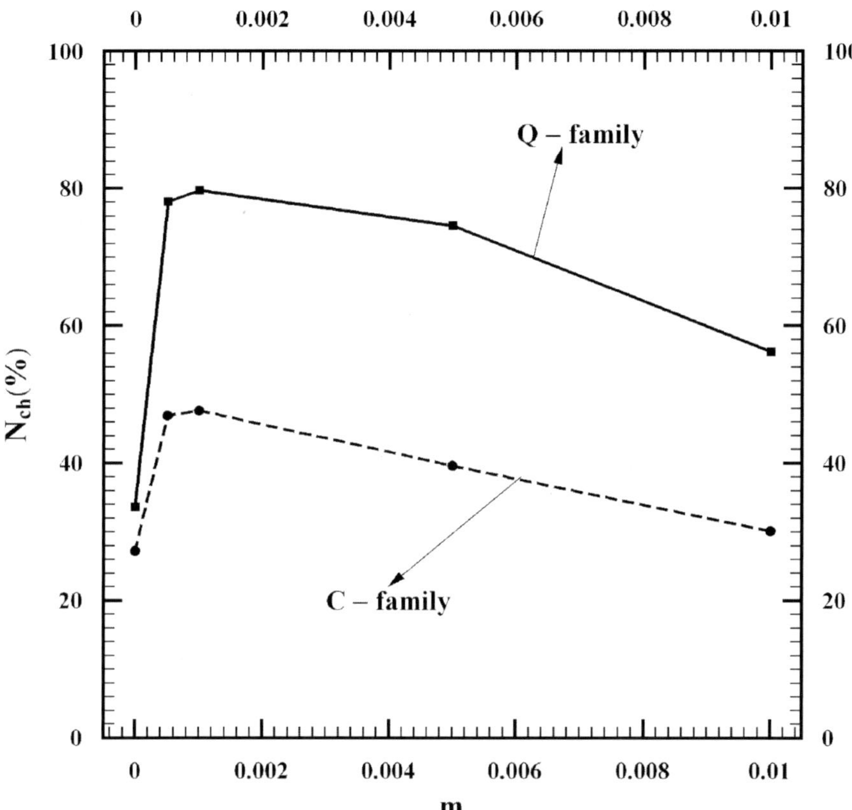

Figure 6. The fractions of mass in chaotic motion in all the models at their snapshots at $t = 150$ as a function of m. Solid line is for the Q family models and dashed line for the C family models. For $m > 0.0010$ the fraction of mass in chaotic motion decreases with m because these systems are self-organized at a rate depending on m.

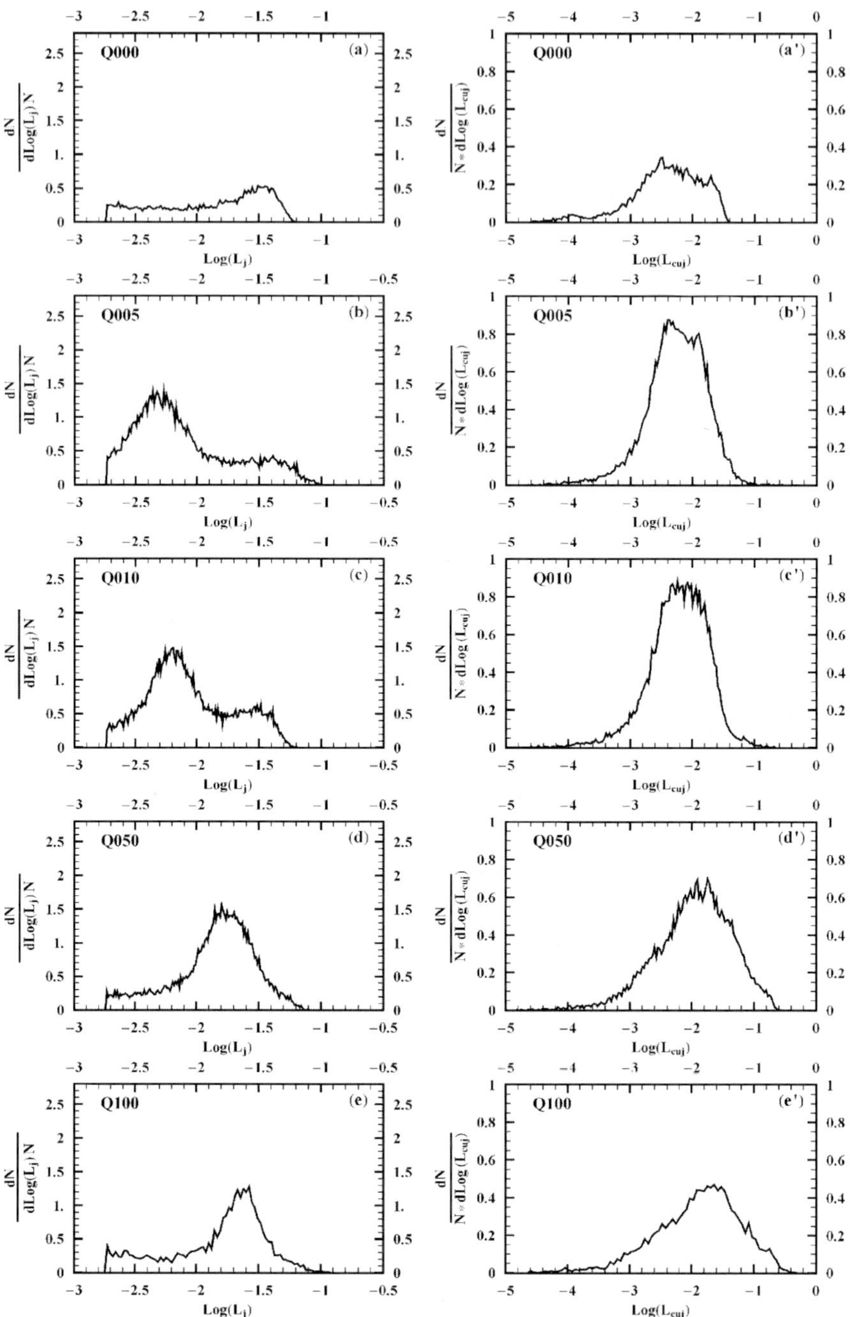

Figure 7. **Left column (a), (b), (c), (d) and (e)**: Distribution of the mass in chaotic motion along the $\log L_j$ axis for the models Q000, Q005, Q010, Q050 and Q100, respectively. In (b), (c), (d) and (e), i.e. for $m \neq 0$, a characteristic maximum appears around a value $\log L_M$ that does not appear in (a). This maximum corresponds to the Box or box-like orbits of the Q000 model that have become chaotic because of the central mass. The values of $\log L_M$ increase with m. **Right column (a'), (b'), (c'), (d') and (e')** : As in the left column, but along the $\log L_{cuj}$ axis. The distributions are now different but their maxima appear at a value $\log L_{cuM}$ which is about the same as $\log L_M$.

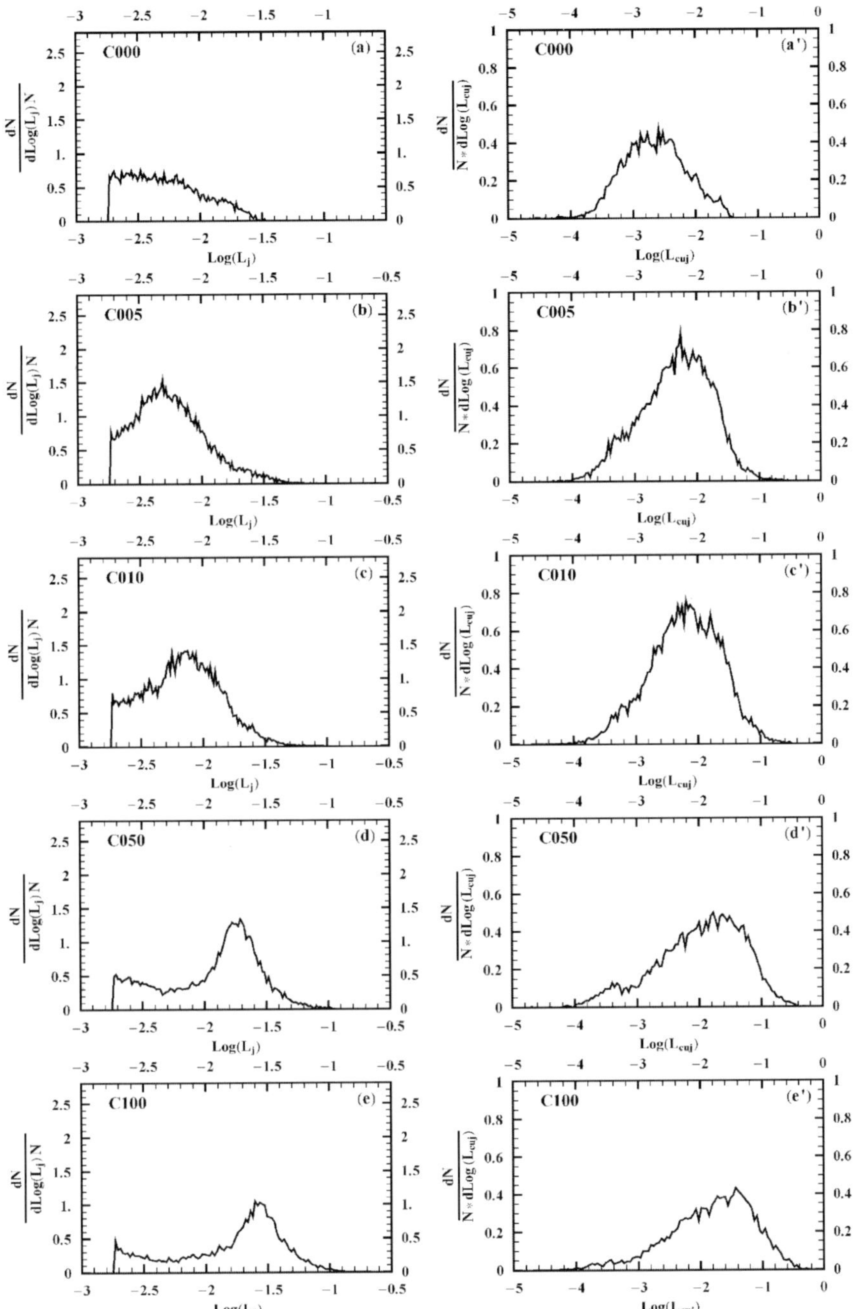

Figure 8.   As in Fig. 7 but for the C family models.

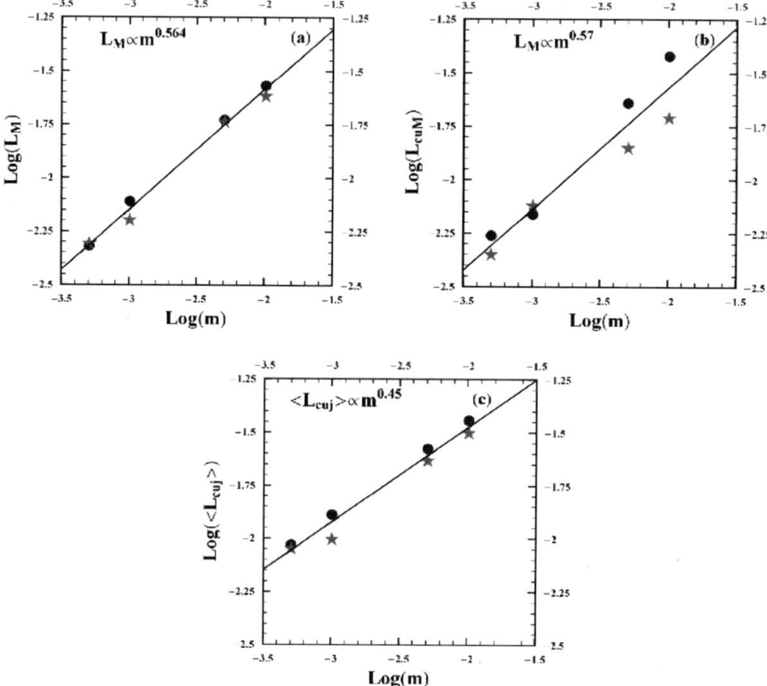

Figure 9. The mean level of the Lyapunov numbers measured either by the values of L_M in **(a)**, or L_{cuM} in **(b)**, or $< L_{cuj} >$ in **(c)** is closely correlated with m obeying approximately a power law with exponent not much different than 0.5. Stars correspond to the Q family models and dots to the C family models.

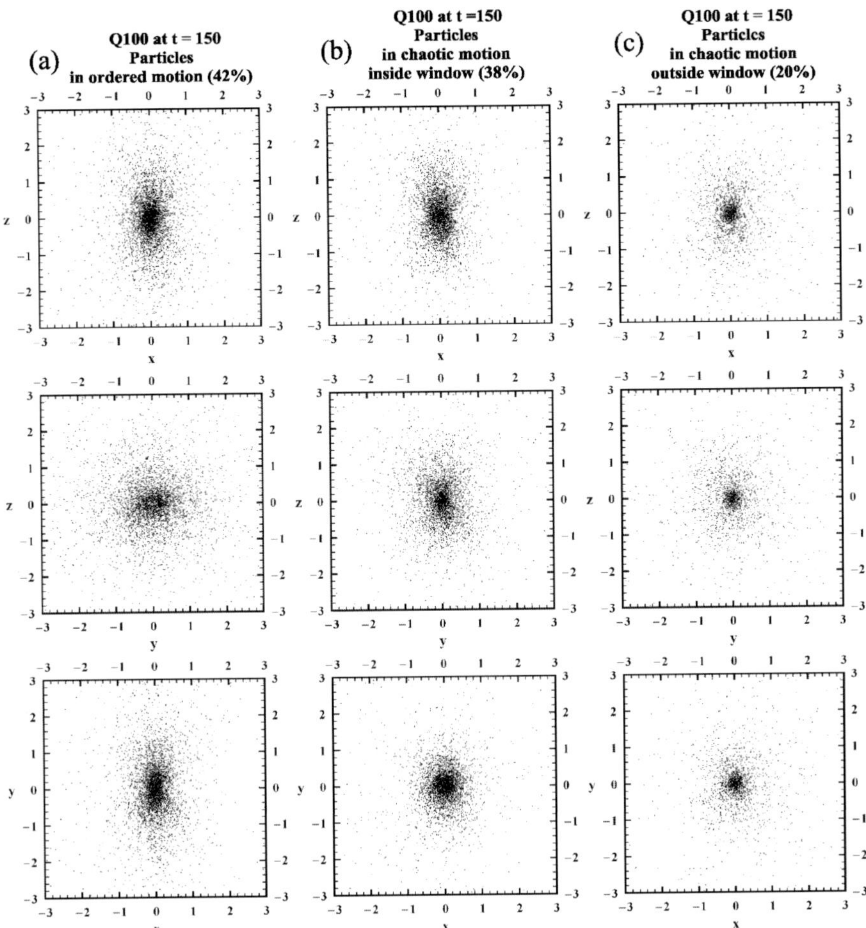

Figure 10. Projection on the planes x-z, y-z, x-y of three different groups of particles in the Q100 model at $t = 150$ as separated at this snapshot. Namely, **column (a)** shows the particles in ordered motion. The majority of their orbits is of SAT type giving an oblate almost spheroidal distribution. **Column (b)** shows the particles in chaotic motion with values of their $\log L_j$ inside a window of width ± 0.3 around the maximum of the distribution in Fig. 7e. These particles still form a bar along the z axis, in contrast to the rest of the mass in chaotic motion (with $\log L_j$ outside the window) shown in **column (c)**, that is almost spherically distributed. Secular evolution of the system is due to the chaotic diffusion of particles in (b).

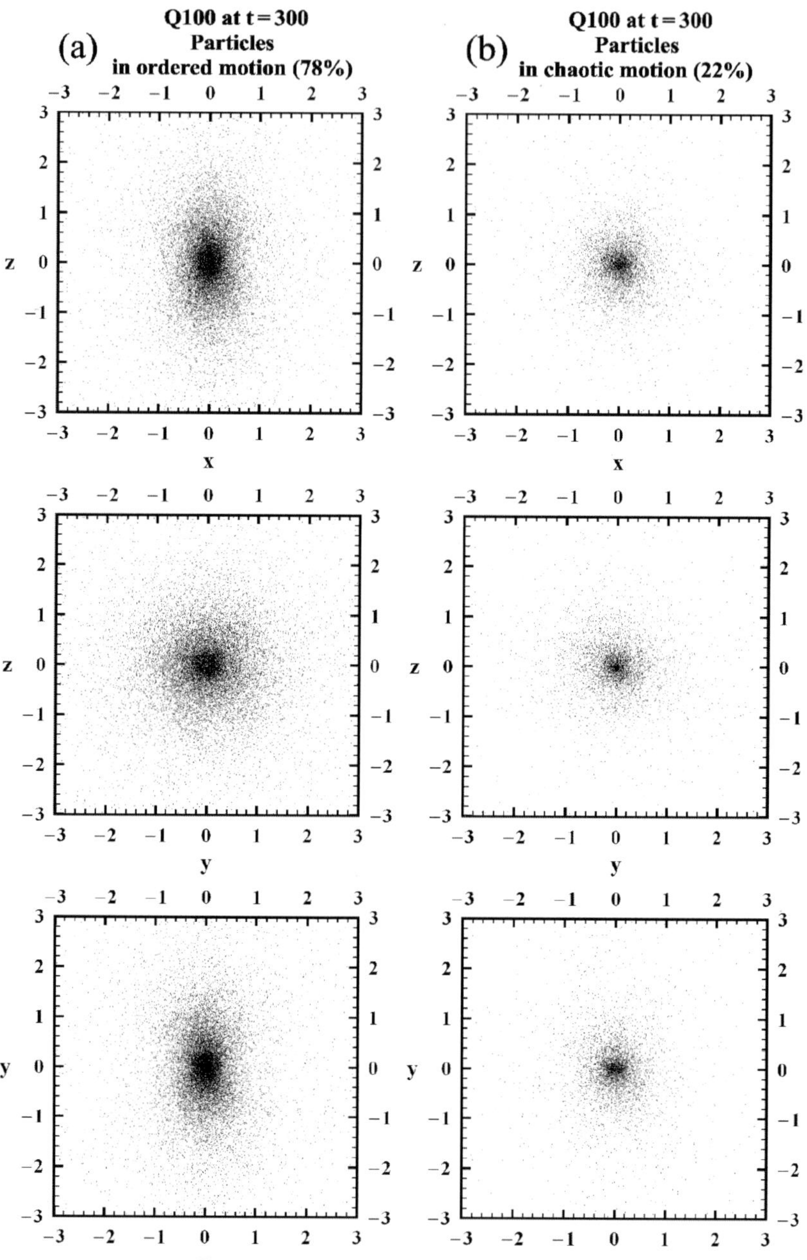

Figure 11. Projections on the planes x-z, y-z, x-y of two different groups of particles in the Q100 model at $t = 300$ as separated at this snapshot. **Column (a)** shows the particles in ordered motion forming an oblate spheroidal configuration formed mainly by SAT orbits, while **column (b)** shows the particles in chaotic motion forming an almost spherical distribution. During the secular evolution from $t = 150$ to $t = 300$, the majority of the chaotic orbits in the column (b) of Fig. 10 are converted to ordered orbits and they are included in column (a) here.

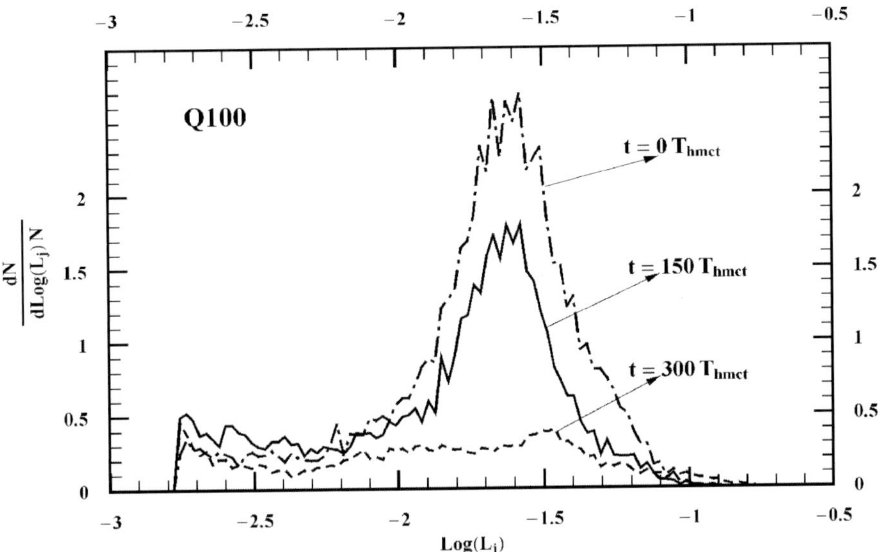

Figure 12. Distributions of the mass in chaotic motion along the $\log L_j$ axis at three different snapshots of the Q100 model, at $t = 0$ (dot-dashed line), at $t = 150$ (solid line) and at $t = 300$ (dashed line). The three distributions are normalized with respect to the mass in the second distribution. The gradual decrease of the maximum shows that the chaotic orbits that are converted to ordered orbits during the secular evolution of the system, belong mainly to this maximum.

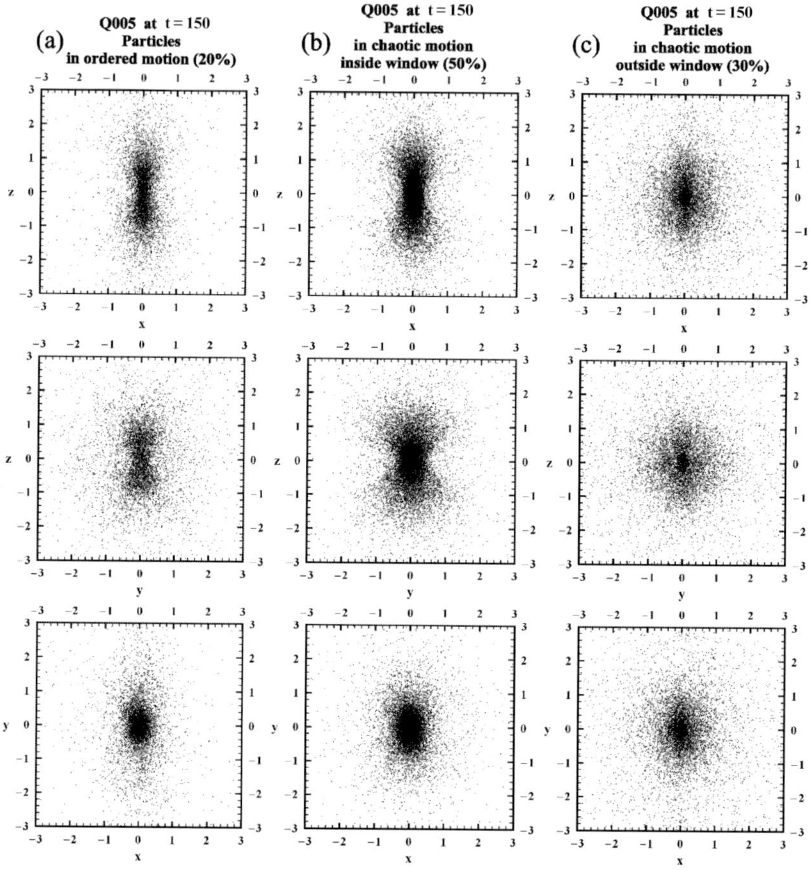

Figure 13. As in Fig. 10, but for the Q005 model at $t = 150$. **Column (a)** shows the particles in ordered motion. They move mainly in HORT orbits and they form a bar strongly elongated along the z axis. **Column (b)** shows the particles in chaotic motion with their $\log L_j$ inside the window ± 0.3 around the value $\log L_M = -2.3$. In principle, this mass can cause secular evolution in the system, but the rate of its chaotic diffusion is too slow for such evolution to be detectable in a Hubble time. The rest of the chaotic mass is shown in **column (c)**. Most of this mass is much more widely diffused than the previous parts.

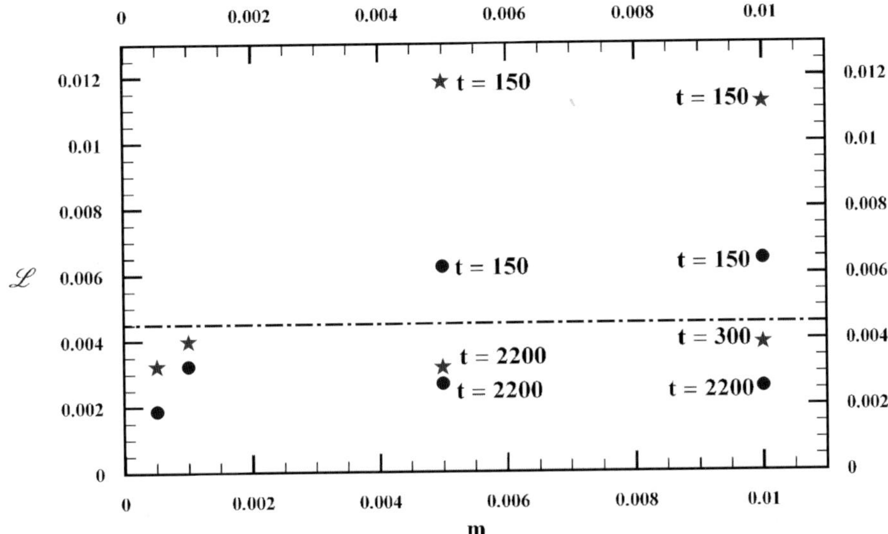

Figure 14. The values of the effective diffusion momentum $\mathscr{L}$ plotted vs m. Stars are for the Q family models and dots for the C family models. The models with secular evolution (Q050, Q100, C050, C100) are plotted twice, i.e. at their intermediate snapshot at $t = 150$ and at a snapshot of their equilibrium as indicated in the figure. Secular evolution is observable as far as $\mathscr{L} \gtrsim 0.0045$, otherwise it is negligible at least in a Hubble time.

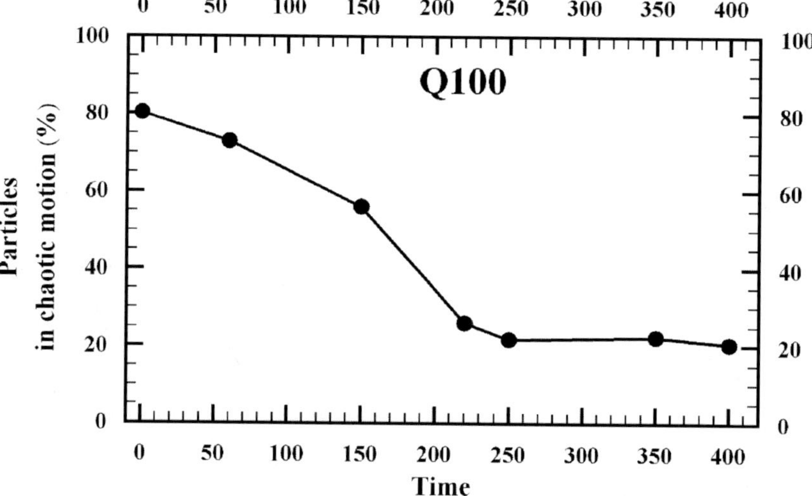

Figure 15. Time evolution of the fraction of mass in chaotic motion in the Q100 model during the phase of its secular evolution. This fraction decreases gradually to the level of about 20% in a period of $t \simeq 250$. Then, it remains at about this level permanently.

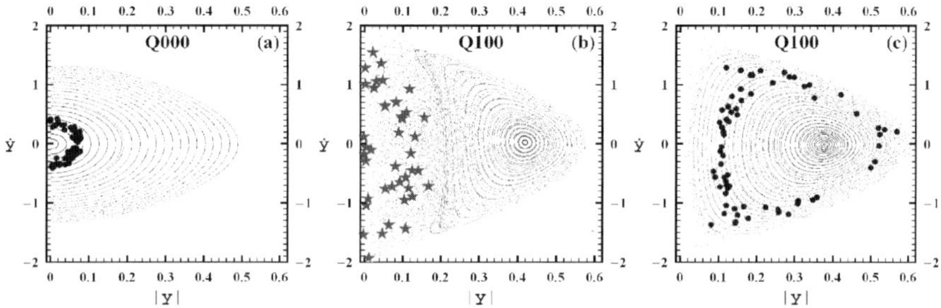

Figure 16. Projections on the plane $(|y|, \dot{y})$ of the 4-dimensional Poincaré section, $(x, y, \dot{x}, \dot{y})$ for $z = 0$ and energy $h = -74$. **(a):** The invariant tori shown are derived by running test particles in the potential of the Q000 and they are quite constant in time. These tori correspond to Box orbits only. The dots upon one of these invariant tori are the Poincaré consequents of the orbit of a real particle in this system of energy $h \simeq -74$, that are collected during a self consistent run of Q000. **(b):** As in (a), but for the potential of the Q100 model fixed at $t = 90$. Comparing (a) and (b) we see that most of the area of Box orbits in (a) is a chaotic area in (b) and that the ordered orbits that are allowed in (b) are of SAT type forming an island of ordered motion around a periodic orbit at ($\simeq 0.4$, 0.0). The Poincaré consequents of the orbit of the real particle in (a), collected during the self-consistent run of the Q100 model for the period from $t = 0$ to $t = 180$, are projected in the chaotic area of this portrait. **(c):** As in (b), but for the potential of the Q100 model fixed at $t = 300$. The island of SAT orbit covers almost the whole portrait except of a small chaotic area near $|y| = 0.0$. The Poincaré consequents of the orbit of the same real particle, collected during the self-consistent run of this model for the period from $t = 180$ to $t = 350$, are now projected inside the island forming approximately a torus.

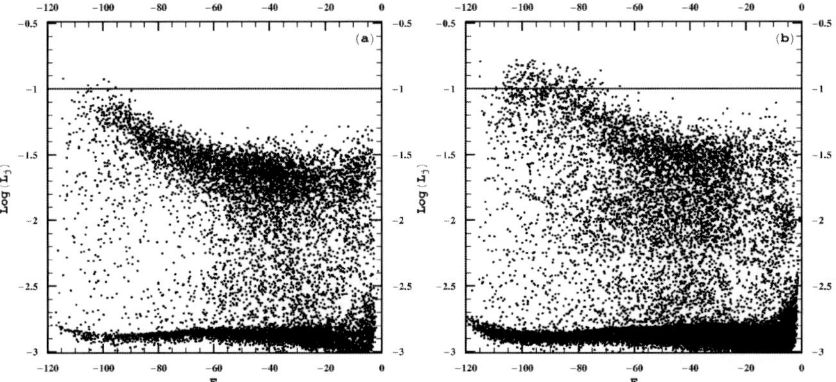

Figure 17.　　**(a)**: Distribution of all the particles of the Q100 model at $t = 150$ on the plane of their binding energies E and their $\log L_j$. **(b)**: As in (a), but for $t = 300$. The mass in chaotic motion in (b) is less than in (a) due to self-organization, but the Lyapunov Numbers in (b) spread to larger values as a compensation of the reduction of entropy.

II

ORAL CONTRIBUTIONS

GAMMA-RAY BURST INTERACTION
WITH DENSE INTERSTELLAR MEDIUM

Maxim V. Barkov,[1] and Gennady S. Bisnovatyi-Kogan[1]

[1]*Space Research Institute (IKI) Russian Academy of Science,*
Moscow, Russia

barmv@mx.iki.rssi.ru

gkogan@mx.iki.rssi.ru

Abstract   Interaction of a cosmological gamma ray burst radiation with a dense interstellar medium of a host galaxy is considered. Gas dynamical motion of the interstellar medium driven by a gamma ray burst is investigated in 2D approximation for different initial density distributions of the host galaxy matter and different total energy of the gamma ray burst. The maximum velocity of the interstellar medium motion is $1.8 \cdot 10^4$ km/s. Light curves of the gamma ray burst afterglow are calculated for a set of nonhomogeneous density distributions, different energy output of gamma ray bursts, and different viewing angles. Spectra of the gamma ray burst afterglows are modeled taking into account the conversion of hard photons (soft X-ray, hard UV) to soft UV and optics photons.

Keywords: gamma ray burst, optical afterglow

1.   Introduction

Although gamma-ray bursts (GRBs) had been discovered more than thirty years ago (Klebesadel, 1973), their origin is still unclear. The most extensive data on the detection of GRBs had been obtained by the Compton Gamma Ray Observatory BATSE experiment (Briggs, 1995; Fishman, 1995; Meegan, 1992). Analysis of the detected GRB sample had shown that their visible distribution on the sky was isotropic, but that there was a significant departure of their (logN – logS) curve from the $N \sim S^{-3/2}$ law corresponding to the spatially uniform source distribution (Briggs, 1995; Kouveliotou, 1994).

Observations of optical afterglows of some GRB, following the identification of these GRB with transient X-ray sources on the satellite Bcppo-SAX,

197

Alexei M. Fridman, et al. (eds.), Astrophysical Disks; Collective and Stochastic Phenomena 197–204
© 2006 *Springer. Printed in the Netherlands*

and a discovery of large (up to $z = 4.5$) redshifts in the spectra of optical transients had confirmed the cosmological origin of long GRB.

The cosmological model suggests that the GRB sources are located in galaxies at distances up to $\sim 10^3$ Mpc. In the framework of this model, the observed fluxes $\sim 10^{-4}$ erg cm^{-2} require a release of enormous amounts of energy ($\sim 10^{51} - 10^{53}$ erg) within a fairly short time interval of the order of several tens of seconds. Such a powerful energy release should have a strong effect on a large volume of surrounding matter in the parent galaxy, and should give rise to formation of the GRB counterpart at other wavelengths. Without specifying the mechanism for the GRB origin, we assume only the existence of a strong flux of the gamma ray radiation, and consider the interaction of this radiation with the interstellar medium on large spatial and temporal scales.

We investigate a response of an interstellar medium of a standard chemical composition to the passage of a short-term powerful pulse of gamma radiation, that is the dynamical behavior, and radiative cooling of matter heated by the gamma rays pulse. The spherical symmetric model was investigated by Bisnovatyi-Kogan and Timokhin (1997). 2D model which allows to study different matter distributions and reaction to the anisotropic GRB, had been studied by Barkov and Bisnovatyi-Kogan (2004a, 2004b), using numerical simulations by PPM method. In what follows we represent the results from Barkov and Bisnovatyi-Kogan (2004a, 2004b).

2.   The main equations

We solve the system of hydrodynamic equations which describe the motion of the matter, together with thermal processes, in the axially symmetric case

$$\frac{\partial \rho}{\partial t} + \nabla(\rho \vec{v}) = 0, \tag{1}$$

$$\frac{\partial(\rho v_r)}{\partial t} + \frac{\partial(\rho v_r^2 + P)}{\partial r} + \frac{1}{r}\frac{\partial(\rho v_r v_\theta)}{\partial \theta} + \frac{2\rho v_r^2 - \rho v_\theta^2 + \rho v_r v_\theta \, ctg \, \theta}{r} = \rho F_\gamma, \tag{2}$$

$$\frac{\partial(\rho v_\theta)}{\partial t} + \frac{\partial(\rho v_r v_\theta)}{\partial r} + \frac{1}{r}\frac{\partial(\rho v_\theta^2 + P)}{\partial \theta} + \frac{3\rho v_r v_\theta + \rho v_\theta^2 \, ctg \, \theta}{r} = 0, \tag{3}$$

$$\frac{\partial}{\partial t}\left(\frac{\rho v^2}{2} + \rho \varepsilon\right) + \nabla\left\{\rho \vec{v}\left(\frac{v^2}{2} + \varepsilon + \frac{P}{\rho}\right)\right\} = \rho H_\gamma - \rho C_\gamma. \tag{4}$$

Here $\rho, P, \varepsilon, v_r$, and v_θ are density, pressure, internal specific energy and two velocity components, respectively. The gamma ray pulse is considered as an instant one having the total energy Γ and the luminosity

$$L = \Gamma \delta \left(t - \frac{r}{c} \right).$$ (5)

We are interested in the behavior of the gas heated by the GRB, which is taken as a fully ionized one. Consider GRBs with a flat spectrum

$$\frac{dL}{dE} = \frac{L}{E_{max}} e^{-E/E_{max}}$$ (6)

The majority of GRB photons have energies larger than the ionization energies of most electrons, so we consider the energy exchange of the GRB with the gas due to the Compton and inverse Compton processes only. The function H_γ in (4) is written as

$$H_\gamma = \frac{L}{4\pi r^2} \frac{\mu_e \sigma_T}{m_u} \frac{E_{max} f_h(E_{max}) - 4kT f_c(E_{max})}{m_e c^2},$$ (7)

where

$$f_c(E_{max}) = \int_0^\infty W(E, E_{max}) q(E) dE,$$

$$f_h(E_{max}) = \frac{1}{E_{max}} \int_0^\infty W(E, E_{max}) s(E) E dE.$$ (8)

We have considered GRB spectra with $E_{max} = 0.6$ MeV and 2 MeV. The functions $s(E)$ and $q(E)$, which include deviations from the Thompson cross section σ_T due to Klein-Nishina corrections (σ_{KN}), are taken from Beloborodov and Illarionov (1995). The functions are normalized so that $s(E) = q(E) = 1$ at $E \ll m_e c^2$. In our cases $f_h = 0.19$; $f_c = 0.33$ for $E = 0.6$ MeV, and $f_h = 0.065$; $f_c = 0.16$ for $E = 2$ MeV.

The radiation force caused by the electron scattering reads

$$F_\gamma = \frac{1}{c} \frac{L}{4\pi r^2} \frac{\mu_e \sigma_T}{m_u} f_f(E_{max}),$$ (9)

where the function

$$f_f(E_{max}) = \frac{1}{\sigma_T} \int_0^\infty W(E, E_{max}) \sigma_{KN}(E) dE$$ (10)

takes into account KN corrections, $f_f = 0.5$ for $E_{max} = 0.6$ MeV and $f_f = 0.32$ for $E_{max} = 2$ MeV. Cooling of the gas is due to different radiative processes (ff, fb, bb).

We use optically thin plasma approximation for description of cooling. It is determined by the function C_γ, calculated by Kirienko (1993) and Raymond, Cox and Smith (1976).

$$C_\gamma = \frac{\Lambda(T) n^2}{\rho}$$ (11)

We approximate analytically the function $\Lambda(T)$ from Kirienko (1993) with a precision not worse than 5%

$$\Lambda(T) = \begin{cases} 0, & T < 10^4 K \\ 10^{-48.8}T^{6.4}, & 10^4 < T < 10^{4.25} \\ 10^{-16.5}T^{-1.2}, & 10^{4.25} < T < 10^{4.5} \\ 10^{-27.48}T^{1.24}, & 10^{4.5} < T < 10^5 \\ 10^{-21.03}T^{-0.05}, & 10^5 < T < 10^{5.4} \\ 10^{-13.6698}T^{-1.413}, & 10^{5.4} < T < 10^{5.86} \\ 10^{-22.8378}T^{0.1515}, & 10^{5.86} < T < 10^{6.19} \\ 10^{-13.1969}T^{-1.406}, & 10^{6.19} < T < 10^{6.83} \\ 10^{-22.2877}T^{-0.075}, & 10^{6.83} < T < 10^{7.5} \\ 10^{-26.6}T^{0.5}, & 10^{7.5} < T \end{cases} \tag{12}$$

It was shown by Barkov and Bisnovatyi-Kogan (2004a) that the heat conductivity may be neglected in this problem.

3.   Numerical results

More than 10 variants had been calculated in the paper of Barkov and Bisnovatyi-Kogan (2004b) for different density distributions and GRBs beaming. We represent here the main results of these calculations.

In the fig. 1 (Barkov and Bisnovatyi-Kogan, 2004b) the evolution of the temperature distribution in the cloud with time is represented, for the GRB exploding in the center of a spherically symmetric uniform cloud with a radius $R = 1.5$ pc, concentration $n_H = 10^5$ cm^{-1}, $\Gamma = 10^{52}$ erg, $E_{max} \geq 1/6$ MeV. It was shown by Barkov and Bisnovatyi-Kogan (2004b), that for $E_{max} \geq 1/6$ MeV, at a big distance from the GRB ($r \geq 0.05$ pc) heating depends only on the GRB energy Γ, and does not depend on E_{max}. The temperature inversion is developed at middle radia ($r = 0.15 \div 0.7$ pc), where the gas is heated up to $T \sim 10^6$ K, and cooling is the most effective. The cooling front is propagating outwards with a superluminal speed, and inward with a subluminal speed (phase velocities). The light curve for sum of a optical and ultraviolet luminosity, observed by the distant observer is represented in fig. 2 (Barkov and Bisnovatyi-Kogan, 2004b).

GRB heating of the cloud is a most intensive in the central parts and leads to a formation of the shock wave propagating outwards. The speed of the shock is about 2×10^8 cm/s for the uniform spherical cloud. It is much less then the speed of light. Therefore, the cloud is heated mainly by the light signal from GRB. Effects connected with the formation of a central shock do not influence the integral light curve, except the radiation in the hard X-ray band produced in the close vicinity of GRB explosion. In the case, when the GRB explosion takes place between two dense clouds, or in the cavity, produced by a strong anisotropic stellar wind, the hydrodynamic effects may be much stronger than

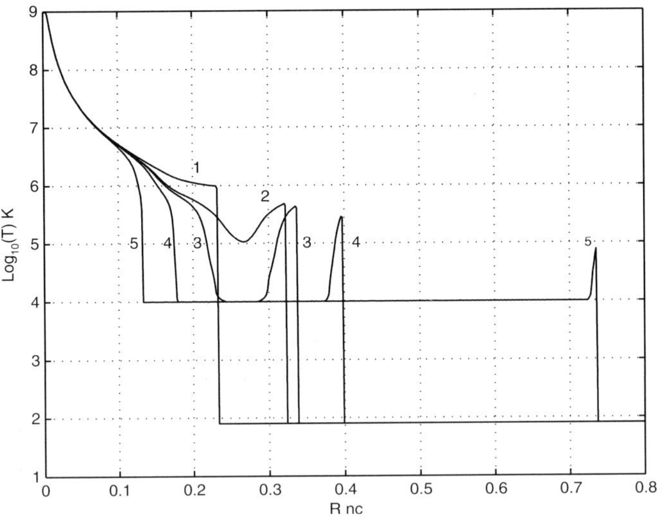

Figure 1.   The evolution of temperature distribution in the cloud with the time is represented for the GRB burst in the center of a spherically symmetric uniform cloud with a radius $R = 1.5$ pc, concentration $n_H = 10^5$ cm^{-1}, $\Gamma = 10^{52}$ erg, $E_{max} \geq 1/6$ MeV. The curves are marked by numbers corresponding to the following time moments after the GRB: 1) 0.76 year, 2) 1.054 year, 3) 1.103 year, 4) 1.30 year, 5) 2.40 year.

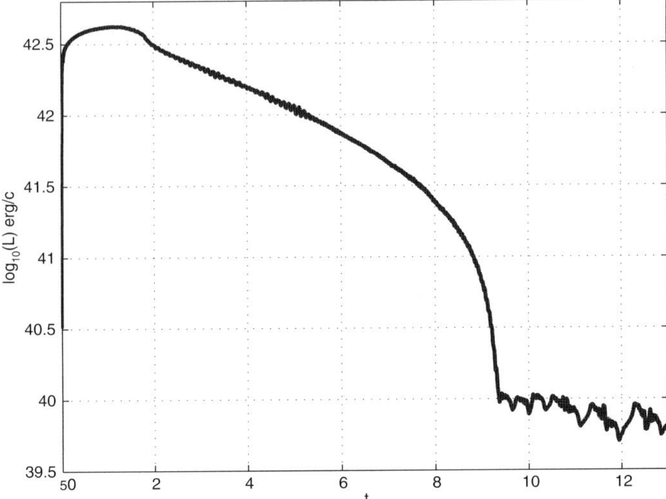

Figure 2.   The light curve for the sum of the optical and ultraviolet luminosity observed by the distant observer, for the same parameters as in fig. 1. Time is given in years.

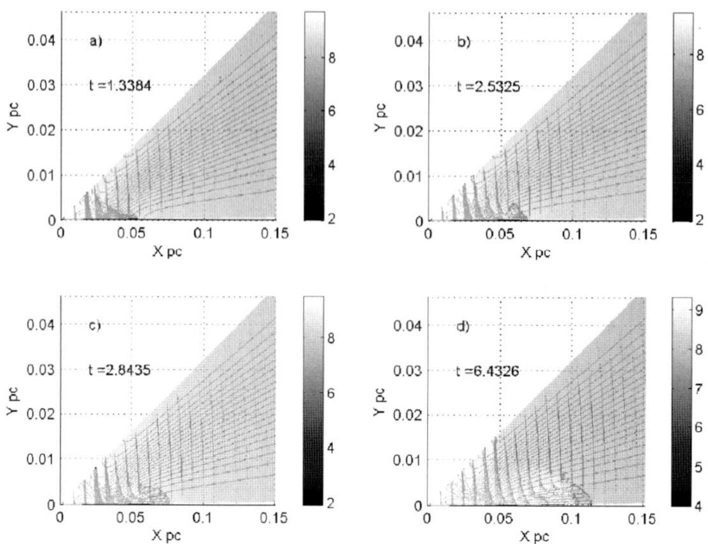

Figure 3.   The evolution of the velocity field is represented for the conical density distribution in the cloud.

in the case of the uniform cloud, and the speed of the shock increases up to $\sim 2 \times 10^9$ cm/s.

In the fig. 3 from Barkov and Bisnovatyi-Kogan (2004a) the evolution of the velocity field is represented for the case of the density distribution in the cloud as $n = 10^5 e^{-2-2\cos(10\theta)}$ cm^{-3}, the cloud radius 1.5 pc, for the energy of the isotropic GRB $\Gamma = 1.6 \times 10^{53}$ erg, and $E_{max} \geq 1/6$ MeV. The calculations had been performed in the region $0 \leq \theta \leq \pi/10$, with the condition $v_\theta = 0$ on the boundaries. The temperature of the heated gas depend only on the distance from the GRB, so the pressure gradient is developed inside the cavity inducing the motion of the matter to the axis of the cone. Collision of the flows at the cone axis produces a cumulative effect, and leads a matter acceleration along the axis up to velocity $\sim 2 \times 10^9$ cm/s (Barkov and Bisnovatyi-Kogan, 2004a). The accelerated matter has a form of the bullet in this case. In the case of the explosion in the space between two spherical clouds, the ejected matter should have a form of an expanding ring.

In the case of anisotropic GRB explosion in the nonuniform gas cloud the observed light curve is different for distant observers at different angular distances from the symmetry axes. Such light curves are represented in fig. 4. from Barkov and Bisnovatyi-Kogan (2004b). Here the anisotropic GRB is considered, with the angular dependence of luminosity inside the beam as $\Gamma(\theta) = 10^{52} e^{-(\vartheta/\vartheta_0)^2}$, $\theta = 0.1$ rad, the total energy of GRB $\Gamma_{tot} = 2.5 \times 10^{49}$ erg, the density distribution $n = 10^5 e^{-(r/r_0)^2}$ cm^{-3}, $r_0 = 0.2$ pc. The explosion

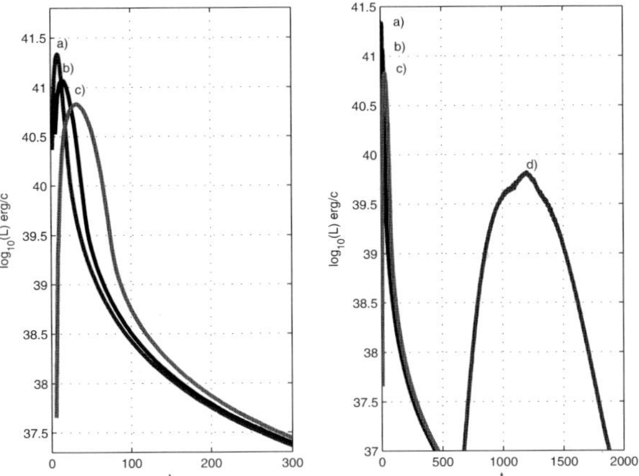

Figure 4. Light curves of the collimated GRB, situated on 1 pc from the center of the molecular cloud. Time is in years. The observer is situated on the line (GRB - MC, case a); on the deviation from this line: by the angle $\alpha = 0.1$ radian (case b), by the angle $\alpha = 0.2$ radian (case c), by the angle $\alpha = \pi/2$ radian (case d).

takes place at a distance 1 pc from the center of the cloud. The shortest optical burst of few days duration, with the largest luminosity $\sim 10^{41.5}$ erg/s is seen by the observer, situated at the symmetry axis on the continuation of the line cloud center – GRB. The observer on the line which is perpendicular to the symmetry axis is observing much longer optical afterglow (~ 1000 days), but with accordingly lower luminosity. All these differences are connected with the kinematic of the light propagation from the nonuniformly and nonsimultaneously heated gas cloud.

4. Discussion

The optical afterglow, connected with the reradiation of the GRB by the dense enough molecular cloud could be observed as an optical transient. Indications that GRB explosions take place in the regions of star formation filled with dense gas clouds Sokolov, 2001; Paczynski, 1999 make this possibility as very probable. Observation of plato in the optical afterglow of GRB 030329 during a month between 64 and 94 days after GRB detection (Ibrahimov et al., 2003) may be connected with such kind of reradiation. The dense molecular clouds have very low temperature, at which dust is formed. Estimations made by Barkov and Bisnovatyi-Kogan (2004b) show, that dust can be evaporated by the GRB pulse in the cloud along the GRB pulse direction, or in the whole cloud by isotropic GRB with $R \sim 1$ pc. In this case it does not influence the light curve of optical, UV and X-ray GRB afterglows. The parts of the cloud outside the GRB beam are not heated, and dust is not evaporated there. Therefore, the observer situated at large angle from the beam axis, would see only the orphan infrared transient source due to the dust reradiation. If the dust cloud is situated at larger distances where it cannot be evaporated by the GRB pulse, the main afterglow radiation is expected in the infrared region for all kind of observers.

References

Barkov M. V., Bisnovatyi-Kogan G. S. Interaction of Cosmological Gamma-ray Burst with dense molecular cloud and formation of directed outbursts Astronomy Reports 48(1):24-35, 2005.

Barkov M. V., Bisnovatyi-Kogan G. S. Afterglow of dence molecular claud after interaction with cosmological gamma ray burst Astronomy Reports in press.

Beloborodov A. M., Illarionov A. F. Compton Heating and Superthermal Electrons in Gamma-loud Active Galactic Nuclei Astrophysical Journal 450:64-69, 1995.

Bisnovatyi-Kogan G. S., Timokhin A. N. Optical counterparts of cosmological gamma-ray bursts: Interaction with matter surrounding the parent galaxy Astronomy Reports 41(4):423-435, 1997.

Briggs M. S. Four Years of BATSE Gamma-Ray Burst Observations Astrophysics and Space Science 231:3-10, 1995.

Fishman G. J., Meegan C. A. Gamma-Ray Bursts Annual Review of Astronomy and Astrophysics 33:415-458, 1995.

Ibrahimov M. A., Asfandiyarov I. M., Kahharov B. B., Pozanenko A. S. and et al., GRB 030329, BVRI photometry. GRB Coordinates Network 2191:1, 2003.

Kirienko A. B. Time-dependent radiative cooling of hot, optically thin interstellar gas Astrophysical Journal 19(1):11-13, 1993.

Klebesadel R. W., Strong I. B. and Olson R. A. Observations of Gamma-Ray Bursts of Cosmic Origin Astrophysical Jornal Letters 182:85, 1973.

Kouveliotou C. BATSE results on observational properties of gamma-ray bursts The Astrophysical Journal Supplement Series 92:637-642, 1994.

Meegan C. A., Fishman G. J., Wilson R. B. et al., Spatial distribution of gamma-ray bursts observed by BATSE Nature 355:143-145, 1992.

Paczynski B. Gamma-Ray Burst - Supernova Relation eprint arXiv:astro-ph 9909048, 1999.

Raymond J., Cox D., Smith B. Radiative cooling of a low-density plasma Astrophysical Journal 204(1):290-292, 1976.

Sokolov V. V., Fatkhullin T. A., Castro-Tirado A. J. and et al., Host galaxies of gamma-ray bursts: Spectral energy distributions and internal extinction Astronomy and Astrophysics 372:438-455, 2001.

MORPHOLOGY OF THE INTERACTION BETWEEN THE STREAM AND COOL ACCRETION DISC IN SEMIDETACHED BINARIES

P. V. Kaigorodov,[1] D. V. Bisikalo,[1] A. A. Boyarchuk,[1] and O. A. Kuznetsov[1,2]

[1] *Institute of Astronomy*
Russian Academy of Sciences
Moscow, Russia

pasha@inasan.rssi.ru

[2] *Keldysh Institute of Applied Mathematics*
Moscow, Russia

Abstract　　It was shown in (Bisikalo et al., 2003) that for realistic parameters of the accretion discs in close binaries ($\dot{\mathfrak{M}} \simeq 10^{-12} \div 10^{-7} M_\odot/\mathrm{yr}$ and $\alpha \simeq 10^{-1} \div 10^{-2}$), the gas temperature in the outer parts of the disc is $\sim 10^4$ K to $\sim 10^6$ K. Our previous gas-dynamical studies of mass transfer in close binaries indicate that, for hot discs (with temperatures for the outer parts of the disc of several hundred thousand K), the interaction between the stream from the inner Lagrange point and the disc is shockless. To study the morphology of the interaction between the stream and a cool accretion disc, we carried out three-dimensional modeling of the flow structure in a binary for the case when the gas temperature in the outer parts of the forming disc does not exceed 13 600 K. The flow pattern indicates that the interaction is again shockless. The computations provide evidence that, as is the case for hot discs, the zone of enhanced energy release (the "hot line") is located beyond the disc, and originates due to the interaction between the circum-disc halo and the stream. It was also shown that during the interaction between the disc and circum-disc halo a sizeable part of matter gets the acceleration in z-direction. This results in formation of a bulge in circum-disc halo with maximum on the phase 0.7.

Keywords:　　accretion: hydrodynamics, accretion disks

205

Alexei M. Fridman, et al. (eds.), Astrophysical Disks; Collective and Stochastic Phenomena 205–216
© *2006 Springer. Printed in the Netherlands*

1.   Introduction

In 1999 – 2003 we developed a 3D gasdynamical model and used it to study the flow patterns in binary systems (Bisikalo et al., 1997a; Bisikalo et al., 1997b; Bisikalo et al., 1998a; Bisikalo et al., 1998b; Bisikalo et al., 1999a; Bisikalo et al., 1999b; Bisikalo et al., 1999c; Bisikalo et al., 2000a; Bisikalo et al., 2000b; Bisikalo et al., 2001a; Bisikalo et al., 2001b; Kuznetsov et al., 2001; Molteni et al., 2001; Boyarchuk et al., 2002). These studies indicate that the flow structure is substantially affected by rarefied gas of the circumbinary envelope. In particular, a self-consistent solution does not include a shock interaction between the stream from the inner Lagrange point L_1 and the forming accretion disc (a "hot spot"). The region of enhanced energy release (the "hot line") is located beyond the disc and is due to the interaction between the envelope and the stream. However, these solutions were obtained for temperatures of the outer parts of the accretion disc of 200 000 – 500 000 K. To check if this behavior is universal, the morphology of the flow must be considered for various disc temperatures.

Analysis conducted in (Bisikalo et al., 2003) has shown that for realistic values of parameters ($\dot{\mathfrak{M}} \simeq 10^{-12} \div 10^{-7} M_{\odot}/\mathrm{yr}$ and $\alpha \simeq 10^{-1} \div 10^{-2}$) the gas temperature in the outer parts of the disc is between $\sim 10^4$K and $\sim 10^6$ K. This implies that cool accretion discs can form in some close binaries.

We will consider the morphology of the interaction between streams of matter and cool accretion discs in semi-detached binary systems (Sections 2 and 3). The basic problem here is whether the interaction between the stream and the disc remains shockless, as was shown for relatively hot discs (Bisikalo et al., 1997a; Bisikalo et al., 1998a; Bisikalo et al., 1998b; Bisikalo et al., 2000a; Boyarchuk et al., 2002). In Section 3 we also describe the formation of a bulge in circum-disc halo with maximum on the phase 0.7. It is shown that the formation of the bulge results from the interaction between the disc and the circum-disc halo, so a substantional part of gas flow accelerates in z-direction. Section 4 presents our main conclusions and a physical basis for the universal nature of the shockless interaction between the stream and disc.

2.   The Model

We described the flow structure in a binary system using a system of gravitational gas-dynamical equations taking into account radiative heating and cooling of the gas for the optically thin case:

$$\begin{cases} \dfrac{\partial \rho}{\partial t} + \mathrm{div}\,\rho\mathbf{v} = 0\,, \\[2em] \dfrac{\partial \rho\mathbf{v}}{\partial t} + \mathrm{div}(\rho\mathbf{v}\otimes\mathbf{v}) + \mathrm{grad}\,P = -\rho\,\mathrm{grad}\,\Phi\,, \\[2em] \dfrac{\partial\rho(\varepsilon + |\mathbf{v}|^2/2)}{\partial t} + \mathrm{div}\,\rho\mathbf{v}(\varepsilon + P/\rho + |\mathbf{v}|^2/2) = \\[1.5em] \qquad = -\rho\mathbf{v}\,\mathrm{grad}\,\Phi + \rho^2 m_p^{-2}\left(\Gamma(T, T_{wd}) - \Lambda(T)\right). \end{cases} \tag{1}$$

Here, as usual, ρ is the density, $\mathbf{v} = (u, v, w)$ the velocity vector, P the pressure, ε the internal energy, Φ the Roche gravitational potential, m_p the proton mass, T_{wd} the temperature of the central object[1] and $\Gamma(T, T_{wd})$ and $\Lambda(T)$ the radiative heating and cooling functions, respectively. The system of gas-dynamical equations was closed with the Clapeyron equation $P = (\gamma - 1)\rho\varepsilon$, where γ is the adiabatic index. We took the parameter γ to be 5/3.

Our main goal here is to study the morphology of the interaction between the stream and the cool accretion disc. According to (Bisikalo et al., 2003) the outer parts of the accretion disc can be cool for $\dot{\mathfrak{M}} \sim 10^{-10} M_\odot/\mathrm{yr}$ at $\alpha \sim 0.1$ and, in particular, in the case of an optically thin disc. The system of equations (1) enables us to carry out three-dimensional modeling of the flow structure in a binary within our formulation of the problem. In the model, the temperature of outer parts of the disc is 13 600 K.

We solved this system of equations using the Roe-Osher method (Boyarchuk et al., 2002; Roe, 1986; Chakravarthy & Osher, 1985), adapted for multi-processing computations via spatial decomposition the computation grid (i.e., partitioning into subregions, with synchronization of the boundary conditions) (Kaigorodov & Kuznetsov, 2002). We considered a semi-detached binary system containing a donor with mass M_2 filling Roche lobe and an accretor with mass M_1. The system parameters were specified to be those of the dwarf nova IP Peg: $M_1 = 1.02 M_\odot, M_2 = 0.5 M_\odot, A = 1.42 R_\odot$.

The modeling was carried out in a non-inertial reference frame rotating with the binary, in Cartesian coordinates in a rectangular three-dimensional grid. Since the problem is symmetrical about the equatorial plane, only half the space occupied by the disc was modeled. To join the solutions, we specified a corresponding boundary condition at the lower boundary of the computation domain. The accretor had the form of a sphere with radius $10^{-2}A$. All matter that ended up within any of the cells forming the accretor was taken to fall onto the star. A free boundary condition was specified at the outer boundaries

---

[1]$T_{wd} = 70\,000$ K for white dwarf

of the disc – the density was constant ($\rho_b = 10^{-8}\rho_{\mathrm{L}_1}$), where ρ_{L_1} is the density at the point L_1, the temperature was $13\,600$ K, and the velocity was equal to zero. The stream was specified in the form of a boundary condition: matter with temperature 5800 K, density $\rho_{\mathrm{L}_1} = 1.6 \times 10^{-8}$ g/cm^3 and velocity along the x axis $v_x = 6.3$ km/s was injected into a zone around L_1 with radius $0.014A$. For this rate of matter input into the system, the model accretion rate was $\sim 10^{-10}M_\odot/$yr.

The size of the computation domain, $1.12A \times 1.14A \times 0.17A$, was selected so that it entirely contains both the disc and stream, including the point L_1. The computation grid with $121 \times 121 \times 62$ cells was distributed between 81 processors, which constituted a two-dimensional 9×9 matrix.

To increase the accuracy of the solution, the grid was made denser in the zone of interaction between the stream and disc, making it possible to resolve well the formed shock wave. The grid was also denser towards the equatorial plane, so that the vertical structure was resolved, even for such a cool disc.

We used the solution obtained for a model without cooling as the initial conditions (Kuznetsov et al., 2001). The model with cooling was computed during approximately five revolutions of the system, until the solution became established. The total time of the computations was ≈ 1000 hours on the MVC1000A computer of the Joint Supercomputer Center (JSC).

3.   Computation results

Figures 1 to 3 present the morphology of gas flows in the binary. Figure 1 shows the density and velocity vector distributions in the equatorial plane of the system (the XY plane), while Fig. 2 present density contours in the frontal (XZ) plane and in the YZ plane containing the accretor and perpendicular to the line connecting the binary components. In spite of the small height of the forming accretion disc, use of the JSC parallel-processing computers made it possible to resolve its vertical structure (the outer parts of the disc were covered by 15 grid cells, and the inner parts by no fewer than 3 cells). The left panel of Fig. 3 gives an enlarged view of the density and velocity vector distributions in the zone of interaction between the stream and the outer edge of the disc (the area in the dashed rectangle in Fig. 1). The right panel of Fig. 3 presents the so-called texture a visualization of the velocity field in the zone of interaction between the stream and disc, constructed using the Line Integral Convolution procedure (LIC) (Cabral & Leedom, 1993).

According to our considerations in (Bisikalo et al., 2000a; Boyarchuk et al., 2002), the gasdynamical flow pattern in a semi-detached binary is formed by the stream from L_1, the disc, a circum-disc halo, and the circumbinary envelope. This subdivision is based on physical differences between these elements of the flow structure: (1) if the motion of the gas is not determined by the

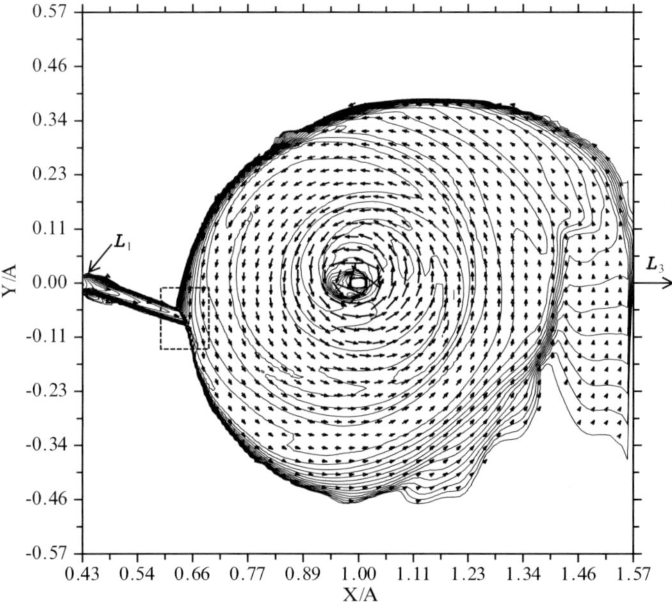

Figure 1.   Contours of constant density and velocity vectors in the equatorial plane XY of the system. The dashed rectangle indicates the zone of interaction between the stream and disc, shown in Figs. 3. The point L_1 and the direction towards L_3 are marked.

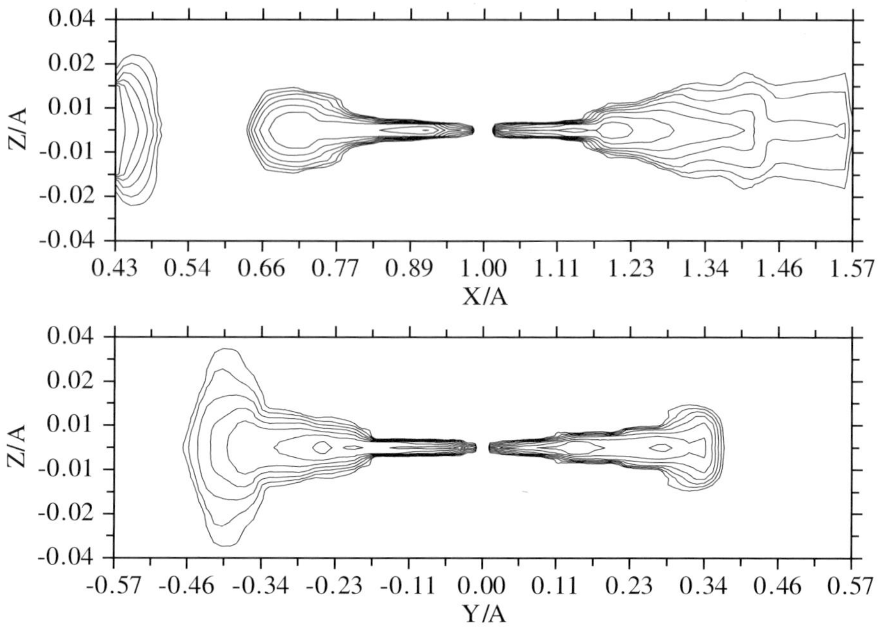

Figure 2.   Density contours in the frontal plane XZ and in the plane YZ of the system.

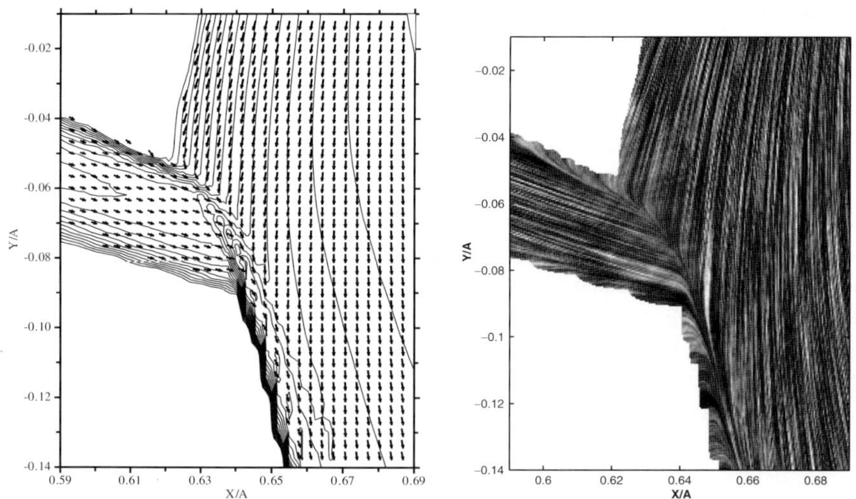

Figure 3.   Contours of constant density and velocity vectors (left panel) and visualization of the velocity field (right panel) in the zone of interaction between the stream and disc (the dashed rectangle in Fig. 1).

gravitational field of the accretor, it forms the circumbinary envelope; (2) if the gas makes one revolution around the accretor, but later mixes with the initial stream, this gas does not become part of the disc, instead forming the circum-disc halo; (3) the disc is formed by that part of the stream that loses its momentum and moves towards the center of gravity after entering the gravitation field of the accretor, rather than interacting with the stream.

In this framework, let us consider the morphology of the flow when the temperature decreases to $13\,600$ K over the entire computation domain due to cooling. Figure 1 indicates that, in this case, the circumbinary envelope is formed primarily in the vicinity of L_3, and does not affect the solution substantially. We can see from Figs. 1 and 2 that the circum-disc halo is compressed up to the disc, and its density increases sharply towards the disc edge.

Figure 3 show that, in the cool-disc case, the interaction between the circum-disc halo and the stream displays all features typical of an oblique collision of two streams. We can clearly see two shock waves and a tangential discontinuity between them. The gases forming the halo and stream pass through the shocks corresponding to their flows, mix, and move along the tangential discontinuity between the two shocks. Further, this material forms the disc itself, the halo, and the envelope.

The solution for the cool case displays the same qualitative characteristics as the solution for the case when the outer parts of the disc are hot: the interaction between the stream and disc is shockless, a region of enhanced energy release is formed due to the interaction between the circum-disc halo and the stream and is located beyond the disc, and the resulting shock is fairly extended, which is particularly important for explaining the observations. However, unlike the solution with a high temperature in the outer regions of the disc (Bisikalo et al., 1997a; Kuznetsov et al., 2001; Boyarchuk et al., 2002), in the cool case, the shape of the zone of shock interaction between the stream and halo is more complex than a simple "hot line". This is due to the sharp increase of the halo density as the disc is approached. Those parts of the halo that are far from the disc have low density, and the shock due to their interaction with the stream is situated along the edge of the stream. As the halo density increases, the shock bends, and eventually stretches along the edge of the disc.

Let us consider the behaviour of the stream after mixing with the matter of circum-disc halo. Figure 3 shows that the oblique collision of flows results in the formation of two shocks diverging at a sharp angle. When passing through the shock the gas pressure increases, so the matter of the gas flow (the mix of the stream and the halo matter) begins to move in the vertical direction, thus the z-component of velocity appears.

The non-zero z-velocity (positive in upper semi-space and negative in lower semi-space) results in the growth of the flow thickness (in z-direction). Rota-

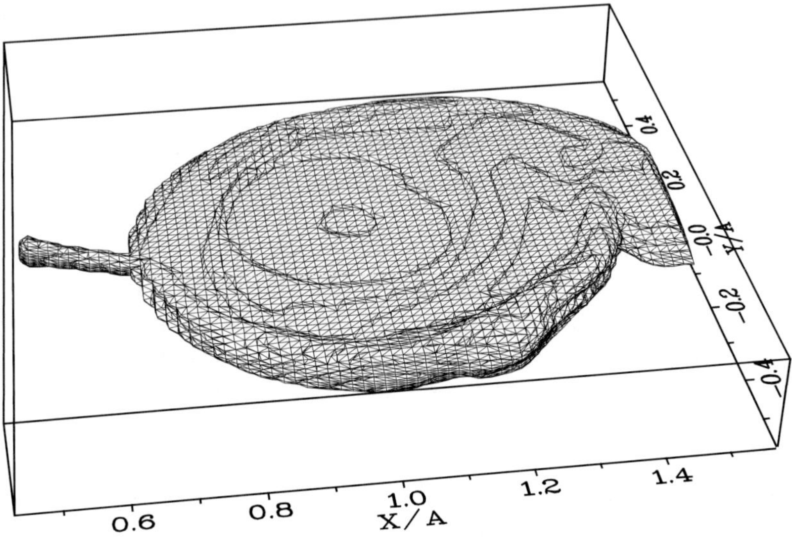

Figure 4.   Bird-eye view of density isosurface at $\rho = 5 \cdot 10^{-11}$ g/cm^3. The z-coordinate is ranged from $-0.17A$ to $0.17A$.

tional movement of matter around the accretor results in a shift the thickenning peak form the place of stream-halo interaction downstream.

When viscosity and other dissipative processes (in particular, radiative cooling) are absent the flow thickness would oscillate near the equilibrium value. In reality the energy gained in the shock dissipates during one revolution so only one maximum and one minimum of thickness are present. The thickening of the flow is seen clearly on the density isosurface (Fig. 4), as well as on Fig. 2 depicting the density distribution in the slice passing through accretor (the YZ plane).

Figure 5 depicts the fragments of flowlines ejected from the vicinity of L_1. It is seen that after leaving an extended mixing zone ("hot line") the flowlines diverge and after that smoothly converge to the equatorial plane.

To investigate the behaviour of flowlines let us introduce a cylindrical coordinate system centered at the accretor ($x = 1.0$, $y = 0.0$, $z = 0.0$) and with azimuthal angle reckoning from L_1 clock-wise (i.e. in the direction opposite to the rotation of the matter in the disc as well as the system rotation), so every point on the flowline is determined by coordinates $(r(t), \varphi(t), z(t))$. The dependence $z(\varphi)$ is drawn on Fig. 6, the flowlines being initiated in the vicinity of L_1. It is seen from this figure that the flowlines begin to rise after entering

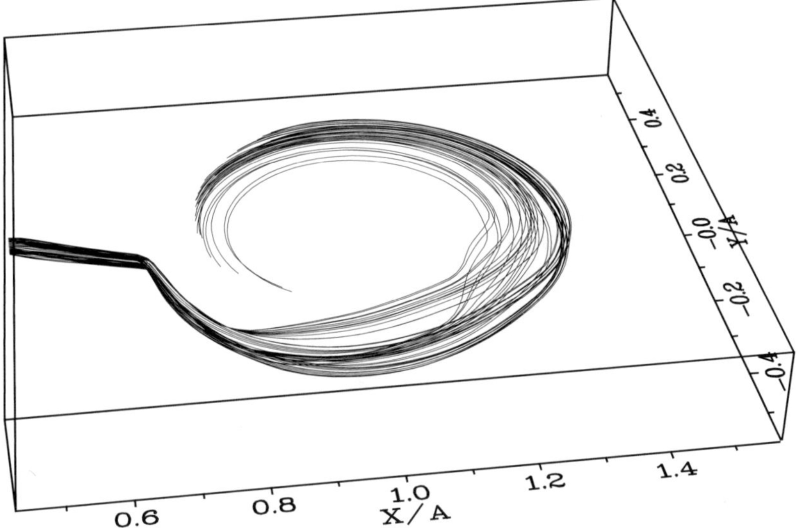

Figure 5. Fragments of flowlines ejected from the vicinity of L_1. The z-coordinate is ranged from $-0.17A$ to $0.17A$.

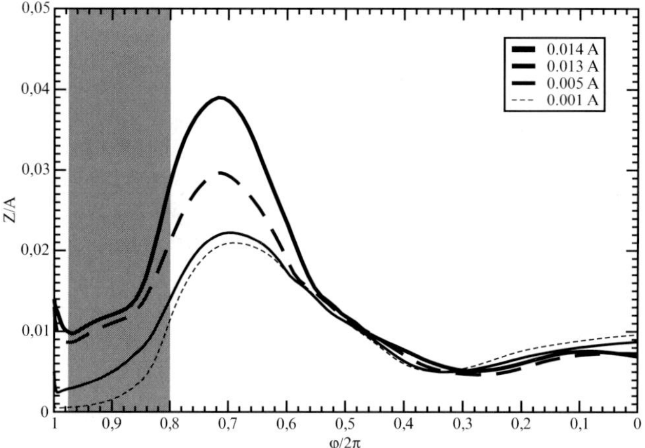

Figure 6. The dependence of $z(\varphi)$ for few flowlines. The initial positions of flowlines are in the vicinity of L_1 with coordinates $(x_{L_1}, 0, z_0)$ (values of z_0 are shown in the figure). The shaded region corresponds to the zone of increasing v_z.

the circum-disc halo due to increasing of the z-velocity (for the majority of flowlines except ones with high initial z-coordinates). The maximum of rise corresponds to the phase ≈ 0.7 and equals to $\approx 0.02 \div 0.04A$ in height. After that flowlines begin to converge to the equatorial plane and its height diminishing up to $\approx 0.005 \div 0.006A$ at phase ≈ 0.3, later the height of the stream reaches the equilibrium value of $\approx 0.007 \div 0.01A$.

Note, that the region of rise of z-velocity (the shaded zone on Fig. 6) coincides with the region where the matter of the stream and circum-disc halo are mixed ("hot line"). The size of this region is rather large and equals to $\sim 63°$ (from 0.8 to 0.975 in phase).

4. Conclusions

Previously, we carried out three-dimensional simulations of the flow structure in close binaries for the case when the temperature of the outer parts of the accretion disc was 200–500 thousand K. Those solutions showed that the interaction between the stream from the inner Lagrange point and the disc was shockless. To determine the generality of the solution, the morphology of the flow for different disc temperatures must be considered. We have presented here the results of simulations for the case when cooling decreases the temperature to 13 600 K over the entire computation domain.

Our analysis of the flow pattern for the cool outer parts of the disc confirms that the interaction between the stream and disc is again shockless. The computations indicate that the solution for the cool disc case displays the same qualitative features as in the case when the outer parts of the disc are hot: the interaction between the stream and disc is shockless, a region of enhanced energy release formed by the interaction between the circum-disc halo and the stream is located beyond the disc, and the shock wave that is formed is fairly extended, and can be considered a "hot line" . The cool solution demonstrates the universal character of our previous conclusions that the interaction between the stream and disc in semidetached binaries is shockless.

The structure of the circum-disc halo manifests a thickening with maximum on the phase ~ 0.7. The full width at half maximum of thickening covers the interval $0.58 \div 0.83$ in phase ($\sim 90°$). The height of thickening is 3–4 times larger than the mean disc height.

Acknowledgments

The work was partially supported by Russian Foundation for Basic Research (projects NN 05-02-16123, 05-02-17070, 05-02-17874, 06-02-16097), by Science Schools Support Program (project N 162.2003.2), by Federal Programme "Astronomy", by Presidium RAS Programs "Mathematical modelling and intellectual systems", "Nonstationary phenomena in astronomy", and by INTAS (grant N 00-491). O.A.K. thanks Russian Science Support Foundation for the financial support.

References

Bisikalo, D.V., Boyarchuk, A.A., Kaigorodov, P.V. and Kuznetsov, O.A.: 2003, *Astron. Zh.* **80**, 879 [2003, *Astron. Rep.* **47**, 809]

Bisikalo, D.V., Boyarchuk, A.A., Kuznetsov, O.A. and Chechetkin, V.M.: 1997, *Astron. Zh.* **74**, 880 [1997, *Astron. Rep.* **41**, 786]

Bisikalo, D.V., Boyarchuk, A.A., Kuznetsov, O.A. and Chechetkin, V.M.: 1997, *Astron. Zh.* **74**, 889 [1997, *Astron. Rep.* **41**, 794]

Bisikalo, D.V., Boyarchuk, A.A., Kuznetsov, O.A. and Chechetkin, V.M.: 1998, *Astron. Zh.* **75**, 706 [1998, *Astron. Rep.* **42**, 621]

Bisikalo, D.V., Boyarchuk, A.A., Chechetkin, V.M., et al.: 1998, *Mon. Not. R. Astron. Soc.* **300**, 39

Bisikalo, D.V., Boyarchuk, A.A., Kuznetsov, O.A. and Chechetkin, V.M.: 1999, *Astron. Zh.* **76**, 270 [1999, *Astron. Rep.* **43**, 229]

Bisikalo, D.V., Boyarchuk, A.A., Kuznetsov, O.A., and Chechetkin, V.M.: 1999, *Astron. Zh.* **76**, 672 [1999, *Astron. Rep.* **43**, 587]

Bisikalo, D.V., Boyarchuk, A.A., Kuznetsov, O.A., and Chechetkin, V.M.: 1999, *Astron. Zh.* **76**, 905 [1999, *Astron. Rep.* **43**, 797]

Bisikalo, D.V., Boyarchuk, A.A., Kuznetsov, O.A., and Chechetkin, V.M.: 2000 *Astron. Zh.* **77**, 31 [2000, *Astron. Rep.* **44**, 26]

Bisikalo, D.V., Harmanec, P., Boyarchuk, A.A., et al.: 2000, *Astron. Astrophys.* **353**, 1009

Bisikalo, D.V., Boyarchuk, A.A., Kilpio, A.A., et al.: 2001, *Astron. Zh.* **78**, 707 [2001, *Astron. Rep.* **45**, 611]

Bisikalo, D.V., Boyarchuk, A.A., Kilpio, A.A. and Kuznetsov, O.A.: 2001, *Astron. Zh.* **78**, 780 [2001, *Astron. Rep.* **45**, 647]

Kuznetsov, O.A., Bisikalo, D.V., Boyarchuk, A.A., et al.: 2001, *Astron. Zh.* **78**, 997 [2001, *Astron. Rep.* **45**, 872]

Molteni, D., Bisikalo, D.V., Kuznetsov, O.A., and Boyarchuk, A.A.: 2001, *Mon. Not. R. Astron. Soc.* **327**, 1103

Boyarchuk, A.A., Bisikalo, D.V., Kuznetsov, O.A., and Chechetkin, V.M.: 2002, *Mass Transfer in Close Binary Stars.* Taylor and Francis, London

Roe, P.L.: 1986, *Ann. Rev. Fluid Mech.* **8**, 337

Chakravarthy, S.R. and Osher, S.: 1985, *AIAA Pap.* No. 85–0363

Kaigorodov, P.V. and Kuznetsov, O.A.: 2002, *Preprint 59, Keldysh Inst. App. Math., Rus. Acad. Sci.*, Moscow

Cabral, B. and Leedom, C.: 1993, in *ACM SIGGRAPH, Computer Graphics Proceedings* **93**, 263

COMPUTER MODELING OF NON-STATIONARY GAS QUASI-KEPLERIAN DISK

Alexander V. Khoperskov, and Sergej S. Khrapov
Volgograd State University, Russia

xss@vlink.ru

khopersk@sai.msu.ru

Abstract   We have considered convective instability in a plane of a disk. In this case nonaxisymmetric perturbations are unstable, and the reason of a convection is connected with a radial non-homogeneity of thermodynamic parameters in quasi-Keplerian disk. On the basis of the linear analysis of stability in WKB-approximation the borders of stability are received. The hydrodynamical model of a non-stationary gas disk in the gravitational field of dot mass is constructed. We have studied nonlinear dynamics of convective unstable perturbations because of radial non-homogeneous entropy, neglecting effects of cooling and viscosity. The opportunity of formation of spiral-cellular structure of convective perturbations with shock waves is shown here.

Keywords: hydrodynamics, accretion disk, instability, convection

1.   Problem of a convection in accretion disks

Within the framework of accretion disk (AD) standard α-model (Shakura & Sunyaev) there is a number of theoretical problems and problems connected with the explanation of the observation data. The latter is concerned with the explanation of low luminosity of X-ray binary system and active galactic nucleus with black holes (see the review (Narayan 2002) and references there). One of the most investigated objects of such type is the source in the centre to Galaxy Sgr A^* (Melia & Falcke). The reduction of luminosity is provided in so-called ADAF-models (advection-dominated accretion flows) (Abramowicz et al. 1995). They are much hotter and Eddington luminosity is reached at smaller rate of accretion.

The beginning of research of ADAF-models was put by the work, (Narayan & Yi 1994), in which self-similar solving of stationary accretion are constructed. The radial advection stream acts as the basic mechanism of energy

217

Alexei M. Fridman, et al. (eds.), Astrophysical Disks; Collective and Stochastic Phenomena 217–230
© 2006 *Springer. Printed in the Netherlands*

transferring, and the transference of the angular moment is provided by α-viscosity. The structure of flow appears to be close to spherical and the rotation of gas essentially differs from Keplerian's law. At the presence of strong viscosity ($\alpha \gtrsim 0.3$) numerical modelling yields results which are similar to ADAF-models (Igumenshchev & Abramowicz 2000). However in calculations with small value of parameter α flows are formed with strong turbulence because of development of convective instability, and this instability essentially changes the spatial structure of current (Stone et al. 1999; Igumenshchev & Abramowicz 2000). Such models are called as "convection-dominanted accretion flows" (CDAF) (Balbus & Hawley 2002).

The major properties of the given models are the generation of convective turbulence, transference of the angular moment to the centre, which compensates a stream of the angular moment outside due to viscosity (or, for example, due to magnetic-rotation instability), together with a stream of thermal energy along radius. Numerical modelling shows the low rate of an accretion and a significant stream of energy outside due to a strong radial convection. The fundamental problem of models CDAF is the presence of a stream of the angular moment into center (Balbus & Hawley 2002). The basic results are obtained in frameworks of axisymmetric models (Igumenshchev & Abramowicz 1999; Stone et al. 1999; Balbus & Hawley 2002). The transition to three-dimensional ADAF-models, apparently, does not change decisions of axisymmetric calculations in qualitative sense (Igumenshchev et al. 2000). It is necessary to note, that the important result concerning with convective transference of the angular moment inside was revealed in the models of the rotating stars with a convective nucleus (Bisnovatyi-Kogan et al. 1979).

The presence of magnetic field (MHD CDAF) essentially can change the properties of flow, and, in particular, the convection can result in a stream of the angular moment as inside, and periphery along radial coordinate, and, apparently, it is connected with the influence of magnetic-rotation instability (Igumenshchev 2002; Balbus & Hawley 2002). The model of gas axisymmetric thick disk with the presence of magnetic field shows, that perturbations with the wave length exceeding a vertical scale of a disk, remain convective unstable (Narayan et al. 2002).

It should be noted that the internal radiative-dominant areas of AD can be unstable concerning to the vertical convection and the disk which is thin enough (Bisnovatyi-Kogan et al. 1979). The nonlinear stage of such convection in thin axisymmetric AD is investigated for the standard model radiative-dominant zone for $r - z$-perturbations in work (Agol et al. 2001).

Apparently, the fundamental problem of the AD is the question on nature of turbulent viscosity (Bisnovatyi-Kogan & Lovelace 2002) and, despite of the great progress, active investigation of turbulence in accreting systems is only at the beginning, and the basic results are still more ahead. There is a point of view, that turbulence in the AD is caused by the development of hydrodynamical instabilities at a nonlinear stage and the analysis of unstable flows is an important problem.

In this article we are going to discuss the consequences of development of convective instability in planes of a nonaxisymmetric disk. In this case the reason of nonaxisymmetric convection is connected with the radial non-homogeneity of thermodynamic parameters in quasi-Keplerian disk. We have analyzed the nonlinear dynamics of convective unstable perturbations because of the radial non-homogeneity of entropy. The opportunity of formation of the convective weak turbulence is shown. Convective intermixing in the plane of a disk can result, on the average, to radial accretion without taking into account the action of viscous forces.

2. Basic equations

The equations describing dynamics enough of a thin gas disk should be written down, in the following form:

$$\frac{\partial \sigma}{\partial t} + \frac{\partial (ru\sigma)}{r\,\partial r} + \frac{\partial (v\sigma)}{r\,\partial \varphi} = 0 \, , \tag{1}$$

$$\frac{\partial u}{\partial t} + u\frac{\partial u}{\partial r} + \frac{v}{r}\frac{\partial u}{\partial \varphi} - \frac{v^2}{r} = -\frac{\partial p}{\sigma\,\partial r} - \frac{\partial \Phi}{\partial r} - D\frac{p}{\sigma}\frac{d\ln\Omega_z}{dr} \, , \tag{2}$$

$$\frac{\partial v}{\partial t} + u\frac{\partial v}{\partial r} + \frac{v}{r}\frac{\partial v}{\partial \varphi} + \frac{uv}{r} = -\frac{\partial p}{\sigma\, r\,\partial \varphi} \, , \tag{3}$$

where σ is surface density, p — surface pressure, u and v are radial and azimuthal components of the velocity, respectively, Φ is potential, and the last component in (2) is connected with the averaging of the equations along vertical coordinate (Gorkavyj & Fridman 1994; Khoperskov & Khrapov 1999) and D depends on the features of vertical distribution of gas in a disk, $\Omega_z^2 = \dfrac{\partial^2 \Phi}{\partial z^2}\Big|_{z=0}$ and in case of Newton's potential for mass M_1 we have $\Omega_z = \Omega_K = \sqrt{GM_1/r^3}$. The equation on pressure should be added to the system of equations (1)–(3)

$$\frac{dp_g}{dt} + \left(1 + 2\frac{\gamma - 1}{\gamma + 1}\right)p_g\nabla\mathbf{v} + 7\frac{\gamma - 1}{\gamma + 1}\left\{\frac{dp_r}{dt} + \frac{9}{7}p_r\nabla\mathbf{v}\right\}$$
$$= 2\frac{\gamma - 1}{\gamma + 1}p\,(\mathbf{v}\nabla)\ln(c\,\Omega_z) \, , \tag{4}$$

where p_g and p_r are gas pressure and radiation pressure, γ is an adiabatic index, and the right part appears from the averaging on z-coordinate and the value of parameter c is expressed through D.

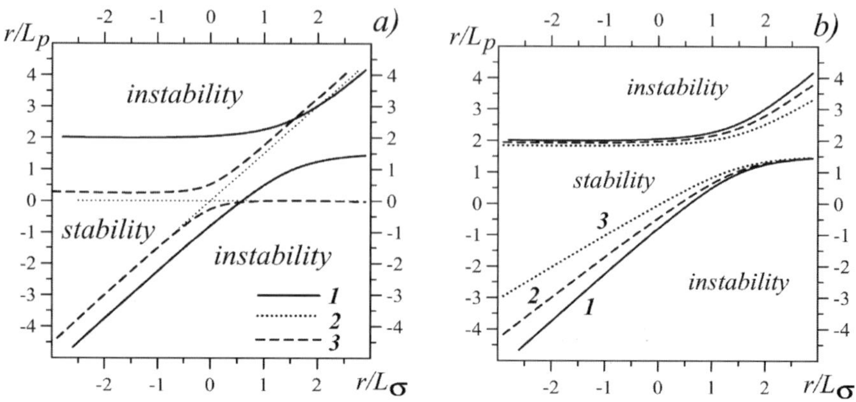

Figure 1. The borders of convective linear instability. *a)* The curve $1 — r/L_\Omega = r/L_z = -3/2$, $\gamma = 5/3$, $\Lambda = 10$; 2 — unrotative atmosphere (8); 3 — for case $r/L_z = 0$. *b)* $1 — \gamma = 5/3$, $2 — \gamma = 4/3$, $3 — \gamma = 1.01$ for $r/L_\Omega = r/L_z = -3/2$, $\Lambda = 10$.

3.   Linear stability analysis

Let's consider a stationary equilibrium non-homogeneous disk without radial motion. The equation (2) gives the balance of radial forces:

$$\frac{v_0^2}{r} - \frac{d\Phi}{dr} - \frac{dp_0}{\sigma_0\,dr} - D\frac{p_0}{\sigma_0}\frac{d\ln\Omega_z}{dr} = 0\,, \qquad (5)$$

where we shall mark equilibrium parameters by an index "0". The two last terms give the small contribution to balance, but the account of pressure gradient $p_0(r)$ is necessary for the development of convection in disk plane. It is convenient to enter specific equilibrium force, for which, taking into account the equation (5), we shall write down

$$g \equiv \frac{d\Phi}{dr} - \frac{v_0^2}{r} + D\frac{p_0}{\sigma_0}\frac{d\ln\Omega_z}{dr} = -\frac{dp_0}{\sigma_0\,dr}\,. \qquad (6)$$

For equilibrium parameters $f = \{\sigma_0, p_0, v_0, \ldots\}$ we shall use the scales of radial non-homogeneity $L_f = (d\ln(f)/dr)^{-1}$.

Let's consider the dynamics of nonaxisymmetric perturbations, taking into account the non-homogeneous distributions $\sigma_0(r)$, $p_0(r)$, $\Omega = v_0(r)/r = \Omega_z(r)$, and present all the functions as $f(r, \varphi, t) = f_0(r) + \tilde{f}(r, \varphi, t)$. We linearize the equations (1) – (4) concerning the perturbing functions $\tilde{f}$ with the account (6). Within the framework of WKB-approximation it is counted that $\tilde{f} = f_1 \cdot \exp\{-i\,\omega t + i\,kr + i\,m\varphi\}$. Thus, we have the system of four linear algebraic equations concerning amplitudes f_1. The condition of existence of untrivial decisions for this system results in the dispersion equation of 4-th degree concerning frequency ω. If we consider separately the radiative-dominant disk $p_r \gg p_g$ or the other case $p_r \ll p_g$, then the dispersion equation results in:

$$\hat{\omega}^4 - \hat{\omega}^2 \cdot \left[\text{æ}^2 + k^2 c_s^2 \cdot \left(1 - 2\,s^2 \frac{r\,d\Omega^2}{\text{æ}^2\,dr} \right) \right] - 2\,s\,\Omega\hat{\omega}kc_T^2\,\Gamma \times$$

$$\frac{d}{dr} \ln \left(\frac{2\Omega\,p_0^{2/\Gamma}}{\text{æ}^2\sigma_0}\,\Omega_z^{\frac{2(1-\Gamma)}{\Gamma}} \right) - \left(\frac{2s\,\Omega kc_T^2}{\text{æ}} \right)^2 \frac{d\,\ln(p_0\Omega_z)}{dr} \cdot \frac{d}{dr} \ln \frac{p_0}{\sigma_0^\Gamma \Omega_z^{\Gamma-1}} = 0,$$

$$(7)$$

where $\hat{\omega} = \omega - m\Omega$, $\Gamma = 1 + 2(\gamma - 1)/(\gamma + 1)$ plays a role of a flat parameter of an adiabatic curve, $s = k_\varphi/\sqrt{k^2 + k_\varphi^2}$ defines the degree of perturbations nonaxisymmetry, $k_\varphi = m/r$. In the case of $p_r \gg p_g$ it is necessary to consider $\gamma = 4/3$. In a limit of $\Omega_z = \text{const}$, the equation (7) was received in work (Morozov & Khoperskov 1990).

In a limit of adiabatic model $\frac{p_0}{\sigma_0^\Gamma \Omega_z^{\Gamma-1}} = \text{const}$, the order of the equation (7) reduced is lowered, as the entropy mode degenerates into $\hat{\omega} = 0$. The equation (7) describes two high-frequency acoustic modes for which it is possible approximately to write down $\hat{\omega}^2 \sim \text{æ}^2 + k^2 c_s^2$, and two low-frequency modes (entropy and vortical).

Formal transition in (7) to the unrotative medium $\Omega = 0$ and $\Omega_z = \text{const}$ gives $\omega^2 = 4\frac{s^2}{\Gamma^2}\frac{1}{L_p}\left(\frac{\Gamma}{L_\sigma} - \frac{1}{L_p} \right)$ for low-frequency waves and for stability it is necessary:

$$\frac{1}{L_p} \left(\frac{\Gamma}{L_\sigma} - \frac{1}{L_p} \right) > 0 . \qquad (8)$$

And it, in accuracy coincides with the condition of convective stability in the non-homogeneous unrotative atmosphere.

Thus, the equation (7) allows to define borders of convective instability with the account of differential rotation ($1/L_\Omega \neq 0$) and the finite thickness of disk in the main approximation ($1/L_z \equiv d\,\ln\Omega_z/dr \neq 0$). Let's write down (7) for low-frequency waves ($|\omega^2| \ll \text{æ}^2$):

$$\Lambda \cdot \nu^2 + \delta \cdot s \cdot \left[\frac{2r}{\Gamma L_p} - \frac{r}{L_\sigma} - 2\frac{\Gamma-1}{\Gamma}\frac{r}{L_z} + \frac{r}{L_\Omega} - 2\frac{r}{L_\text{æ}} \right] \cdot \nu$$

$$+ s^2 \frac{\delta^2}{\Gamma^2} \cdot \left[\frac{r}{L_p} - \Gamma\frac{r}{L_\sigma} - (\Gamma-1)\frac{r}{L_z} \right] \cdot \left[\frac{r}{L_p} + \frac{r}{L_z} \right] = 0, \qquad (9)$$

where $\nu = \hat{\omega}/\Omega$, $\Lambda = \frac{\text{æ}^2}{\Omega^2} + \frac{k^2 c_s^2}{\Omega^2} \cdot \left[1 + 6s^2 \right]$, $\delta = 2kc_s^2/\Omega^2\,r)$. Taking into account the estimation for half-thickness of disk $h \sim c_s/\Omega$, parameter Λ can

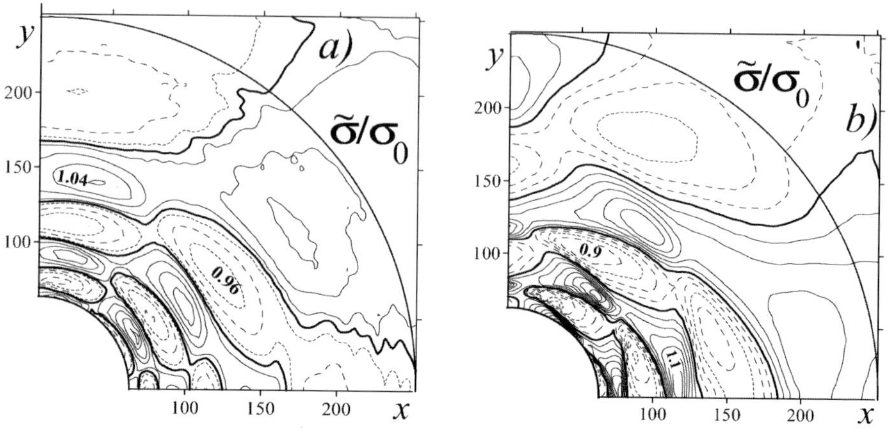

Figure 2.   Contours of surface density σ/σ_0 at two different moments of time $t_1 = 15000$ (a), $t_2 = 41000$ (b).

accept values from $\Lambda \simeq 1$ in a long-wave limit, up to $\Lambda \lesssim 100$ for perturbations whose wave length is comparable to thickness of the disk. For many stationary models of AD it is possible to accept the power characteristic of equilibrium parameters of a disk on radial coordinate (Shakura & Sunyaev): $r/L_f = \mathrm{const}$. The condition $\mathrm{Im}(\nu) > 0$ gives the unstable solutions, considered in work (Morozov & Khoperskov 1990) in a case $r/L_z = 0$.

In fig. 1 on a plane of parameters r/L_σ and r/L_p the borders of convective instability determined from a condition $\mathrm{Im}(\nu) = 0$, for base model $r/L_\Omega = r/L_z = -3/2$, $\gamma = 5/3$, $\Lambda = 10$ are represented. Here for comparison there are borders of stability for the unrotative medium for which, the balance is provided only by the external force and the pressure gradient, and for model $r/L_z = 0$. As we see, the rotation and the finite thickness of disk appreciablly change the conditions of convective instability. And, depending on values of L_σ and L_p, the zones of stability on planes $(r/L_\sigma, r/L_p)$, can both be increased and decreased. For equilibrium distributions with $r/L_\sigma < 0$ and $r/L_p < 0$ differential rotations and final thickness of disks are stabilizing factors.

Influence of parameter γ on the condition of convective instability is shown in fig. 1a. At any values of r/L_σ with diminution of γ critical value $|r/L_p|$ becomes less. This effect is corresponded to criterion (8). The parameter Λ characterizes a spatial structure of perturbations. Large values of the parameter $\Lambda \gg 1$ are reached for short-wave waves in radial direction $k \gtrsim \Omega/c_s$ and Λ is higher for perturbations with large azimuthal number m. It is necessary to emphasize, that from the point of view of the equations (9), (7) the most unstable waves are the extremely nonaxisymmetric perturbations $s = 1$ (as $\mathrm{Im}\,\omega \propto s$), for which the made approximations are broken certainly. The

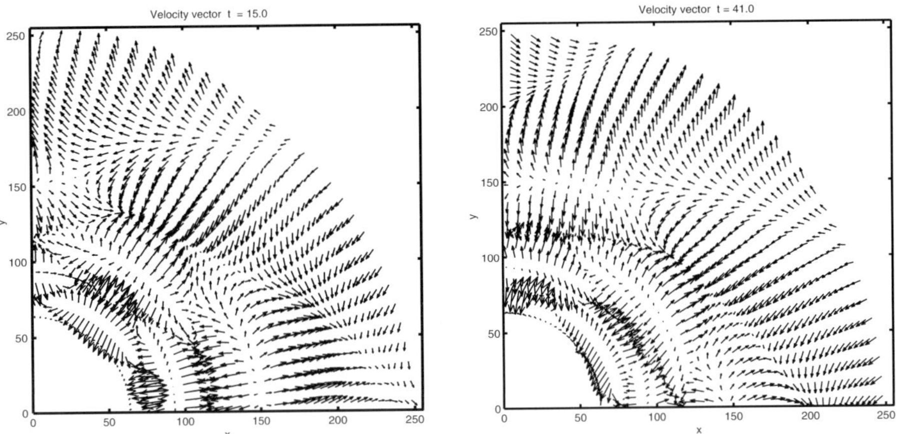

Figure 3. Fields of velocity in a disk plane for the model represented on fig. 2: (a) — $t_1 = 15000$, (b) — $t_2 = 41000$. The length of arrows is proportional to the logarithm of the velocity modulus.

borders of convective instability in disk plane depend on the non-axisymmetry degree of perturbations only through parameter $\Lambda(\Theta)$. The parameter Λ influences on model with $r/L_p > 0$ most of all. Thus models with $r/L_p > 0$ can be convective unstable only concerning small-scale disturbations.

The presence of extreme case (8) affirms the fact, that the physical mechanism causing the growth of perturbations with the time, is similar to classical convective instability at presence of a gradient entropy, which is co-directed to external force. The Archimed's force of buoyancy leads to the convective motion. However, in our case the effects of rotation play the important role. The basic question demanding deep analyze and studying is the influence of strong differential rotation on convective cells at a nonlinear stage.

4. Nonlinear stage of a radial convection

Let's consider the problem on influence of strong differential rotation on convective cells. For solving the equations of hydrodynamics we use the method TVD-E, having limited by $\Omega_z = \text{const}$.

Our model has free parameters: $\delta_p = r/L_p$, $\delta_\sigma = r/L_\sigma$, γ, $\mathcal{M} = r\Omega/c_s$. At the initial moment of time we set the power characteristics of density and pressure ($r/L_p = \text{const}, r/L_\sigma = \text{const}$). Dimensionless coordinates and time: $t = 1$ — a cycle time on radius $r = 1$ should be used. On the external border of settlement area r_{ex} conditions of free course of substance are used. On the internal border r_{in} the conditions of solid wall are used, considering, that the disk reaches the surface of accreting stars in case of a neutron star or

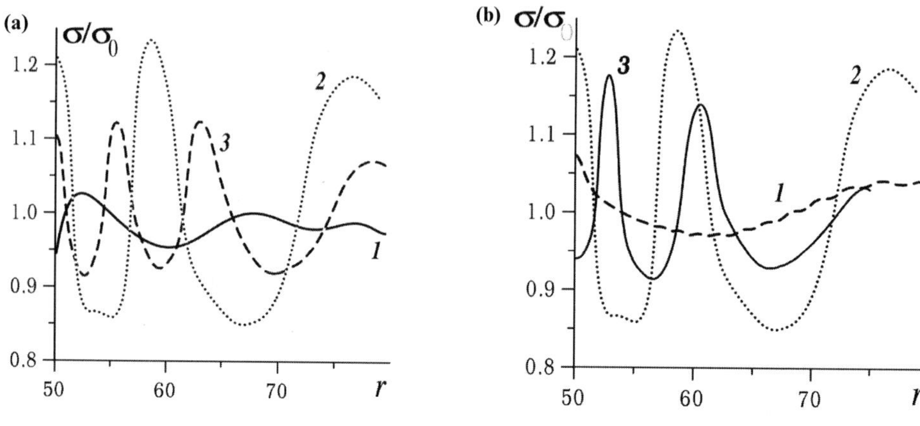

Figure 4. The radial distributions of relative density σ/σ_0. (a) $\mathcal{M} = 10$. The perturbations differ by azimuthal structure: $1 - m = 4$, $2 - m = 8$, $3 - m = 12$. (b) For fixed $m = 8$ models differ by the Mach number $\mathcal{M}$: $1 - \mathcal{M} = 5$, $2 - \mathcal{M} = 10$, $3 - \mathcal{M} = 15$.

the white dwarf. Let's consider only a limit $p_{gas} \gg p_{rad}$, that is carried out at $r_{in} \gg r_G$ ($r_G = 2\,GM/c^2$ — gravitational radius).

We are going to analyze the dynamics of the waves which are differentiated by azimuthal number m, depending on initial spatial structure of perturbations. For the formation of the certain harmonic m the sector of a disk on a corner φ is considered only. In this case along azimuthal coordinate periodic boundary conditions are used.

Structure of a convection at a nonlinear stage. If the equilibrium condition which is determined by functions $p_0(r)$ and $\sigma_0(r)$ provides stability of a disk according to (7) ($\mathrm{Im}(\omega) = 0$) the fact is that the increasing of perturbations in due course does not occur in numerical models. Control calculation in case of $r/L_p = -2$, $r/L_\sigma = -1$ shows, that on an extent $t \le 10^5$ at the presence of initial perturbations with initial amplitude $\lesssim 2\%$ their further increase does not occur.

Let's consider the model with $r/L_p = -3/2$ and $r/L_\sigma = -1/2$, $\gamma = 5/3$ which gets into the unstable area according to (9). Regardless of amplitude of initial perturbation, we obtain typical spiral-cellular wave structure.

In fig. 2 there are contours of ratio of density $\sigma(r, \varphi)$ to equilibrium value $\sigma_0(r)$ for two moments of time $t_1 = 15\,000$ and $t_2 = 41\,000$. At the initial stage typical convective cells (see fig. 2 a) with small relative amplitude of surface density $|\sigma - \sigma_0|/\sigma_0 \lesssim 0,05$ are formed. With time the increasing of amplitude and the complication of spatial structure because of the differential rotation of disk take place (see fig. 2 b).

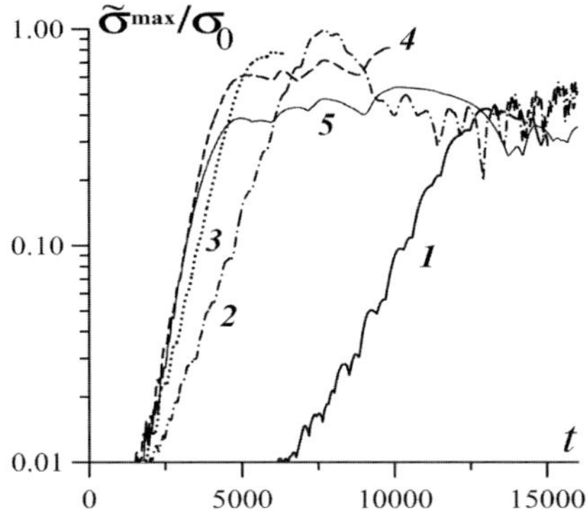

Figure 5. Dependences of the maximal value of the perturbed density $\tilde{\sigma}^{max}/\sigma_0$ for a convective cell from time for various models: (a) $\mathcal{M} = 10$, $1 - m = 8$, $2 - m = 12$, $3 - m = 20$. (b) For fixed m models differ by number of the Mach $\mathcal{M}$: $1 - \mathcal{M} = 10$, $2 - \mathcal{M} = 20$, $3 - \mathcal{M} = 40$.

The law of rotation differs from Keplerian Ω_K just a little. With the increasing of $\mathcal{M}$ these deviations are decreasing. Since the full radial component of velocity u and the perturbation of azimuthal velocity $\tilde{v} = v - v_0$ are rather small in comparison with the equilibrium velocity of rotation $v_0 \simeq r\Omega_K$ it is more convenient to consider only the perturb components but not a full field of velocities. In fig. 3 the vector field of velocity perturbations is represented in a plane of a disk. The field of velocities demonstrates vortical character of flow.

The spatial structure of perturbation can be characterized by radial wave number k and by azimuthal number m, which are independent within the framework of the linear analysis. The radial structure of unstable perturbations at the nonlinear stage is defined by parameters of model ($\mathcal{M}$, m, δ_p, δ_σ, γ), and we can change azimuthal number m, varying the initial perturbations along angle φ. In fig. 4 it is visible, that with the increasing of azimuthal number the perturbations become more small-scale in the radial direction as well. There are similar effect in the case of increasing number of the Mach ($\mathcal{M}$).

It follows from the linear analysis, that the increment of instability is proportional $\text{Im}\,\omega \propto m$, and as a whole this fact is affirmed at an initial stage of evolution of perturbations. In fig. 5 time dependences of amplitude of relative density for the chosen convective cell for various m and $\mathcal{M}$ are shown. During the typical times $t^{(sat)} \simeq 350\mathcal{M}$ the increasing of perturbations up to much nonlinear stage, close to saturation, takes place. The disturbance amplitude with small azimuthal number arises slowly.

At the non-linear stage of instability the spiral shock waves (SW) are formed in a disk, and it is caused by supersonic flowing of gas onto the convective cells. In fig. 6 the structure of shock waves is shown. The distributions of $(\text{div}\,\vec{v})^2$

most evidently demonstrates positions of fronts of shock waves in fig. 6. Radial structure of parameters of gas at the fixed values of an azimuthal angle φ are represented in fig. 7. The shock wave is formed on the back edge of the spiral density perturbation, and the wave of underpressure is observed on the front edge.

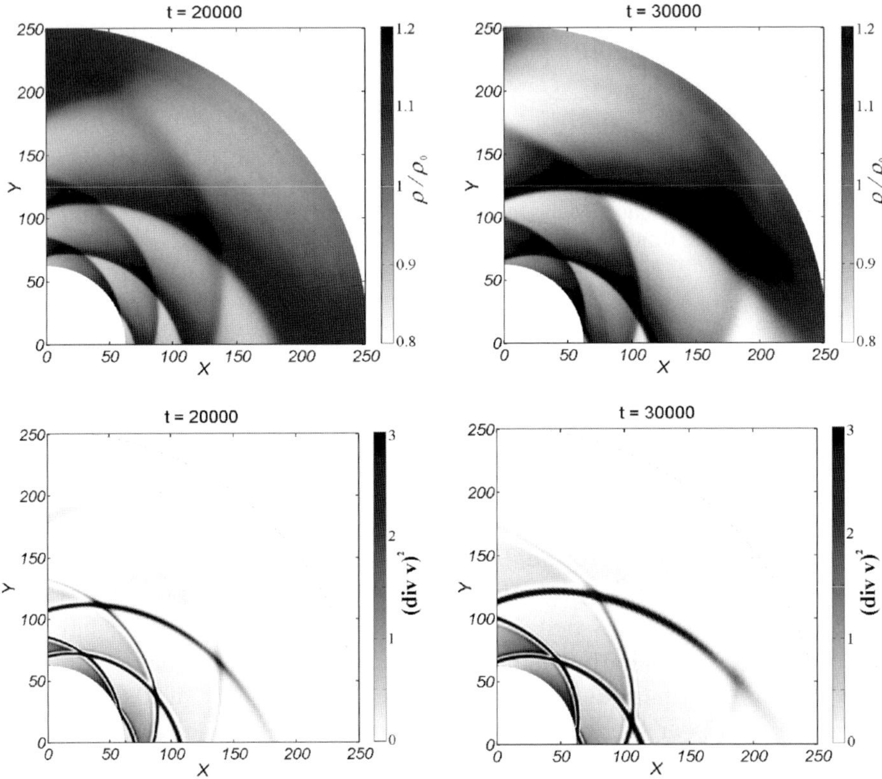

Figure 6. Contours of the relative density and $(\mathrm{div}\,\vec{v})^2$ for the two different moments of the time.

Is viscosity necessary for an accretion? The result of convection in disk plane is gas falling onto the gravitational center, on average.

In fig. 8 time dependences of the current of mass $\dot{M} = r \int_0^{2\pi} \sigma u\, d\varphi$ are represented on three various radiuses. Non-stationary character of accretion is connected with the absence of the stationary solution because of used boundary conditions in radius r_{in}. The condition of solid wall on inner boundary results in the accumulation of mass in a disk that has an effect for other parameters as well (fig. 9). The current of the angular moment is directed outside, that

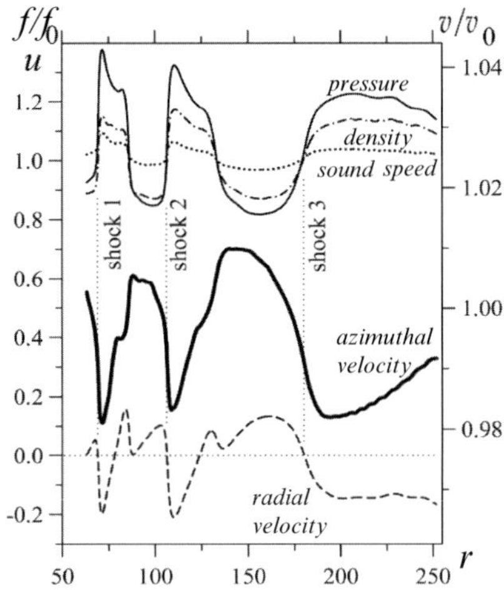

Figure 7. The radial dependences σ/σ_0, p/p_0, c_s/c_{s0}, u (left axis), v/v_0 (right axis) for the azimuthal angle $\varphi = 0$ at time $t = 2 \cdot 10^4$. The positions of shock waves are indicated.

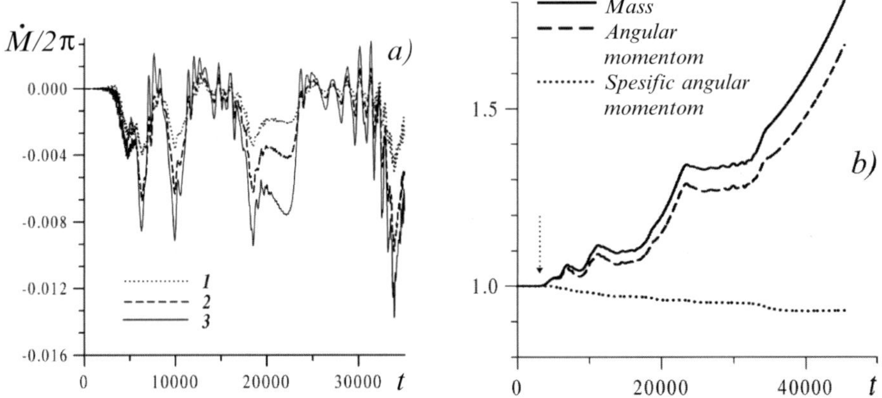

Figure 8. *a*) The dependences $\dot{M}$ from time through three various circles of the fixed radius. *b*) The time dependences of integral values (mass M, angular moment $\mathcal{L}$ and specific angular moment $l = \mathcal{L}/M$) in calculated region of numerical model.

appears to be the important distinctive feature from models CDAF (Balbus & Hawley 2002). Such received direction of the current affirms, that rotation is a source of free energy (8).

In fig. 9 the radial distributions of relative sound speed in a disk at the different time moments are shown. Monotonous increase of c_s in due course points out to the heating of a disk as a result of gravitational energy release.

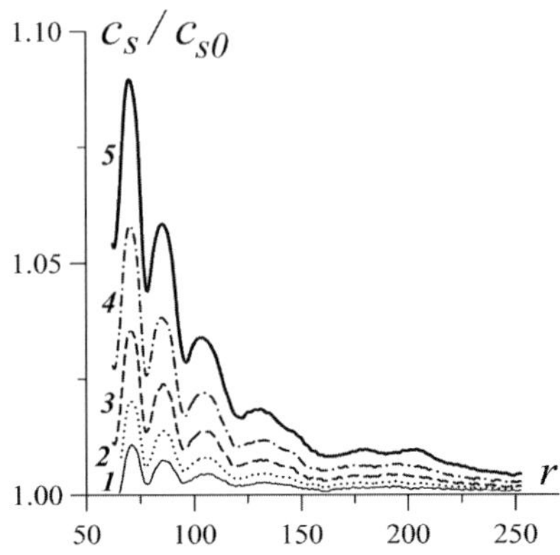

Figure 9.   The dependence $c_s(r)/c_{s0}(r)$ at the different time moments (the value c_s is received as the result of averaging on a angle φ, and c_{s0} is initial equilibrium value): 1 — $t_1 = 30\,000$, 2 — $t_2 = 35\,000$, 3 — $t_3 = 40\,000$, 4 — $t_4 = 45\,000$, 5 — $t_5 = 50\,000$.

The increasing of temperature in the internal part of a disk is higher, than in periphery.

In summary, it should be noted, that the received results, made in frameworks of quasi-Keplerian disk, are kept, on the whole, in case of rotation laws with $r/L_\Omega > -3/2$. In the appendix to the gas subsystems of disk galaxies, the values of parameter r/L_Ω lay from 0 (close to solid-body rotation in the central zone) up to -1, and the curve of plateau-type is characteristic for the majority of galaxies in the most part of a disk.

5.   Discussion of results

We have studied nonlinear dynamics of convective unstable perturbations because of radial non-homogeneity of entropy, neglecting effects of cooling and viscosity. Unlike numerous works on studying of convection in a plane $r - z$, we have shown principled opportunity of development of convective instability in thin quasi-Keplerian disk, where vertical motion (if they are present) are not a reason of the convection.

The opportunity of formation of the weak convective turbulence in characteristic times $\sim(10^3 \div 10^4)\,\tau$ in the central zone of disk (τ — Keplerian period on radius $3r_g = 6\,GM/c^2$) is shown. The rate of convection generation is increased for the peripheral region of disk. The convective intermixing in disk plane can result in average radial current of mass without taking into account action of viscous forces.

The convective instability considered here in a plane of a disk is not connected with the fact of gas rotation, as in case of fluid motion between two rotating cylinders (Taylor flow), that results in the formation of Taylor vortex. The opportunity of convection development in the disk plane is kept and in the case of solid-body rotation $L_\Omega^{-1} = \frac{d\Omega}{\Omega\,dr} = 0$, and it can have interest for models of gas disks of galaxies. The convection conditions can be fulfilled for the curve of rotation of galactic gas disks such as a plateau $V = r\Omega = \text{const}$. In contrast to the convection in the appendix to stars (Brun & Toomre 2002), the degree of differential rotation in the AD are much higher. The physical reason of such instability is the decrease of specific entropy with radius. The role of external force which is required for convection development is played by the value $g \simeq \frac{\partial\Phi}{\partial r} - r\Omega^2$, and in the case of $\frac{ds_0}{dr} < 0$ it is necessary $g > 0$. It should be noted, that instability is possible even at $g < 0$ in the case of $\frac{ds_0}{dr} > 0$.

Acknowledgments

This work was supported by the Russian Foundation for Basic Research through the grants RFBR 04-02-96500 and 04-02-16518.

References

Abramowicz M.A., Chen X., Kato S., Lasota J.-P., Regev O., 1995, ApJ, **438**, L37.

Agol E., Krolik J., Turner N.J., Stone J.M. 2001, ApJ, **558**, 543.

Balbus S.A., Hawley J.F., 2002, ApJ, **573**, 749.

Bisnovatyi-Kogan G.S., Blinnikov S.I., Kostiuk N.D., Fedorova A.V. 1979, Soviet Astron., **23**, 432.

Bisnovatyi-Kogan G.S., Blinnikov S.I. 1977, AA, **59**, 111.

Bisnovatyi-Kogan G.S., Lovelace R.V.E. 2001, New Astr. Reviews, 2001, **45**, 663.

Brun A.S., Toomre T. 2002, ApJ, **570**, 865.

Igumenshchev I.V., Abramowicz M.A. 1999, MN, **303**, 309.

Igumenshchev I.V., Abramowicz M.A., 2000, ApJS, **130**, 463.

Igumenshchev I.V., Abramowicz M.A., Narayan R. 2000, ApJ, 537, 27L.

Igumenshchev I.V. 2002, ApJ, **577**, 31L.

Gor'kavyj N.N., Fridman A.M. Physics of planetary rings: Celestial mechanics of continuous medium. M.: Nauka, 1994.

Khoperskov A.V., Khrapov S.S., 1999, Astr. Rep., **43**, 216.

Melia F., Falcke H., 2001, Ann. Rev. A&A, **39**, 309.

Morozov A.G., Khoperskov A.V. 1990, SvAL, **16**, 244.

Narayan R., Yi I., 1994, ApJ, **428**, 13.

Narayan R. astrp-ph/0201260.

Narayan R., Quataert E., Igumenshchev I.V., Abramowicz M.A. 2002, ApJ, **577**, 295.

Shakura N.I., Sunyaev R.A., 1973, A& A, **24**, 337.

Stone J.M., Pringle J.E., Begelman M.C., 1999, MNRAS, **310**, 1002.

3D STRUCTURE OF GASEOUS DISKS IN SPIRAL GALAXIES

I. G. Kovalenko
Department of Physics, Volgograd State University
Universitetskij prosp. 100, Volgograd 400062, Russia

igk@vlink.ru

M. A. Eremin
Department of Physics, Volgograd State University
Universitetskij prosp. 100, Volgograd 400062, Russia

emk@vlink.ru

V. V. Korolev
Department of Physics, Volgograd State University
Universitetskij prosp. 100, Volgograd 400062, Russia

ppvito@vlink.ru

Abstract   We examine the structure and dynamics of a gaseous disk in a spiral galaxy in the vicinity of a spiral arm. We show numerically that traditional assumptions of a unique galactic shock wave are inadequate and should be oriented to a compages of shocks. In a vertical plane we find that relaxation of interstellar gas to a steady state behind the primary galactic shock wave occurs through a sequence of multiple vertical shocks which forces gas circulation in the vertical plane.

Keywords: spiral galaxies, hydrodynamical instability, shock waves

1.   Introduction

Considering large-scale perturbations in a thin disk it seems natural to abstract oneself from a cross-wise structure of a disk and to treat perturbations as in-plane 2-dimensional ones.

The most widespread method for exclusion of the vertical direction is averaging over vertical coordinate z assuming fast relaxation to hydrostatic

Alexei M. Fridman, et al. (eds.), Astrophysical Disks; Collective and Stochastic Phenomena 231–240
© 2006 *Springer. Printed in the Netherlands*

equilibrium. An idea of this method came from the theory of shallow water for incompressible fluid and is actually based on a standard expedient of separation of motions into fast and slow motion in the theory of waves.

For incompressible fluid the condition of fast relaxation is fulfilled as soon as disk thickness $2h$ is much smaller than a size of perturbation λ. After averaging the equations of motion in a plane agree with equations for 2-dimensional compressible fluid within change of 3-dimensional, bulk, adiabatic index γ_3 to 2-dimensional, surface, index γ_2 (Landau & Lifshitz, 1986). Applying this procedure of averaging to a compressible gaseous disk embedded in an external square potential well of stellar disk one gets (Churilov & Shukhman, 1981).

$$\gamma_2 = \frac{3\gamma_3 - 1}{\gamma_3 + 1} \, . \tag{1}$$

As mentioned above an expression (1) is correct solely in the limit of instant relaxation to a hydrostatic equilibrium.

A situation de facto with astrophysical disks is usually more intricate inasmuch as (i) the matter of disks is compressible gas and the rate of relaxation to hydrostatic equilibrium is limited and can be comparable with rates of other processes; (ii) an unperturbed state of disk is generally fast supersonic rotation and horizontal velocity is much greater than the vertical one.

A shortcoming of traditional local hydrostatic equilibrium approximation was first pointed out by Fridman, Khoruzhij and Libin in Appendix to the book (Fridman & Gor'kavyi, 1999; 1994 in a Russian edition).

Indeed, a condition of fast relaxation is fulfilled unless and until the period of vertical oscillations $T_z \sim \Omega_z^{-1}$ significantly exceeds the time of vertical relaxation t_z, where $\Omega_z \sim (\partial^2\psi/\partial z^2|_{z=0})^{1/2}$ is the frequency of disk oscillations across an equatorial plane $z = 0$, and ψ is the gravitational potential. This is usually not true of astrophysical disks where one has $T_z \sim t_z$ since $T_z \sim \Omega_z^{-1}$ and $t_z \sim h/c_s \sim \Omega_z^{-1}$. Here h is the disk half-thickness and c_s is the adiabatic sound speed. Thus, one obviously has to take into account the effects of inertia of vertical motions and local disequilibrium of a disk.

The primary aim of the present article is to demonstrate that 3-dimensionality of thin disks can lead to non-trivial, quite surprising effects. We discuss those effects which "cannot be observed" in principle in 2-dimensional models.

2.   A Vertical Structure of Disks: Beyond the Hydrostatic Model

To account for the effects of deviation from equilibrium Fridman, Khoruzhij and Libin suggested a procedure based on solution of integral equations. Kovalenko and Lukin (1999) put forward alternatively a moment expansion technique to exclude z-coordinate through expansions of all variables over z. Their idea involves integration of hydrodynamical equations, multiplied successively by $z^n/n!$, over definitional domain $z \in [h_-, h_+]$, and reduction to a system of coupled equations for the moments.

For the velocity one uses conventional power expansions

$$\mathbf{v}_{||} = \sum_{n=0}^{\infty} \mathbf{v}_{||n} \frac{z^n}{n!} \,, \qquad w = \sum_{n=0}^{\infty} w_n \frac{z^n}{n!} \,, \tag{2}$$

where total velocity is split into in-plane and transversal components, $\mathbf{v} = (\mathbf{v}_{||}, w)$, the moments of which $\mathbf{v}_{||n} = (u_n, v_n)$ and w_n are considered to be functions of surface coordinate $\mathbf{r}_{||} = (x, y)$ and time t.

Density and pressure are expanded into integral moments:

$$\sigma_n = \int_{h_-}^{h_+} \rho \frac{z^n}{n!}\, dz, \qquad \mathcal{P}_n = \int_{h_-}^{h_+} p \frac{z^n}{n!}\, dz \,. \tag{3}$$

Dynamics in pure 2-dimensional disk can be expressed by unique dimensionless parameter, Mach number

$$M = \frac{u_0}{c_0} \,. \tag{4}$$

The effects of 3-dimensionality of a disk can be allowed for by introduction of two next-order non-dimensional parameters

$$K = \frac{w_1 h}{c_0} = w_1 \left(\frac{\sigma_2}{\gamma_3 \mathcal{P}_0} \right)^{1/2}, \qquad L = \frac{h}{h_{eq}} = \left(\frac{2\sigma_2 \psi_2}{\mathcal{P}_0} \right)^{1/2} . \tag{5}$$

The parameter K characterizes intensity of vertical pulsations expressed in terms of longitudinal short-length sound speed

$$c_0 = \left(\frac{\gamma_3 \mathcal{P}_0}{\sigma_0} \right)^{1/2}, \tag{6}$$

and the second parameter, L, presents the ratio of local, h, and equilibrium, h_{eq}, half-thicknesses of a disk defined as

$$h = \left(\frac{\sigma_2}{\sigma_0} \right)^{1/2}, \qquad h_{eq} = \left(\frac{\mathcal{P}_0}{2\sigma_0 \psi_2} \right)^{1/2} . \tag{7}$$

Given the instant flow fields, parameters K and L provide a convenient tool to trace zones in a disk where 3-dimensional effects are significant. In the hydrostatic approximation $K \approx 0$ and $L \approx 1$. In an unbalanced case K and L deviate from their equilibrium values markedly.

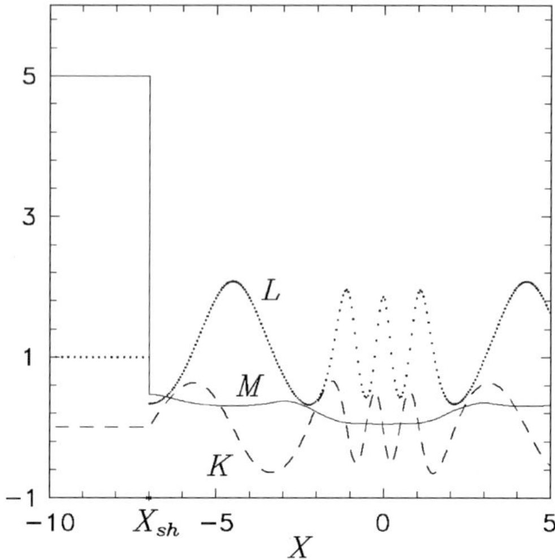

Figure 1.   A sketch of vertical oscillations of a disk behind the front of a galactic shock wave according to (Kovalenko & Lukin, 1999). The Mach number M is shown by solid line, dimensionless vertical velocity K and relative half-thickness of disk L are shown by dashed and dotted lines respectively. Variability of spatial period of oscillations is due to non-uniformity of the flow within the potential well of a spiral arm in the area $-5 < x < 5$.

Probably the most critical zones where the effects of 3-dimensionality approve themselves are shock waves. Indeed, gas passing through the shock front loses its hydrostatic balance instantly and begins oscillating. What should happen just behind the galactic shock front schematically illustrates Fig. 1.

Here we consider in detail a fine vertical structure of the galactic shock wave.

3.   The Numerical Model

An interest to the processes inspirated by galactic shock waves in a vertical direction was arisen by Walters & Cox (2001) not far ago who considered galactic shocks as a potential effective mechanism of mass and energy transfer from the disk plane to gaseous halo.

In the present article we would like to discuss, quite the contrary, possible reciprocal influence of shocks on a disk plane through the vertical direction.

We study the process of relaxation to a steady state of the vertical flow in the vicinity of a spiral arm as well as ultimate steady state configurations.

We consider interstellar matter as a perfect gas with an adiabatic index $\gamma_3 = 5/3$. No other effects such as self-gravitation, viscosity or magnetic fields are taken into account.

An ultimate state is prepared in two steps. Initially the potential is specified as pure disk potential

$$\psi_{disk}(z) = \begin{cases} \psi_0 (z/z_{cutoff})^2, & |z| \le z_{cutoff}; \\ \psi_0, & |z| \ge z_{cutoff}. \end{cases} \tag{8}$$

Gas moves at a constant supersonic speed u along the plane of a disk. In the vertical direction gas is hydrostatically balanced. A constant temperature along z is preset. The scale height of gaseous disk is then that of for isothermal disk: $h = c_s z_{cutoff} / \sqrt{\psi_0}$.

To ensure accurate relaxation of a shock front to a steady state position, the circle potential well ψ_{spir} is slowly (compared with the typical hydrodynamical time) building up to its maximum value

$$\psi_{spir}(x, z) = \begin{cases} \psi_1 \cos^2(\pi\sqrt{x^2 + z^2}/2d), & x^2 + z^2 \le d^2; \\ 0, & x^2 + z^2 \ge d^2; \end{cases} \tag{9}$$

This potential mimics the vertical cross-section of a spiral arm.

Total potential is added by both constituents:

$$\psi(x, z) = \psi_{disk}(z) + \psi_{spir}(x, z) \,. \tag{10}$$

The parameters of the flow can be expressed in the dimensionless form in terms of density and sound speed taken in an equatorial plane and of an unperturbed 1/5 half-thickness of gaseous disk

$$\rho_0 = 1 \,, \qquad c_{s0} = 1 \,, \qquad h = 1 \,, \qquad \psi_0 = 25 \,, \qquad z_{cutoff} = 5 \,,$$

$$\psi_1 = -6 \,, \qquad d = h \,, \qquad M_0 = 3 \div 5 \,.$$

Parameters taken in that way correspond to the typical values $c_s = 10^4$ K, $h = 150$ pc and $z_{cutoff} = 750$ pc.

The computational domain is taken as square with side length equal to 200. Extended height of the domain allows us to minimize the influence of boundaries. The boundary conditions at the entrance left side $x = -50$ are fixed, reflecting boundary is taken at the equatorial plane $z = 0$, and 'free-efflux' conditions are prescribed at the exit right $x = 150$ and upper $z = 200$ boundaries.

3-dimensional model is computed in a uniform cube with the same parameters as for the 2-dimensional square with the difference that bending of a spiral arm in the x-y-plane is taken into account. The potential $\psi_{spir}(x, y, z)$ has now a segment $0 < y < 100$ which is slowly reshaping into the concave circular arc with the curvature radius $R_c = 50$. In the case of oblique incidence gas has non-zero inflow y-velocity v.

An explicit version of Versatile Advection Code is used on the uniform Cartesian grid that consists of 600 x 600 in 2D or 100 x 100 x 100 in 3D calculations. Grid cells are taken as squares (cubes) with linear sizes equal to 50 (300) pc.

4. Basic Results

Figure 2 demonstrates evolution of disk in a vertical plane for the case of inflow Mach number $M_0 = 3$. As the spiral potential well deepens a shock wave forms just behind the arm. Unbalanced postshock gas executes vertical oscillations as it is predicted by theory but due to nonlinear amplitude relaxation to equilibrium occurs through formation of serial shocks.

The maximum number of shocks arises at intermediate times. One can distinguish up to 7 fan-shaped shock fronts. After several hydrodynamical times distant shocks slowly dissipate, and only 2-3 shock fronts survive.

Figures 3, 4 and 5 illustrate behavior of parameters K and L and various moments of density and pressure.

At large Mach numbers shocks become pressed against an equatorial plane, and their number reduces (Fig. 6).

All experimental findings are cumulated in Table 1. From Table 1 one can notice a curious fact that larger Mach numbers reduce intensity of the primary shock front. Indeed, multiplying M_0 by factor sin(angle of inclination) we get an effective Mach number $M_{norm} = 1.9$ for $M_0 = 3$ and $M_{norm} = 1.5$ for $M_0 = 5$. This effect can never be noticed in 2-dimensional hydrodynamics since the latter predicts exactly the opposite result.

Table 1. Geometrical parameters of multiple vertical shocks for different models

Inflow Mach number	phase (i= intermediate, ss =steady state)	number of shocks	angles of inclination (deg)	distances between shocks (in x-dir)
3	i (Fig. 2a)	6	104, 35, 33, 25, 16, 10	15, 22, 22, 22, 22
3	ss (Fig. 2b)	3	40, 24, 22	9, 17
5	i (Fig. 6a)	4	31, 15, 15, 12	20, 40, 36
5	ss (Fig. 6b)	3	18, 11, 13	20, 38

In a 3-dimensional case we try to improve the model by considering arched shape of a spiral arm in the plane of a disk and oblique inflow of gas at some pitch angle.

The initial state of simulation repeats the structure of flow in the 2D model. The spiral arm potential well has the shape of a uniform cylinder infinitely elongated along y-axis. A new point is that we now allow y-velocity to be non-zero. After that the potential well slowly protrudes downstream acquiring $\supset$-form. For details of ultimate state preparation we refer the reader to the article by Chernin et al. in the present volume.

Weak bending of the spiral arm in a plane of a disk does not alter the character of the flow qualitatively. One may still distinguish secondary shock waves

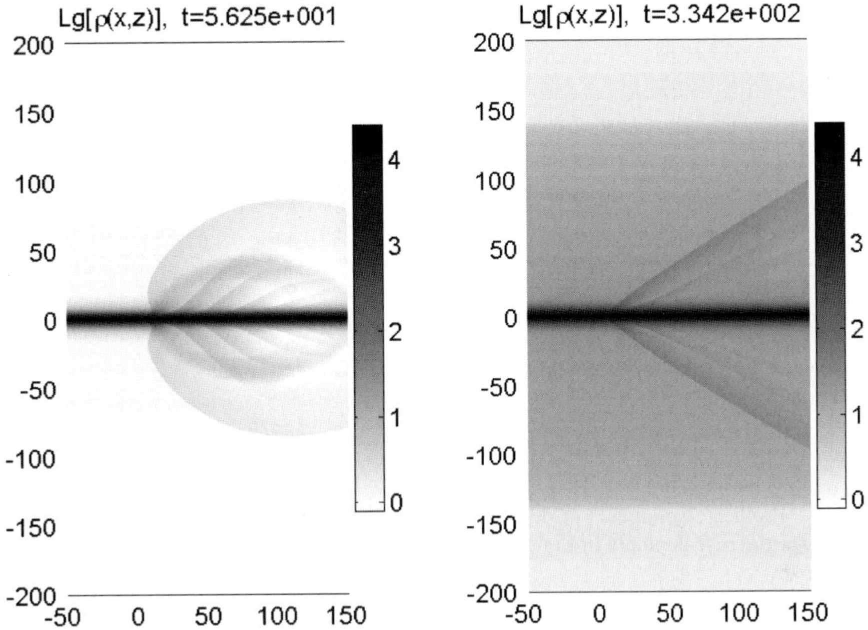

Figure 2. Vertical structure of disk with multiple shock fronts behind a spiral arm at the phase of relaxation (left frame) and after relaxation to a steady state (right frame). Gray-scale distribution of density is depicted. The potential well of the spiral arm is a circle of radius $d = 5$ with the centre at $x = 0$, $z = 0$. Time is given in units $t = 2 \cdot 10^8$ years. Distances are given in half-thicknesses of gaseous disk h (=150 pc). Inflow Mach number is $M_0 = 3.0$.

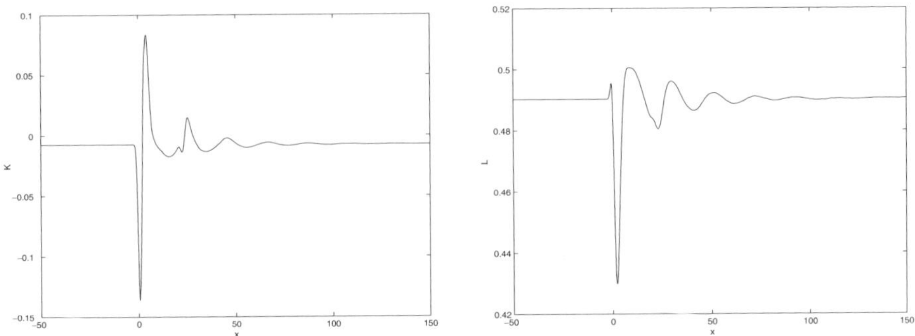

Figure 3. Profiles of dimensionless vertical velocity K and relative half-thickness L calculated for the case depicted in Fig. 2 ($M_0 = 3$).

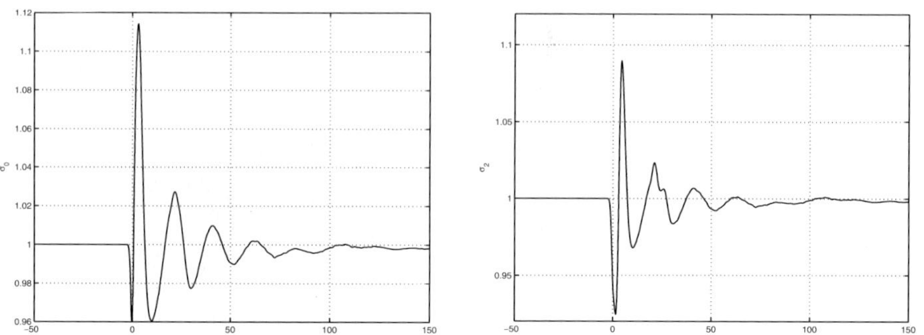

Figure 4. Profiles of moments of density σ_0 and σ_2, normalized by their initial values, and calculated for the case depicted in Fig. 2 ($M_0 = 3$).

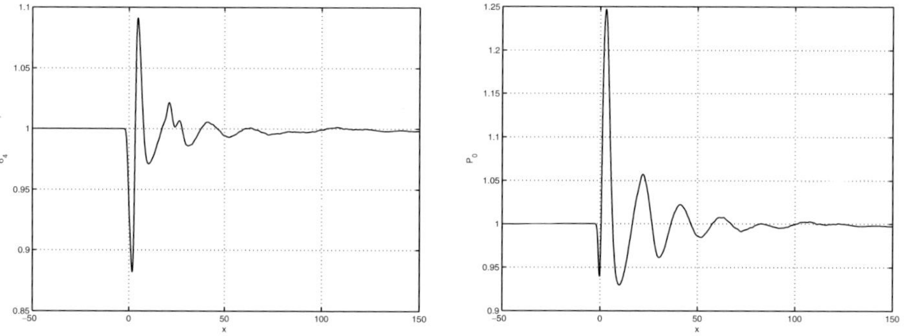

Figure 5. Profiles of moments of density σ_4 and pressure $\mathcal{P}_0$, normalized by their initial values, and calculated for the case depicted in Fig. 2 ($M_0 = 3$).

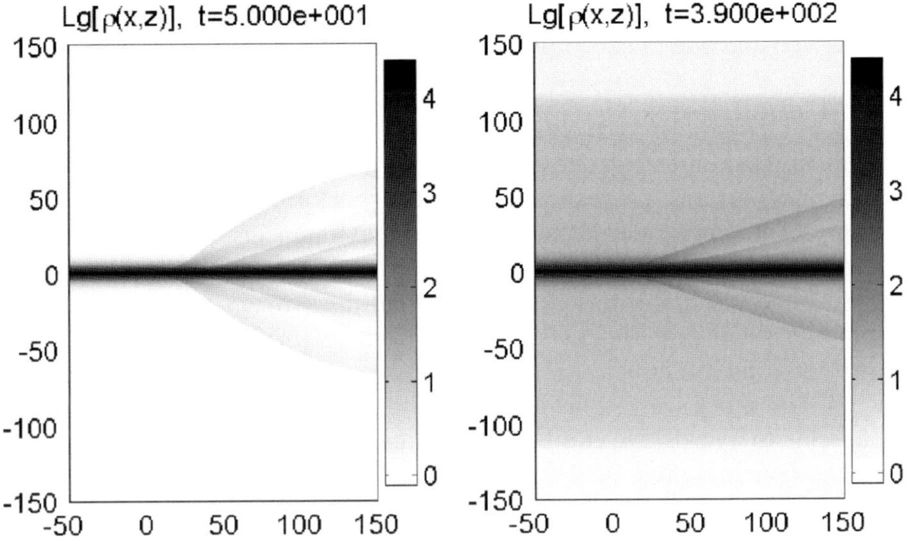

Figure 6.   The same as in Fig. 2 but for $M_0 = 5$.

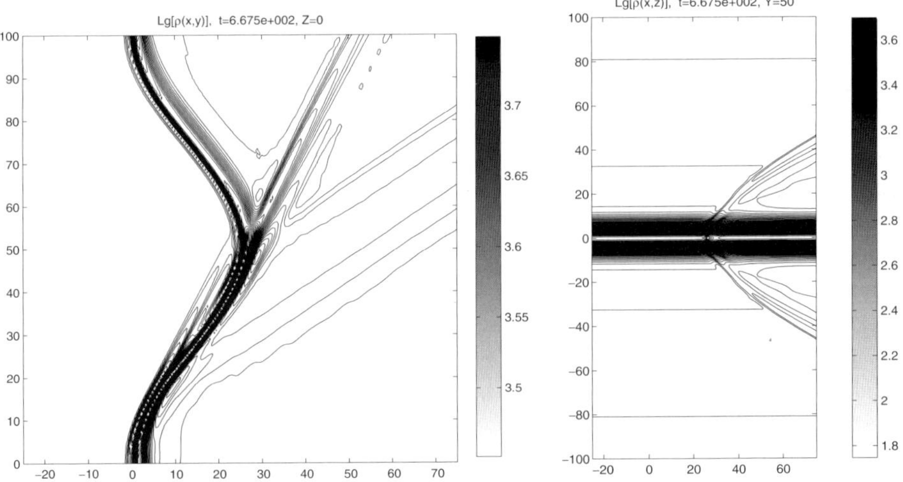

Figure 7.   Horizontal (left frame) and vertical (right frame) structure of a galactic shock in a steady state. Isolines of density are shown. The spiral arm is presented by a tube with circular vertical cross-section with radius $d = 1$. Gas flows north-east (left frame). Time is given in units $t = 2 \cdot 10^8$ years. Distances are given in half-thicknesses of gaseous disk h (=150 pc). Inflow x-velocity is $u_0 = M_{0x} = 3.0$, and y-velocity is $v_0 = M_{0y} = 2.0$.

which now become much weaker due to the effects of axial defocusing on a concave well and due to loss of linear resolution on 3-dimensional grid.

5.   Conclusions

We distinguish three basic results.

1. The gaseous disk unbalanced behind the shock front oscillates in vertical direction. The amplitude of oscillations is maximum just behind the shock front and slowly damps downstream.

2. Secondary shock fronts can form behind the primary front (Figs. 2, 6). Positions of secondary shocks shift away from the primary shock as M increases. At Mach number ~ 3 six or even seven secondary shocks may be detected. The shock waves play the role of strong dissipative factor forcing the disk to relax to hydrostatic equilibrium within several oscillations.

3. The opening angle of the shock front and the intensity of the primary shock *decrease* as the Mach number grows (Table 1). This fact contradicts 2-dimensional hydrodynamics and points out the necessity of taking into account vertical direction in simulations of thin gaseous disks.

Acknowledgments

Dr. P. Grosbøl has noted that the lattice of secondary waves behind the primary galactic shock wave with distances between shocks 300-1200 pc (last column in Table 1) can be related to the system of fine linear dust structures clearly seen near and between arms in face-on spiral galaxies.

The work was supported in part by the Russian Foundation for Basic Research (project 04-02-96500) and by the Russian Ministry of Education (grant E02-11.0-39).

References

Chernin, A. D., Korolev, V. V. and Kovalenko, I. G.: 2006, present volume, p. 321.

Churilov, S. M. and Shukhman, I. G.: 1981, *Astron. Tsirkular* N 1157, 1.

Fridman, A. M. and Gor'kavyi, N. N.: 1999, *Physics of Planetary Rings*. Springer-Verlag, Berlin and Heidelberg.

Kovalenko, I. G. and Lukin D. V.: 1999, *Astron. Lett.* **25**, 215.

Landau, L. D. and Lifshitz, E. M.: 1986, *Hydrodynamics*. Nauka, Moscow.

Walters, M. A. and Cox D. P.: 2001, *ApJ* **549**, 353.

HYDRODYNAMICAL TURBULENCE IN ACCRETION DISCS

O. A. Kuznetsov
Keldysh Institute of Applied Mathematics, Moscow, Russia
and
Institute of Astronomy Russ. Acad. Sci., Moscow, Russia
kuznecov@spp.keldysh.ru

Abstract   Our 2D numerical simulations confirm a possibility of excitation of hydrodynamic turbulence in the accretion disk due to shear instability under finite-amplitude perturbations and at high Reynolds number. These results disproves the well-known claims about the impossibility to excite the turbulence in accretion disks by hydrodynamic shear instability. Our estimations have shown that the development of turbulence results in the value of accretion rate corresponding to the α_{SS}-coefficient in the range $0.08 \div 0.15$.

Keywords: accretion, accretion discs – hydrodynamics – instabilities – turbulence

1.   Introduction

The release of gravitational energy in the process of accretion is one of the most powerful energy sources in the Universe. The mass accretion in the disk requires expulsion of angular momentum to the exterior of the disc. If angular momentum of the fluid element is conserved

$$\partial_t \mathfrak{L} + \mathbf{v} \cdot \nabla \mathfrak{L} = -\frac{r}{\rho} \cdot \partial_\varphi P - r \cdot \partial_\varphi \Phi \approx 0,$$

as it takes place in axially symmetric disk with negligible viscosity, the matter could reach only the so called circularization radius $r_{circ} = \mathfrak{L}^2/2GM$. Here $\mathfrak{L}$ is vertical component of angular momentum $\mathbf{r} \times \mathbf{v}$, and partial derivatives are written as $\partial_t \equiv \partial/\partial t$, $\partial_\varphi \equiv r^{-1}\partial/\partial\varphi$, $\partial_r \equiv \partial/\partial r$, M is the mass of central object, G is gravitational constant. A lot of mechanisms were proposed to explain the angular momentum transfer, which is required for explanation of observational data, let us list the main of them: tidal interaction

Alexei M. Fridman, et al. (eds.), Astrophysical Disks; Collective and Stochastic Phenomena 241–258
© 2006 *Springer. Printed in the Netherlands*

(Papaloizou & Pringle, 1977; Lin et al., 1990; Dgani et al., 1994); spiral shocks (Michel, 1984; Sawada et al., 1986; Spruit, 1987; Syer & Narayan, 1993); convection (Bisnovatyi-Kogan & Blinnikov, 1976; Paczyński, 1976; Shakura et al., 1978; Lin et al., 1980); disc wind and jets (Blandford & Paine, 1982; Lin et al., 1994; Cao & Spruit, 2002; Lynden-Bell, 2003); different disk instabilities: hydromagnetic (Velikhov, 1959; Chandrasekhar, 1981; Balbus & Hawley, 1991; Winters et al., 2003); parametric (Goodman, 1993; Lubow et al., 1993); baroclinic (Cabot, 1984; Li et al., 2000; Klahr et al., 2003); over-reflection (Fridman et al., 2003) etc.; as well as angular momentum transport via wave propagation (Lin & Papaloizou, 1979; Lubow, 1981; Vishniac & Diamond, 1989). Angular momentum transfer was also considered in reviews by Papaloizou & Lin (1995) and Balbus (2003).

Nevertheless, the most natural way of angular momentum transport is a viscous dissipation which gives

$$\partial_t \mathfrak{L} + \mathbf{v} \cdot \nabla \mathfrak{L} = -\frac{r}{\rho} \cdot \partial_\varphi P - r \cdot \partial_\varphi \Phi + r^{-1} \partial_r \left(\nu r^3 \partial_r (r^{-2} \mathfrak{L}) \right), \qquad (1)$$

where ν is kinematic viscosity coefficient. In the thin stationary disk with Keplerian angular velocity it follows from (1) the expression for the radial velocity $v_r = -{}^3/_2 \nu / r$. Molecular viscosity with coefficient $\nu_{mol} = c_s \cdot \lambda$ (c_s is a speed of sound, $\lambda = 1/n\sigma$ is a mean free path, n is a number of particles in 1 cm^3, $\sigma \simeq 10^{-16}$ cm^2 is a cross-section of impact between particles) results in a very low accretion rates

$$\dot{\mathfrak{M}} = 2\pi \cdot r \cdot H \cdot \rho \cdot |v_r|, \qquad\qquad H = c_s \cdot \Omega_K^{-1},$$

$$\dot{\mathfrak{M}} \simeq 10^{-18} \left(\frac{T}{10^5 \text{ K}} \right) \left(\frac{r}{10^5 \text{ km}} \right)^{3/2} \left(\frac{M}{M_\odot} \right)^{-1/2} M_\odot/\text{yr},$$

which is in contradiction with observations (here H is a disc thickness, Ω_K is angular Keplerian velocity of the disc rotation, T is a disc temperature). On the other hand the flow in the disk has extremely large Reynolds number $\mathfrak{Re} = r\Omega_K H/\nu_{mol}$:

$$\mathfrak{Re} \simeq 10^9 \left(\frac{n}{10^{15} \text{ cm}^{-3}} \right) \left(\frac{r}{10^5 \text{ km}} \right).$$

Accordingly, Lynden-Bell (1969), Shakura (1972), and Lynden-Bell & Pringle (1974) suggested a possible origin of turbulence due to similar mechanism as those occurring in laboratory flows. Shakura (1972) and Shakura & Sunyaev (1973) introduced the so called turbulent viscosity and replaced viscosity coefficient as $\nu_{mol} = c_s \cdot \lambda \rightarrow \nu_{turb} = \alpha_{SS} \cdot c_s \cdot H$, where α_{SS} is an heuristic constant. According to

$$\frac{H}{\lambda} \simeq 10^7 \left(\frac{n}{10^{15} \text{ cm}^{-3}} \right) \left(\frac{T}{10^5 \text{ K}} \right)^{1/2} \left(\frac{r}{10^5 \text{ km}} \right)^{3/2} \left(\frac{M}{M_\odot} \right)^{-1/2},$$

the turbulent viscosity coefficient is many order of magnitude larger than molecular one, and provides reasonable values for accretion rate.

How does the turbulence develop in accretion disc? It is known that flow in the pipe (Poisseille flow) can show turbulence developing under some conditions. The main aim of this paper is to find the turbulence development in the accretion disc considering it as a (curved) pipe.

2. System of equations and results for inertial frame

Let us describe briefly the basic equations. We start with 3D hydrodynamical non-viscous equations:

$$\partial_t \rho + \nabla(\rho \mathbf{u}^{\|}) + \partial_z(\rho w) = 0,$$

$$\partial_t(\rho \mathbf{u}^{\|}) + \nabla(\rho \mathbf{u}^{\|} \otimes \mathbf{u}^{\|} + P) + \partial_z(\rho \mathbf{u}^{\|} w) = -\rho \cdot \nabla \Phi,$$

$$\partial_t(\rho w) + \nabla(\rho w \mathbf{u}^{\|}) + \partial_z(\rho w^2 + P) = -\rho \cdot \partial_z \Phi,$$

$$\partial_t E + \nabla(\mathbf{u}^{\|}(E + P)) + \partial_z(w(E + P)) = -\rho \cdot \mathbf{u}^{\|} \cdot \nabla \Phi - \rho \cdot w \cdot \partial_z \Phi$$

with equation of state of perfect gas with ratio of heat capacities γ

$$P = (\gamma - 1)\left(E - \frac{1}{2}\rho|\mathbf{u}^{\|}|^2 - \frac{1}{2}\rho w^2\right).$$

Here $\mathbf{u}^{\|} = (u, v)$ – vector of horizontal velocity, w – vertical of component of velocity, $\Phi = -GM/\sqrt{r^2 + z^2}$ – gravitational potential, $\nabla = (\partial_x, \partial_y)$, $(\mathbf{u}^{\|} \otimes \mathbf{u}^{\|})_{i,j} = u_i^{\|} u_j^{\|}$ $(i, j = 1, 2)$.

We can apply a standard procedure of averaging for thin discs to obtain the 2D hydrodynamical equations. Let us assume that horizontal velocity and gradient of gravitational potential does not depend on z-coordinate: $\mathbf{u}^{\|} = \mathrm{const}(z)$ (i.e. $|\mathbf{u}^{\|}| \gg |\frac{1}{2}\partial_{zz}\mathbf{u}^{\|} \cdot H^2|$, here we took into account the symmetry of the disc w.r.t. $z = 0$ plane), $\nabla \Phi = \mathrm{const}(z)$ (i.e. $H^2 \ll r^2$), and vertical component of velocity gives a negligible income to kinetic energy: $\mathbf{u}^{\|2} + w^2 \approx \mathbf{u}^{\|2}$. The accuracy of this approach can be estimated as $\mathcal{O}(H^2/r^2)$. Integrating from $-\infty$ to ∞ and taking into account that $\int_{-\infty}^{\infty} \partial_z(\ldots)\,\mathrm{d}z = 0$ we get (hereinafter $\mathbf{u} \equiv \mathbf{u}^{\|}$):

$$\partial_t \sigma + \nabla(\sigma \mathbf{u}) = 0,$$

$$\partial_t(\sigma \mathbf{u}) + \nabla(\sigma \mathbf{u} \otimes \mathbf{u} + \Pi) = -\sigma \nabla \Phi,$$

$$\partial_t \Upsilon + \nabla((\Upsilon + \Pi)\mathbf{u}) = -\sigma \mathbf{u} \nabla \Phi$$

with equation of state $\Pi = (\gamma - 1)(\Upsilon - \frac{1}{2}\sigma|\mathbf{u}|^2)$. Here $\sigma = \int_{-\infty}^{\infty} \rho\,\mathrm{d}z$, $\Pi = \int_{-\infty}^{\infty} P\,\mathrm{d}z$, $\Upsilon = \int_{-\infty}^{\infty} E\,\mathrm{d}z$, $\Phi = -GM/r$.

For numerical purposes we use the polar coordinates (r, φ), $u \equiv u_r$, $v \equiv u_\varphi$, but rewrite the 2D hydrodynamical equations in pseudo-Cartesian form (see, e.g., Bisnovatyi-Kogan & Pogorelov, 1997):

$$\partial_t \sigma + \partial_r(\sigma u) + \partial_\varphi(\sigma v) = -\sigma \cdot \frac{u}{r}, \tag{2}$$

$$\partial_t(\sigma u) + \partial_r(\sigma u^2 + \Pi) + \partial_\varphi(\sigma u v) = -\sigma \cdot \partial_r \Phi - \sigma \cdot \frac{u^2 - v^2}{r}, \tag{3}$$

$$\partial_t(\sigma v) + \partial_r(\sigma u v) + \partial_\varphi(\sigma v^2 + \Pi) = -\sigma \cdot \frac{2uv}{r}, \tag{4}$$

$$\partial_t \Upsilon + \partial_r((\Upsilon + \Pi)u) + \partial_\varphi((\Upsilon + \Pi)v) = -\sigma \cdot u \cdot \partial_r \Phi - \frac{(\Upsilon + \Pi)u}{r}. \tag{5}$$

We will exploit the non-dimensional variables using arbitrary values r_0, ρ_0, $\sqrt{GM/r_0}$ and $\sqrt{r_0^3/GM}$ as scales for distance, density, velocity and time. Dimensionless rotational period at $r = 1$ is equal to $P_0 = 2\pi$.

We consider a piece of the Keplerian disc $0.8 < r < 1.2$, $0 < \varphi < \pi/3$ with initial values $u_0 = 0$, $v_0 = \Omega_K r = r^{-1/2}$, $\sigma_0 = 1$, $\Pi_0 = c_0^2 \sigma_0/\gamma$, $c_0 = 0.5$ and disturb it on the lower edge $\varphi = 0$ as $\sigma = \sigma_0(1 + A\sin(kr))$, $\Pi = \Pi_0(1 + A\sin(kr))$ with $k = 160$ (it corresponds to the wavelength $\lambda = 2\pi/k = 0.04$) and amplitude of perturbations 50% (i.e. $A = 0.5$). Hereinafter this variant will be referred as run A. Figure 1 shows the entropy distribution for this run.

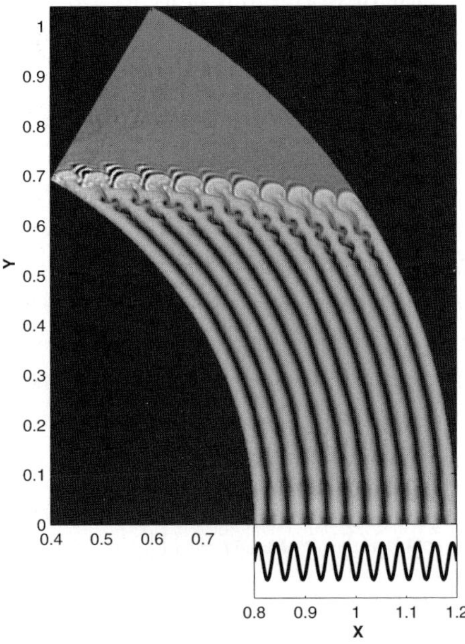

Figure 1.   Entropy distribution for run A.

It is seen that the flow is laminar. Small deviations from laminar flow appear to correspond the so called "convective" not "absolute" instabilities. These terms were described by Landau & Lifshitz (1986) as

> "...In one case, despite of the motion of a wave package, the perturbations grow with time unrestrictedly at any point of the flow; such instability with respect to infinitesimal perturbations we call *absolute*. In another case the package is propagating so quickly, that in each fixed point of a space the perturbation tends to zero as $t \to \infty$; such instability we call *convective*. For Poisseille flow, apparently, the second case takes place...
>
> It is necessary to note, that the distinction between both cases has the relative character in the sense that it depends on a choice of the reference frame, in relation to which the instability is examined: convective instability becomes absolute in reference frame driven "together with a package", and the absolute instability becomes convective in reference frame quickly "running away" from the package. At any given a case, however, the physical sense of this distinction is established by existence of the preferred reference frame, with respect to which the instability should be examined – reference frame, in which the walls of the pipe are in rest. Moreover, as the real pipes may have large, but final length, the perturbation arising somewhere may, in principle, be carried out from the pipe before a true breaking up of the laminar flow."

In run A we have unperturbed (initial) azimuthal velocity ranging from 1.12 on the left edge to 0.91 on the right edge so the flow is supersonic everywhere ($c_0 = 0.5$). Following Landau & Lifshitz (1986) ideas let us consider this flow in rotating coordinate system.

3. System of equations and results for rotating frame

Let us transfer to the coordinate system which is rotating with angular velocity $\Omega_1 = \Omega_K(r_1) = 0.76$, where $r_1 = 1.2$ denotes the right edge of the disc. The Coriolis and centrifugal forces should be added in the equations so they now read as

$$\partial_t(\sigma u) + \ldots = -\sigma \cdot \partial_r \Phi_1 - \sigma \cdot \frac{u^2 - v^2}{r} + 2\Omega_1 \cdot v \cdot \sigma,$$

$$\partial_t(\sigma v) + \ldots = -\sigma \cdot \frac{2uv}{r} - 2\Omega_1 \cdot u \cdot \sigma,$$

$$\partial_t \Upsilon + \ldots = -\sigma \cdot u \cdot \partial_r \Phi_1 - \frac{(\Upsilon + \Pi)u}{r}, \qquad \Phi_1 = -\frac{GM}{r} - \frac{1}{2}\Omega_1^2 r^2.$$

We consider a piece of the Keplerian disc $0.8 < r < 1.2$, $0 < \varphi < \pi/6$ with the same initial values (except v_0, now $v_0 = (\Omega_K - \Omega_1)r = r^{-1/2} - 0.76r$) and the same perturbations as in run A. Hereinafter this variant will be referred as run B. In this run the unperturbed (initial) azimuthal velocity ranging from 0.51 on the left edge to 0 on the right edge so the flow is subsonic almost everywhere. Figures 2–3 show the entropy distribution for run B. Figure 3 also

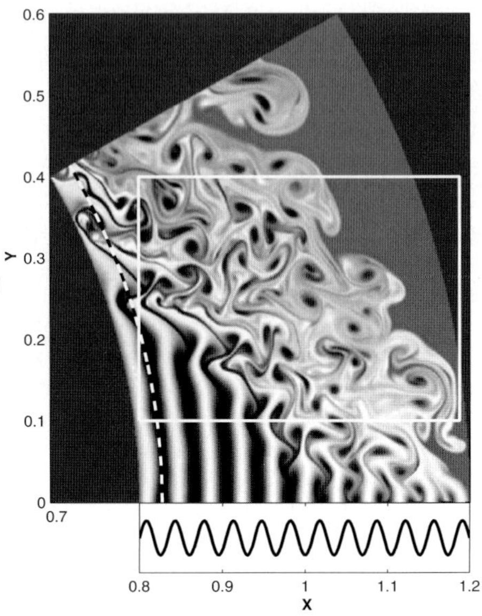

Figure 2.   Entropy distribution for run B. Bold black-white line divides the supersonic and subsonic regions for unperturbed (initial) flow.

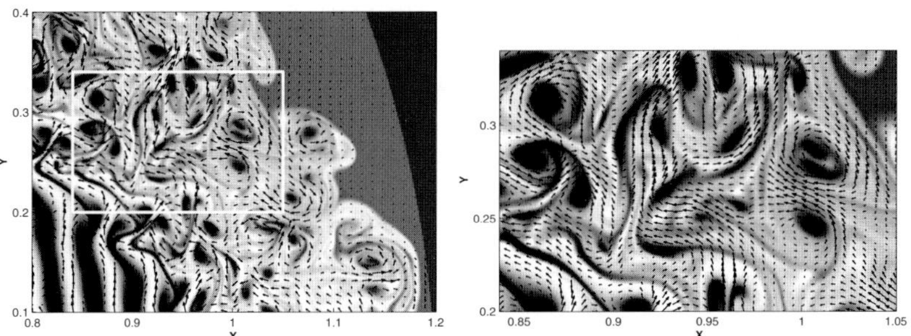

Figure 3.   Entropy distribution and perturbed velocity field for run B.

show the perturbed velocity field $\tilde{\mathbf{v}} = \mathbf{v} - \mathbf{v}_0$. It is seen that the flow contains both cyclonic and anticyclonic vortices. In Fig. 3 the so called dipole vortices (modons) are also seen (see Larichev & Reznik, 1976; Fridman, 1988; Dolotin & Fridman, 1991 for the discussion on the modons origin). Figure 4 shows the visualization of perturbed velocity field using Line Integral Convolution method for two different moments of time. In Figs. 2–4 we can see that the flow is turbulent except left lower corner (supersonic and transonic zone) and right upper corner (zone of unperturbed flow).

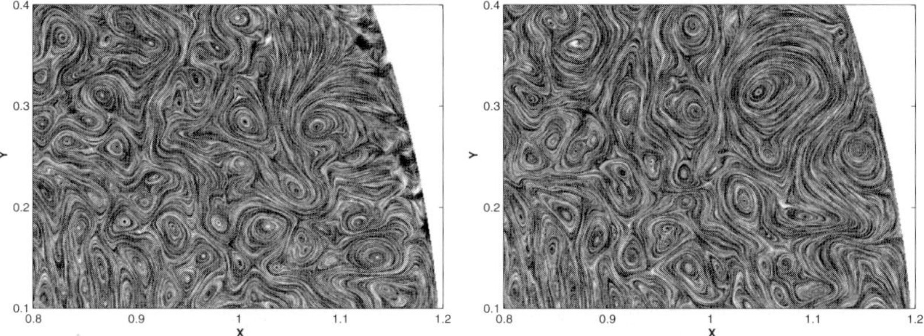

Figure 4. Visualization of perturbed velocity field for run B using the Line Integral Convolution method, for two moments of time.

4. Balbus & Hawley analysis

Results of run B are partly in contradiction with analysis made by Balbus et al. (1996), Hawley et al. (1999). Let as repeat briefly their arguments. Let us start from Euler equations in inertial reference frame (here $-\Omega_K r^2$ is the gravitational force)

$$\partial_t u + u \cdot \partial_r u + v \cdot \partial_\varphi u + \frac{1}{\sigma} \cdot \partial_r \Pi = -\Omega_K^2 r + \frac{v^2}{r},$$

$$\partial_t v + u \cdot \partial_r v + v \cdot \partial_\varphi v + \frac{1}{\sigma} \cdot \partial_\varphi \Pi = -\frac{uv}{r},$$

and let us introduce the velocity deviations from unperturbed flow $\mathbf{u}_0 = (0, r\Omega_K)$, $v = \Omega_K r + \tilde{v}$, $u = \tilde{u}$. We have now

$$\partial_t \tilde{u} + \tilde{u} \cdot \partial_r \tilde{u} + (\Omega_K r + \tilde{v}) \cdot \partial_\varphi \tilde{u} = \left[\frac{2\Omega_K r + \tilde{v}}{r}\right] \tilde{v} - \frac{1}{\sigma} \cdot \partial_r \Pi, \quad (6)$$

$$\partial_t \tilde{v} + \tilde{u} \cdot \partial_r \tilde{v} + (\Omega_K r + \tilde{v}) \cdot \partial_\varphi \tilde{v} = -\left[\frac{(\Omega_K r^2)' + \tilde{v}}{r}\right] \tilde{u} - \frac{1}{\sigma} \cdot \partial_\varphi \Pi, \quad (7)$$

and for the radial and azimuthal components of kinetic energy we have, after multiplication of (6),(7) by $\sigma\tilde{u}$ and $\sigma\tilde{v}$, respectively, and taking into account of continuity equation (2):

$$\partial_t \left(\tfrac{1}{2}\sigma\tilde{u}^2\right) + \tilde{u} \cdot \partial_r \left(\tfrac{1}{2}\sigma\tilde{u}^2\right) + (\Omega_K r + \tilde{v}) \cdot \partial_\varphi \left(\tfrac{1}{2}\sigma\tilde{u}^2\right) =$$

$$= \left[\frac{2\Omega_K r + \tilde{v}}{r}\right] \sigma\tilde{u}\tilde{v} - \tilde{u} \cdot \partial_r \Pi - \left(\partial_r \tilde{u} + \frac{\tilde{u}}{r} + \partial_\varphi \tilde{v}\right) \left(\tfrac{1}{2}\sigma\tilde{u}^2\right), \qquad (8a)$$

$$\partial_t \left(\tfrac{1}{2}\sigma\tilde{v}^2\right) + \tilde{u} \cdot \partial_r \left(\tfrac{1}{2}\sigma\tilde{v}^2\right) + (\Omega_K r + \tilde{v}) \cdot \partial_\varphi \left(\tfrac{1}{2}\sigma\tilde{v}^2\right) =$$

$$= -\left[\frac{(\Omega_K r^2)' + \tilde{v}}{r}\right] \sigma\tilde{u}\tilde{v} - \tilde{v} \cdot \partial_\varphi \Pi - \left(\partial_r \tilde{u} + \frac{\tilde{u}}{r} + \partial_\varphi \tilde{v}\right) \left(\tfrac{1}{2}\sigma\tilde{v}^2\right).$$

$$(8b)$$

After linearizing, we obtain

$$\frac{d}{dt}\left(\tfrac{1}{2}\sigma\tilde{u}^2\right) = \left[2\Omega_K\right]\sigma\tilde{u}\tilde{v} + \ldots, \tag{9}$$

$$\frac{d}{dt}\left(\tfrac{1}{2}\sigma\tilde{v}^2\right) = -\left[\frac{\ae^2}{2\Omega_K}\right]\sigma\tilde{u}\tilde{v} + \ldots, \tag{10}$$

where $d/dt = \partial_t + \tilde{u}\cdot\partial_r + (\Omega_K r + \tilde{v})\cdot\partial_\varphi$ – Lagrangian time derivative, $\ae^2 = 2\Omega_K(\Omega_K r^2)'/r = 2\Omega_K(2\Omega_K + \Omega_K' r)$. In Balbus et al. (1996) these equations are commented as:

> "...These simple relations are very general. The non-inertial rotational forces couple to the flow as though they were background velocity gradients. In each of equations (8), the first term on the right represents the interaction between the mean flow and the fluctuations. *Note that it enters equations (9) and (10) with opposite signs* for Rayleigh-stable flows. ... Equation (10) presents an interesting difficulty: since $d(r^2\Omega)/dr > 0$ in astrophysical disks, and viscosity acts only as a sink, outward turbulent transport ($\langle\rho u_r u_\varphi\rangle > 0$) would leave angular momentum fluctuations without a source from the mean flow."

So we can see that this analysis refers to Rayleigh criterion only, meantime a more thorough non-linear and non-axisymmetrical analysis is desirable.

5.　　Nonlinear analysis in local coordinate frame

Let us rewrite the Euler equations in Cartesian coordinates in rotating frame:

$$\partial_t u + (\mathbf{u}\cdot\nabla)\cdot u + \frac{1}{\sigma}\cdot\partial_x\Pi = -\partial_x\Phi_1 + 2\Omega_1\cdot v,$$

$$\partial_t v + (\mathbf{u}\cdot\nabla)\cdot v + \frac{1}{\sigma}\cdot\partial_y\Pi = -\partial_y\Phi_1 - 2\Omega_1\cdot u.$$

It is seen that $\Pi = \Pi_0 = \mathrm{const}$, $u = u_0 = -y/r^{3/2} + \Omega_1 y$, $v = v_0 = x/r^{3/2} - \Omega_1 x$ give the unperturbed solution. As usual, let us introduce the deviations from unperturbed solution $u = u_0 + \tilde{u}$, $v = v_0 + \tilde{v}$ to obtain

$$\partial_t\tilde{u} + (\mathbf{u}_0\cdot\nabla)\cdot\tilde{u} + (\tilde{\mathbf{u}}\cdot\nabla)\cdot u_0 + (\tilde{\mathbf{u}}\cdot\nabla)\cdot\tilde{u} + \frac{1}{\sigma}\cdot\partial_x\Pi = 2\Omega_1\cdot\tilde{v},$$

$$\partial_t\tilde{v} + (\mathbf{u}_0\cdot\nabla)\cdot\tilde{v} + (\tilde{\mathbf{u}}\cdot\nabla)\cdot v_0 + (\tilde{\mathbf{u}}\cdot\nabla)\cdot\tilde{v} + \frac{1}{\sigma}\cdot\partial_y\Pi = -2\Omega_1\cdot\tilde{u}.$$

These equations include u_0, v_0, $\partial_x u_0$, etc., defining as follows:

$$u_0(x,y) = -\frac{y}{r^{3/2}} + \Omega_1\cdot y, \qquad\qquad v_0(x,y) = \frac{x}{r^{3/2}} - \Omega_1\cdot x,$$

$$\partial_x u_0(x,y) = \tfrac{3}{2}\cdot\frac{xy}{r^{7/2}}, \qquad\qquad \partial_y u_0(x,y) = -\frac{1}{r^{3/2}} + \tfrac{3}{2}\cdot\frac{y^2}{r^{7/2}} + \Omega_1,$$

$$\partial_x v_0(x, y) = \frac{1}{r^{3/2}} - {}^3/_2 \cdot \frac{x^2}{r^{7/2}} - \Omega_1, \qquad\qquad \partial_y v_0(x, y) = -{}^3/_2 \cdot \frac{xy}{r^{7/2}}.$$

Let us shift the origin of coordinate system to an arbitrary point $(x_\star, y_\star)$ and introduce the polar coordinates $(\mathfrak{r}, \phi)$ ($\mathfrak{r} = 0$ corresponds to $(x_\star, y_\star)$): $x \to x_\star + \mathfrak{r} \cdot \cos \phi$, $y \to y_\star + \mathfrak{r} \cdot \sin \phi$. We are going to bring a local analysis with $\mathfrak{r} \ll r_\star$ so we can adopt $u_0(x, y)$, $v_0(x, y)$, $\partial_x u_0(x, y)$, etc. in the form $u_0(x_\star, y_\star)$, $v_0(x_\star, y_\star)$, $\partial_x u_0(x_\star, y_\star)$, etc. Moreover, with no loss of generality, we can put $y_\star = 0$ (this corresponds to a choice of $\phi = 0$ direction coinciding with $(0, 0) - (x_\star, y_\star)$ line) then $u_0(x_\star, y_\star) = \partial_x u_0(x_\star, y_\star) = \partial_y v_0(x_\star, y_\star) = 0$, and non-zero coefficients are

$$v_0 = \sqrt{\frac{GM}{r_\star}} - \Omega_1 \cdot r_\star, \quad \partial_y u_0 = -\frac{1}{r_\star^{3/2}} + \Omega_1, \quad \partial_x v_0 = -{}^1/_2 \cdot \frac{1}{r_\star^{3/2}} - \Omega_1,$$

where $r_\star = x_\star$. The equations in the new coordinate system read:

$$\partial_t \tilde{u} + X \cdot \partial_{\mathfrak{r}} \tilde{u} + Y \cdot \partial_\phi \tilde{u} = \left(\frac{Y}{\mathfrak{r}} + 2\Omega_1 - B \right) \tilde{v} + A \cdot \tilde{u} - \frac{1}{\sigma} \cdot \partial_{\mathfrak{r}} \Pi,$$

$$\partial_t \tilde{v} + X \cdot \partial_{\mathfrak{r}} \tilde{v} + Y \cdot \partial_\phi \tilde{v} = - \left(\frac{Y}{\mathfrak{r}} + 2\Omega_1 - C \right) \tilde{u} - A \cdot \tilde{v} - \frac{1}{\sigma} \cdot \partial_\phi \Pi,$$

where $\partial_{\mathfrak{r}} \equiv \partial / \partial \mathfrak{r}$, $\partial_\phi \equiv \mathfrak{r}^{-1} \partial / \partial \phi$, $V_\star = v_0$ and

$$X = \tilde{u} + V_\star \sin \phi, \qquad Y = \tilde{v} + V_\star \cos \phi, \qquad A = {}^3/_2 r_\star^{-3/2} \cos \phi \sin \phi,$$

$$B = \Omega_1 + r_\star^{-3/2} \left({}^3/_2 \sin^2 \phi - 1 \right), \qquad C = \Omega_1 + r_\star^{-3/2} \left({}^3/_2 \cos^2 \phi - 1 \right).$$

Multiplying these equations with $\tilde{u}$ and $\tilde{v}$, and combining to continuity equation we obtain the non-linear version of equations (8):

$$\frac{d}{dt} \left({}^1/_2 \sigma \tilde{u}^2 \right) = \left[\frac{\tilde{v} + V_\star \cos \phi}{\mathfrak{r}} + 2\Omega_1 - B \right] \sigma \tilde{u}\tilde{v} + \left[A \right] \sigma \tilde{u}^2 + \dots,$$

$$\frac{d}{dt} \left({}^1/_2 \sigma \tilde{v}^2 \right) = - \left[\frac{\tilde{v} + V_\star \cos \phi}{\mathfrak{r}} + 2\Omega_1 - C \right] \sigma \tilde{u}\tilde{v} - \left[A \right] \sigma \tilde{v}^2 + \dots.$$

It is seen that production of $\mathfrak{r}$- and ϕ-components of perturbed kinetic energy is not sign-definite more due to angular dependence. Moreover, the equations contain nonlinear terms $\sim \mathfrak{r}^{-1}$ corresponding to strong production of turbulence at small scales as it usually occurs in 2D case (see Batchelor, 1992).

6. Geostrophic approximation

Figure 5 shows the pressure distribution and perturbed velocity field for run B. It is seen that both cyclonic and anticyclonic vortices corresponds to *minimum* of pressure. This is in contradiction, in particular, to the standard rotational flow model (e.g., the Earth atmosphere) where cyclones produce low pressure (cloudy sky) and anticyclones produce high pressure (clear sky). To resolve let us come back to the non-linear analysis. Let as adopt that the vortex is steady-state and axially symmetrical: $\partial_t = \partial_\phi = \tilde{u} = 0$. Averaging and taking into account

$$\overline{\cos\phi} = \overline{\sin\phi} = \overline{\cos\phi\,\sin\phi} = 0, \qquad \overline{\sin^2\phi} = \overline{\cos^2\phi} = {}^1\!/_2,$$

we obtain

$$\sigma^{-1}\partial_{\mathfrak{r}}\Pi = \frac{\tilde{v}^2}{\mathfrak{r}} + \zeta\tilde{v}, \tag{11}$$

where $\zeta = 2\Omega_1 - B$ ($\zeta \approx +1$ for our calculations). Introducing the Rossby number

$$\mathfrak{Ro} = \frac{|\tilde{v}|}{2\Omega_1\mathfrak{r}},$$

we can rewrite (11) as

$$\sigma^{-1}\partial_{\mathfrak{r}}\Pi = \mathfrak{Ro} \cdot 2\,\Omega_1|\tilde{v}| + \zeta\tilde{v}.$$

The values of Rossby numbers for different systems are given in Tab. 1. Low values corresponds to the so-called geostrophic approximation when non-linear term disappears. In this case $\partial_{\mathfrak{r}}\Pi \propto \tilde{v}$ and $\tilde{v} > 0$ (cyclones) corresponds to $\partial_{\mathfrak{r}}\Pi > 0$ (low pressure in the center) while $\tilde{v} < 0$ (anticyclones) corresponds

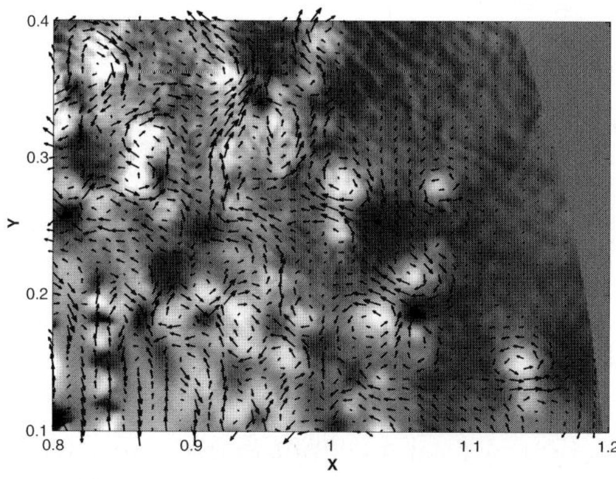

Figure 5. Pressure distribution and perturbed velocity field for run B. White color corresponds to the *minimum* of pressure.

Table 1. The values of Rossby numbers for different systems.

Earth's atmosphere	$\mathfrak{r} = 1000\,\text{km}$	$\tilde{v} = 10\,\text{m/s}$	$\Omega_\oplus = 7 \cdot 10^{-5}\,\text{s}^{-1}$	$\mathfrak{Ro} = 0.07$
Our calculations	$\mathfrak{r} < 0.05$	$\tilde{v} \simeq c_0 = 0.1 \div 0.5$	$\Omega_1 = 0.76$	$\mathfrak{Ro} > 1.3 \div 6.6$
Jupiter's Red Spot	$\mathfrak{r} = 30\,000\,\text{km}$	$\tilde{v} = 300\,\text{m/s}$	$\Omega_{2\!\!\!+} = 10^{-4}\,\text{s}^{-1}$	$\mathfrak{Ro} = 0.03$
Bath drain	$\mathfrak{r} = 10\,\text{cm}$	$\tilde{v} = 1\,\text{cm/s}$	$\Omega_\oplus = 7 \cdot 10^{-5}\,\text{s}^{-1}$	$\mathfrak{Ro} = 700$

to $\partial_r \Pi > 0$ (high pressure in the center). It is seen that our case is far from geostrophic approximation due to non-linear interaction. If we adopt the mean size of vortex as $\mathfrak{r} \simeq 0.05$ then it is seen that all cyclones $\tilde{v} > 0$ and almost all anticyclones $\tilde{v} < -0.05$ produce the minimum of pressure.

7.   The α_{SS} coefficient

Let us combine simple formulas

$$\nu = \alpha_{SS} \cdot c_s^2 \cdot \Omega_K^{-1} \qquad\qquad v_r = -{}^3\!/_2 \frac{\nu}{r}$$

to get

$$\alpha_{SS} = {}^2\!/_3 \frac{|v_r| \cdot \Omega_K \cdot r}{c_s^2}.$$

Figure 6 shows distribution of α_{SS} coefficient for zones where accretion takes place, i.e. where $v_r < 0$ (zones with $v_r > 0$ or decretion zones are shown by white color). Azimuthal averaging over all zones of accretion gives the distribution $\alpha_{SS}(r)$ with mean value $\bar{\alpha}_{SS} = 0.085$.

8.   Impact of grid resolution, Mach number, amplitude of perturbations, etc.

Apart from runs A and B we also conducted eight other run varying grid resolution, Mach number, amplitude of perturbation etc. Run C was conducted on the grid 1600×1600 with low temperature initial condition $c_0 = 0.2$. Runs D–I are variations of run C: runs D–F were conducted with extended computational domain in azimuthal direction to 2, 3, and 5 times, respectively. Run G was conducted with even lower initial temperature $c_0 = 0.1$. Runs H and I were conducted with decreased (30%) and increased (70%) amplitude of perturbation. Run J will be discussed below.

Figure 7 shows entropy and α_{SS} coefficient distributions for run C. It is seen that there is developed turbulence in subsonic zone and accretion occurs there.

Figures 8–10 show entropy and α_{SS} coefficient distributions for runs D–F and leads to the same conclusion. Nevertheless here we can see that the flow becomes turbulent in the part of supersonic zone as well.

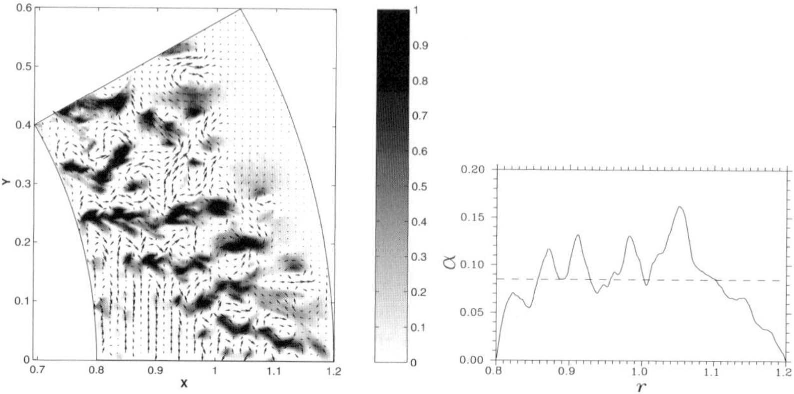

Figure 6.   Left: Distribution of α_{SS} coefficient for run B. White color corresponds to zones of decretion.   Right: Azimuthally averaged distribution of α_{SS} coefficient for run B. Dashed line shows the mean value $\bar{\alpha}_{SS} = 0.085$.

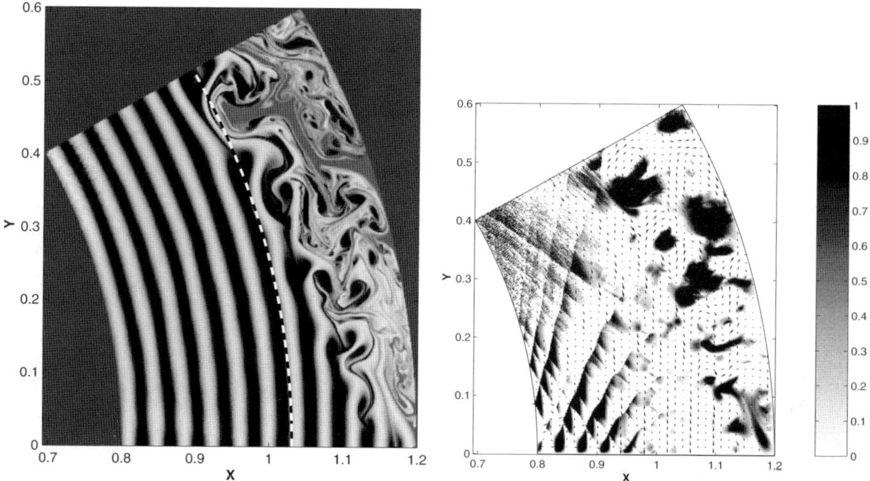

Figure 7.   Left: Entropy distribution for run C. Bold black-white line divides the supersonic and subsonic regions for unperturbed (initial) flow.   Right: Distribution of α_{SS} coefficient for run C.

Results for runs G–I show that very low temperature ($c_0 = 0.1$) and small value of the amplitude of perturbations ($A = 0.3$) produce a weaker turbulence, while large amplitude of perturbations ($A = 0.7$) produce a stronger turbulence. In Fig. 11 (run I with large amplitude of perturbations) we also can see that the flow becomes turbulent in the part of supersonic zone.

As about the value of α_{SS} coefficient for different run, it is seen that for all turbulent runs the value of α_{SS} lies in the range $0.05 \div 0.15$, or excluding very

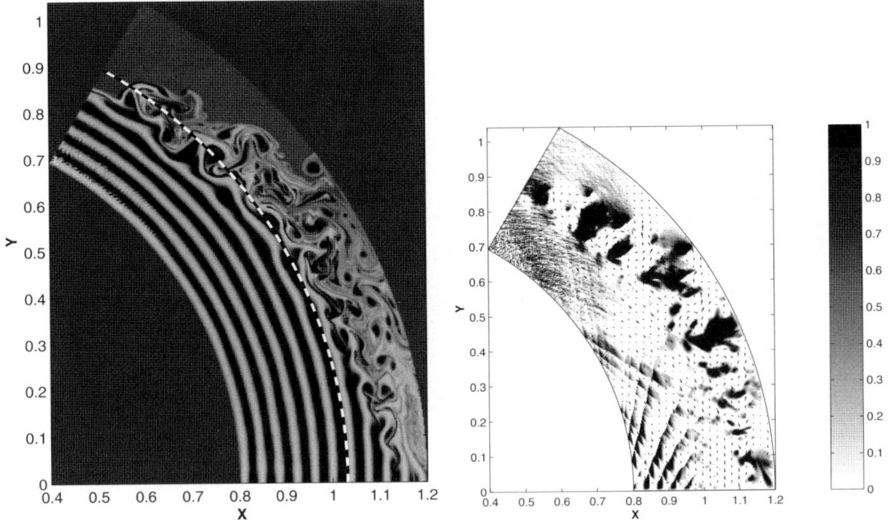

Figure 8. Left: Entropy distribution for run D. Bold black-white line divides the supersonic and subsonic regions for unperturbed (initial) flow. Right: Distribution of α_{SS} coefficient for run D.

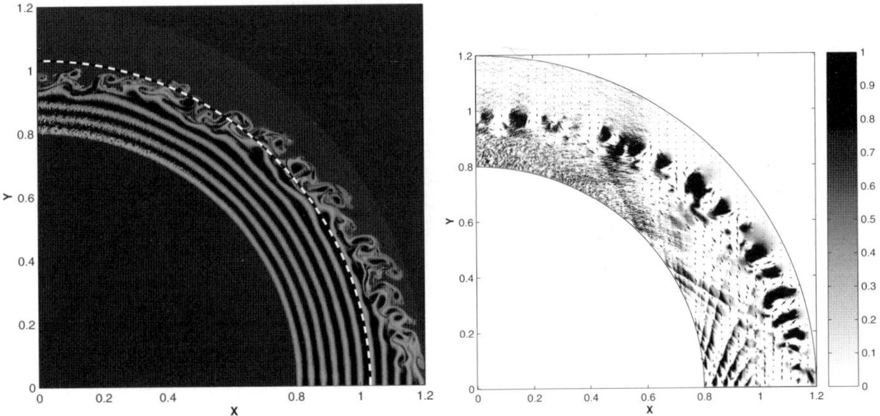

Figure 9. Left: Entropy distribution for run E. Bold black-white line divides the supersonic and subsonic regions for unperturbed (initial) flow. Right: Distribution of α_{SS} coefficient for run E.

cool and low-amplitude runs G, H, it lies in the range $0.08 \div 0.15$. This result is in the good agreement with theoretical and observational estimations of α_{SS} for accretion discs in dwarf novae during outburst (see Smak, 1984; Cannizzo, 2001; Ak et al., 2002).

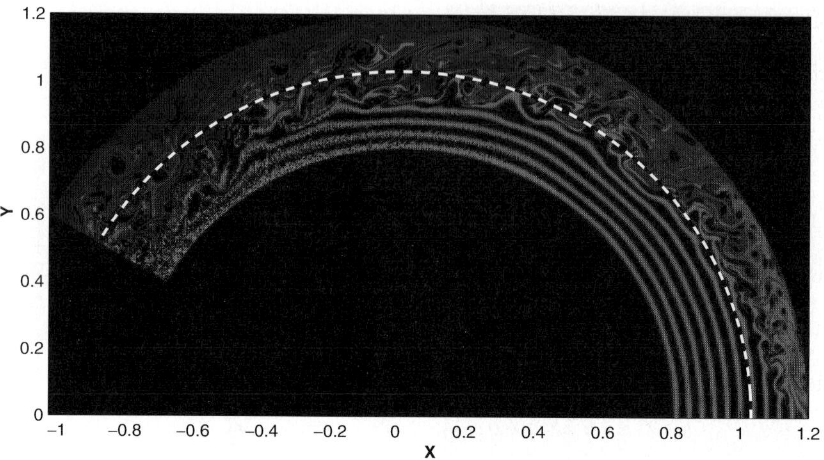

Figure 10. Entropy distribution for run F. Bold black-white line divides the supersonic and subsonic regions for unperturbed (initial) flow.

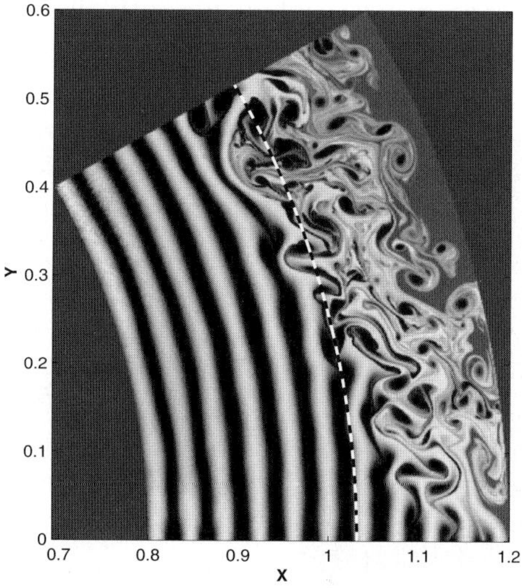

Figure 11. Entropy distribution for run I.

9. Discussion

So, our 2D inviscid, non-linear, compressible hydrodynamical simulations indicate that finite-amplitude perturbations lead to development of the pure hydrodynamical turbulence in accretion discs. To obtain the turbulence in the simulations the choice of special (comoving) reference frame should be done. The simulations have been performed up to the time, when quasi-stationary

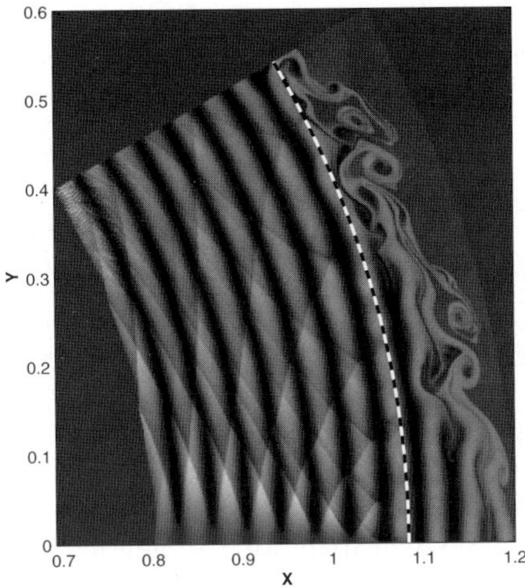

Figure 12.   Entropy distribution for run J.

turbulent state was reached with almost constant average characteristics. What is the reason of the development of turbulence? Without pretending on generality of our conclusions we would like to stress some possible ways.

First of all the development of turbulence in our calculations appears not to depend on the violation of Rayleigh criterion. To check we conducted an additional run J with gravitational potential $\Phi = -1/r^3$. The unperturbed distribution of angular velocity in this case is $\Omega_K \propto r^{-5/2}$ so $\partial_r(\Omega_K r^2) < 0$ and Rayleigh criterion is violated. The results for this run is shown in Fig. 12. It is seen that turbulence development is not stronger than in the case of Newtonian potential producing Rayleigh-stable flow (in fact, the turbulence is even weaker due to diminishing of subsonic zone).

Another channel of turbulence development is an instability dealing with the kink (sharp bend) of rotational curve (Nezlin et al., 1986; Baev et al., 1987). Of course, initial rotational curve is smooth in our case, but as turbulent develops it becomes non-smooth and contains kinks can lead to instabilities which in turns enforce the turbulence.

One more possible reason for development of turbulence is highly non-geostrophic character of the flow (i.e the Rossby number is not small or, in other words, the Rossby radius $|\tilde{u}|/2\Omega_1$ is large in comparison with the size of vortex). In this case the perturbations describe by the so called Fridman equation with vector non-linearity (Dolotin & Fridman, 1991; Fridman, 1989). The solution of this equation includes dipole vortices (modons) which also appear in our calculations (see Fig. 3). These dipole vortices can enforce the development of turbulence.

The turbulence also can develop due to existence of local regions of non-Keplerian angular velocities as it was pointed out by (Li et al., 2000; Lovelace et al., 1999).

And, finally, one more way to turbulence development is the so called transient growth of small scale perturbations (Fridman, 1989; Ioannou & Kakouris, 2001; Lominadze et al., 1988; Chagelishvili et al., 2003). This mechanism provides transient growth of energy of perturbations despite their asymptotic stability, and may lead in between to the value of the amplitude, where nonlinear effects make the system unstable.

10. Conclusions

Summarizing, we can assert that our results deals with non-linear finite-amplitude hydrodynamical shear instability which leads to the development of turbulence. We can draw the following conclusions from our study:

- Pure hydrodynamical turbulization of accretion disk is possible.

- The development of turbulence deals with non-linear finite-amplitude hydrodynamical shear instability.

- The value of α_{SS} coefficient is $0.05 \div 0.15$.

Acknowledgments

The work was partially supported by Russian Foundation for Basic Research (projects NN 05-02-16123, 05-02-17070, 05-02-17874, 06-02-16097), by Science Schools Support Program (project N 162.2003.2), by Federal Programme "Astronomy", by Presidium RAS Programs "Mathematical modelling and intellectual systems", "Nonstationary phenomena in astronomy", and by INTAS (grant N 00-491). Author thanks Russian Science Support Foundation for the financial support as well as D.V. Bisikalo, G.S. Bisnovatyi-Kogan, V.M. Chechetkin, A.M. Fridman, A.V. Koldoba and D. Molteni for useful discussions.

References

Papaloizou, J. and Pringle, J.E.: 1977, *MNRAS* **181**, 441

Lin, D.N.C, Papaloizou, J.C.B. and Savonije, G.J.: 1990, *ApJ* **365**, 748

Dgani, R., Livio, M. and Regev, O.: 1994, *ApJ* **436**, 270

Michel, F.C.: 1984, *ApJ* **279**, 807

Sawada, K., Matsuda, T. and Hachisu, I.: 1986, *MNRAS* **219**, 75

Spruit, H.C.: 1987, *A&A* **184**, 173

Syer, D. and Narayan, R.: 1993, *MNRAS* **262**, 749

Bisnovatyi-Kogan, G.S. and Blinnikov, S.I.: 1976, *Sov. Astron. Lett.* **2**, 191

Paczyński, B.: 1976, *Comments on Astrophysics* **6**, 95

Shakura, N.I., Sunyaev, R.A. and Zilitinkevich, S.S.: 1978, *A&A* **62**, 179

Lin, D.N.C. and Papaloizou, J.: 1980, *MNRAS* **191**, 37

Blandford, R.D. and Paine, D.G.: 1982, *MNRAS* **199**, 883

Lubow, S., Papaloizou, J.C.B. and Pringle, J.E.: 1994, *MNRAS* **268**, 1010

Cao, X. and Spruit, H.C.: 2002, *A&A* **385**, 289

Lynden-Bell, D.: 2003, *MNRAS* **341**, 1360

Velikhov, E.P.: 1959, *Zh. Exper. Theor. Phys.* **36**, 1398

Chandrasekhar, S.: 1981, *Hydrodynamic and Hydromagnetic Stability.* Dover, New York

Balbus, S.A. and Hawley, G.F.: 1991, *ApJ* **376**, 214

Winters, W.F., Balbus, S.A. and Hawley, J.F.: 2003, *MNRAS* **340**, 519

Goodman, J.: 1993, *ApJ* **406**, 596

Lubow, S.H., Pringle, J.E. and Kerswell, R.R.: 1993, *ApJ* **419**, 758

Cabot, W.: 1984, *ApJ* **277**, 806

Li, H., Finn, J.M., Lovelace, R.V.E. and Colgate, S.A.: 2000, *ApJ* **533**, 1023

Klahr, H.H. and Bodenheimer, P.: 2003, *ApJ* **582**, 869

Fridman, A.M., Boyarchuk, A.A., Bisikalo, D.V., Kuznetsov, O.A., Khoruzhii, O.V., Torgashin, Yu.M. and Kilpio, A.A.: 2003, *Phys. Lett. A* **317**, 181.

Lin, D.N.C and Papaloizou, J.C.B.: 1979, *MNRAS* **186**, 799

Lubow, S.H.: 1981, *ApJ* **245**, 274

Vishniac, E.T. and Diamond, P.: 1989, *ApJ* **347**, 435

Papaloizou, J.C.B. and Lin, D.N.C.: 1995, *ARA&A* **33**, 505

Balbus, S.A.: 2003, *ARA&A* **41**, 555

Lynden-Bell, D.: 1969, *Nature* **223**, 690

Shakura, N.I.: 1972, *AZh* **49**, 921 (*SvA* **16**, 756)

Lynden-Bell, D. and Pringle, J.E.: 1974, *MNRAS* **168**, 603

Shakura, N.I. and Sunyaev, R.A.: 1973, *A&A* **24**, 337

Bisnovatyi-Kogan, G.S. and Pogorelov, N.V.: 1997, *Astron. Astroph. Trans.* **12**, 263

Landau, L.D. and Lifshitz, E.M.: 1986, *Hydrodynamics.* Nauka, Moscow

Larichev, V.D. and Reznik, G.M.: 1976, *Dokl. Akad. Nauk USSR* **231**, 1077

Fridman, A.M.: 1988, *Astron. Tsirk.*, No. 1532; No. 1533

Dolotin, V.V and Fridman, A.M.: 1991, *Zh. Exper. Theor. Phys.* **99**, 3

Balbus, S.A., Hawley, J.F. and Stone, J.M.: 1996, *ApJ* **467**, 76

Hawley, J.F., Balbus, S.A. and Winters, W.F.: 1999, *ApJ* **518**, 394

Batchelor, G.K., 1992, *The Theory of Homogeneous Turbulence.* Cambridge Univ. Press, Cambridge

Smak, J.: 1984, *Acta Astron.* **34**, 161; 1998, *Acta Astron.* **48**, 677; 1999, *Acta Astron.* **49**, 391

Cannizzo, J.: 2001, *ApJ* **556**, 847; **561**, L175

Ak, T., Ozkan, M.T. and Mattei, J.A.: 2002, *A&A* **389**, 478

Nezlin, M.V., Polyachenko, V.L., Snezhkin, E.N., Trubnikov, A.S. and Fridman, A.M.: 1986, Sov. *Astron. Lett.* **12**, 213

Baev, P.V., Makov, Y.N. and Fridman, A.M.: 1987, *Sov. Astron. Lett.* **13**, 406

Fridman, A.M.: 1989, *Sov. Astron. Lett.* **15**, 487

Lovelace, R.V.E., Li, H., Colgate, S.A. and Nelson, A.F.: 1999, *ApJ* **513**, 805

Ioannou, P.J. and Kakouris, A., 2001, *ApJ* **550**, 931

Lominadze, D.G., Chagelishvili, G.D. and Chanishvili, R.G., 1988, *Sov. Astron. Lett.* **14**, 364

Chagelishvili, G.D., Zahn, J.-P., Tevzadze, A.G. and Lominadze, J.G., 2003, *A&A* **402**, 401

2D-SIMULATIONS OF SUBCRITICAL AND SUPERCRITICAL ACCRETION DISKS AROUND BLACK HOLES

Toru Okuda
Hakodate College, Hokkaido Univ. of Education
Hachiman-Cho 1-2, Hakodate 040-8567, Japan

Abstract   If Shakura-Sunyaev disk with the α-model for the viscosity is considered, very luminous accretion disks around black holes are thermally unstable and we except that these disks may exhibit various time-dependent behaviors of the luminosity and the mass outflow. For the observational appearances of such luminous disks, we examine subcritical and supercritical accretion disks around the black holes by two-dimensional radiation hydrodynamical calculations.

Keywords: black hole physics, accretion disks, hydrodynamics, instabilities

1.   Introduction

Disk accretion is an essential process for such dynamic phenomena as energetic X-ray sources, active galactic nuclei, and protostars. Since the early works by Pringle & Rees (1972) and Shakura & Sunyaev (1973), a great number of studies have been devoted to problems of the disk accretion onto gravitating objects. The standard accretion disk model (Shakura & Sunyaev, 1973) with geometrically thin disk have proven to be particularly successful in applications to cataclysmic variables. The standard model shows that very luminous disks become thermally unstable if the α-model for the viscosity is used. Under these conditions the standard model itself is broken down because the basic assumptions used will be invalid. And so it is is not definitely clear what's the final fate of such unstable disk. The accretion disks with supercritical ($\dot{m} \geq 1$) and subcritical ($1 \geq \dot{m} \geq 0.06$) accretion rates belong to the above category, where $\dot{m}$ is the accretion rate normalized to the Eddington critical rate M_{E} given by

$$L_{\mathrm{E}} = \eta \dot{M}_{\mathrm{E}} c^2 \tag{1}$$

Alexei M. Fridman, et al. (eds.), Astrophysical Disks; Collective and Stochastic Phenomena 259–274
© 2006 *Springer. Printed in the Netherlands*

where L_E is the Eddington luminosity, η the conversion efficiency of gravitational energy to radiation, and c the speed of light. Here we take η as 1/16.

For these luminous disks, the slim disk model has been proposed by Abramowicz et al. (1988), where strong advective cooling depresses the thermal instability and stable disk may be obtained. The thermal instability of the standard disk is interpreted by the relation between the accretion rate $\dot{m}$ and the surface density Σ at a fixed radius r. When the curve of $\dot{m} = \dot{m}(\Sigma)$ has a characteristic S-shape with three branches (lower, middle, and upper), the disks in the upper and lower branches are stable against thermal instability but in the middle branch it is unstable. If the disk is under the unstable middle branch, it is expected that the disk exhibits a limit-cycle behavior of luminosity. Taking account of advective cooling, Abramowicz et al. (1988) calculated the $\dot{m} - \Sigma$ curve at $r = 5r_g$ which shows the S-shape, from the transonic solutions of a black hole with $M_* = 10M_\odot$ and the viscosity parameter $\alpha = 10^{-3}$. Based on the slim disk model, for a typical case at the subcritical luminosity, Honma et al. (1991) and Szuszkiewicz & Miller (1998) carried time-dependent calculations of transonic accretion disks around a black hole with $\dot{m} = 0.06$ and $\alpha = 0.1$ and showed the limit-cycle behaviors of the disk luminosity. Recently, to understand the bursting behavior of the microquasar GRS 1915+105, Watarai & Mineshige (2003) also calculated time evolution of a luminous accretion disk around a black hole with the same parameter and concluded that the basic properties of the bursts of GRS 1915+105 can be explained by the limit-cycle oscillation model.

A unified $\Sigma - \dot{m}$ picture of accretion flows at a fixed radius r in the case of low viscosity was obtained (Abramowicz et al., 1995), where four classes of solutions, that is, Shakura-Sunyaev disk (SSD), Shapiro-Lightman-Eardley disk (SLE) (Shapiro et al., 1976), slim disk, and advection-dominated accretion flow (ADAF) are included. Gu & Lu (2000) predicted that the thermal instability of a radiation pressure-supported SSD can possibly trigger two different kinds of behavior of the flow, namely, the limit cycle and the SSD-ADAF transition. They show that, for low values of viscosity $\alpha \leq 0.5$, only the limit-cycle behavior can occur, while for large values of $\alpha \geq 0.5$, either of the limit-cycle behavior and the SSD-ADAF transition can occur.

However, two-dimensional calculations in relation to the above studies have not been sufficiently performed. Therefore we need to confirm these results for the luminous accretion disks, also by 2D hydrodynamical simulations. Here we examine some cases of the subcritical and supercritical accretion disks around the black holes by 2D radiation hydrodynamical calculations.

2.   Numerical Methods

2.1   Basic Equations

A set of relevant equations consists of six partial differential equations for density, momentum, and thermal and radiation energy. These equations include the full viscous stress tensor, heating and cooling of the gas, and radiation transport. The radiation transport is treated in the gray, flux-limited diffusion approximation (Levermore & Pomraning, 1981). We use spherical polar coordinates (r, ζ, φ), where r is the radial distance, ζ is the polar angle measured from the equatorial plane of the disk, and φ is the azimuthal angle. The gas flow is assumed to be axisymmetric with respect to Z-axis $(\partial/\partial\varphi = 0)$ and the equatorial plane . In this coordinate system, the basic equations for mass, momentum, gas energy, and radiation energy are written in the following conservative form (Kley, 1989):

$$\frac{\partial\rho}{\partial t} + \mathrm{div}(\rho\boldsymbol{v}) = 0, \tag{2}$$

$$\frac{\partial(\rho v)}{\partial t} + \mathrm{div}(\rho v\boldsymbol{v}) = \rho\left[\frac{w^2}{r} + \frac{v_\varphi^2}{r} - \frac{GM_*}{(r-r_\mathrm{g})^2}\right] - \frac{\partial p}{\partial r} + f_r + \mathrm{div}\boldsymbol{S}_r + \frac{1}{r}S_{rr}, \tag{3}$$

$$\frac{\partial(\rho rw)}{\partial t} + \mathrm{div}(\rho rw\boldsymbol{v}) = -\rho v_\varphi^2\tan\zeta - \frac{\partial p}{\partial\zeta} + \mathrm{div}(r\boldsymbol{S}_\zeta) + S_{\varphi\varphi}\tan\zeta + f_\zeta, \tag{4}$$

$$\frac{\partial(\rho r\cos\zeta v_\varphi)}{\partial t} + \mathrm{div}(\rho r\cos\zeta v_\varphi\boldsymbol{v}) = \mathrm{div}(r\cos\zeta\boldsymbol{S}_\varphi), \tag{5}$$

$$\frac{\partial\rho\varepsilon}{\partial t} + \mathrm{div}(\rho\varepsilon\boldsymbol{v}) = -p\,\mathrm{div}\boldsymbol{v} + \Phi - \Lambda, \tag{6}$$

and

$$\frac{\partial E_0}{\partial t} + \mathrm{div}\boldsymbol{F_0} + \mathrm{div}(\boldsymbol{v}E_0 + \boldsymbol{v}\cdot P_0) = \Lambda - \rho\frac{(\kappa+\sigma)}{c}\boldsymbol{v}\cdot\boldsymbol{F_0}, \tag{7}$$

where ρ is the density, $\boldsymbol{v} = (v, w, v_\varphi)$ are the three velocity components, G is the gravitational constant, M_* is the central mass, p is the gas pressure, ε is the specific internal energy of the gas, E_0 is the radiation energy density per unit volume, and P_0 is the radiative stress tensor. The subscript "0" denotes the value in the comoving frame and that the equations are correct to the first order of $\boldsymbol{v}/c$ (Kato et al., 1998). We adopt the pseudo-Newtonian potential (Paczyńsky & Wiita, 1980) in equation (3), where r_g is the Schwarzschild radius. The force density $\boldsymbol{f}_\mathrm{R} = (f_r, f_\zeta)$ exerted by the radiation field is given by

$$\boldsymbol{f}_\mathrm{R} = \rho\frac{\kappa+\sigma}{c}\boldsymbol{F_0}, \tag{8}$$

where κ and σ denote the absorption and scattering coefficients and $\boldsymbol{F_0}$ is the radiative flux in the comoving frame. $S = (\boldsymbol{S}_r, \boldsymbol{S}_\zeta, \boldsymbol{S}_\varphi)$ denote the viscous stress tensor. $\Phi = (S\,\nabla)\boldsymbol{v}$ is the viscous dissipation rate per unit mass.

The quantity Λ describes the cooling and heating of the gas, i.e., the energy exchange between the radiation field and the gas due to absorption and emission processes,

$$\Lambda = \rho c \kappa (S_* - E_0), \tag{9}$$

where S_* is the source function and c is the speed of light. For this source function, we assume local thermal equilibrium $S_* = aT^4$, where T is the gas temperature and a is the radiation constant. For the equation of state, the gas pressure is given by the ideal gas law, $p = R_G \rho T / \mu$, where μ is the mean molecular weight and R_G is the gas constant. The temperature T is proportional to the specific internal energy, ε, by the relation $p = (\gamma - 1)\rho\varepsilon = R_G \rho T / \mu$, where γ is the specific heat ratio. To close the system of equations, we use the flux-limited diffusion approximation (Levermore & Pomraning, 1981) for the radiative flux:

$$\boldsymbol{F_0} = -\frac{\lambda c}{\rho(\kappa + \sigma)}\,\mathrm{grad}\,E_0, \tag{10}$$

and

$$P_0 = E_0 \cdot T_{\mathrm{Edd}}, \tag{11}$$

where λ and T_{Edd} are the *flux-limiter* and the *Eddington Tensor*, respectively, for which we use the approximate formulas given in Kley (1989). The formulas fulfill the correct limiting conditions in the optically thick diffusion limit and the optically thin streaming limit, respectively.

Table 1.   Model parameters

Model	$\dot{m}$	α	$R_{\mathrm{max}}/r_{\mathrm{g}}$
1	0.06	0.1	1.2×10^4
2	0.2	0.8	1.2×10^4
3	3	0.001	440
4	10	0.001	5×10^4

For the kinematic viscosity, ν, we adopt a modified version (Papaloizou & Stanley, 1986) of the standard α-model. The modified prescription for ν is given by

$$\nu = \alpha\, c_{\mathrm{s}}\, \min\left[H_{\mathrm{p}}, H\right], \tag{12}$$

where c_{s} is the local sound speed, H the disk height, and $H_{\mathrm{p}} = p/|\mathrm{grad}\,p|$ the pressure scale height on the equatorial plane.

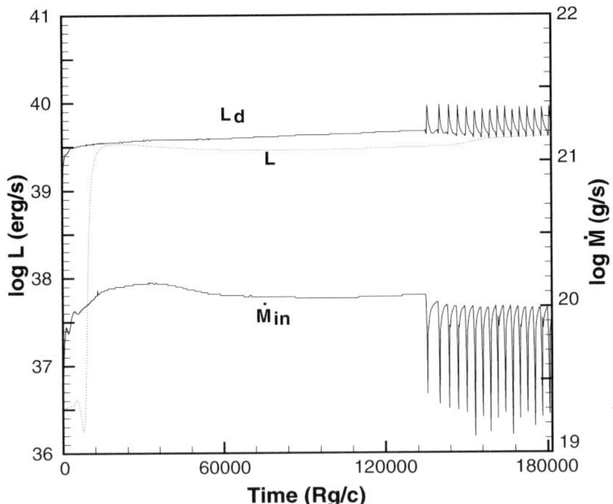

Figure 1.   Time evolution of total luminosity L, disk luminosity L_d, mass-inflow rate $\dot{M}_{in}$ swallowed into the black hole through the inner boundary for model 1.

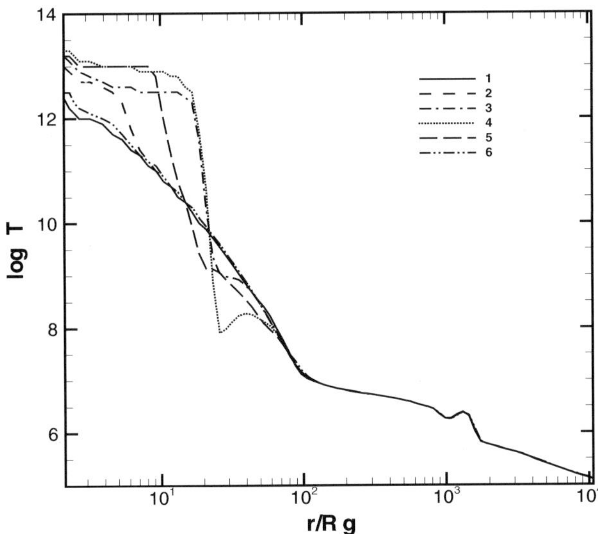

Figure 2.   Temperature profiles on the equatorial plane at $t \sim 1.6 \times 10^5 r_g/c$ for model 1. The numbers $1 - 6$ in legend denote six successive time number during one cycle of the QPO-like phenomena of the disk luminosity.

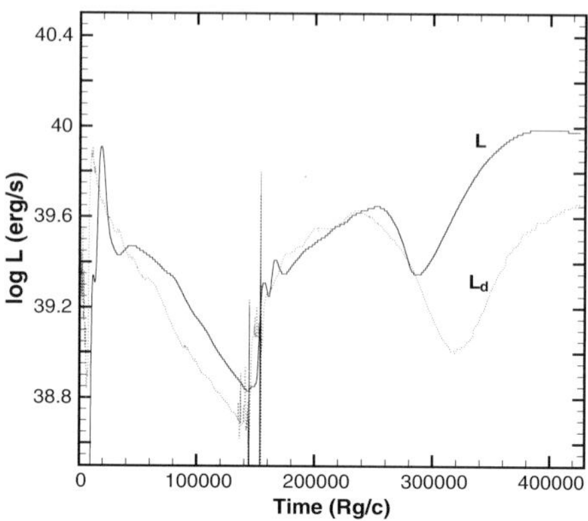

Figure 3.   Time evolution of total luminosity L and disk luminosity L_d for model 2.

2.2   Numerical Scheme

The set of partial differential equations (2)–(7) is numerically solved by a finite-difference method under adequate initial and boundary conditions. The numerical schemes used are basically the same as that described by Kley (1989) and Okuda et al. (1997). The methods are based on an explicit-implicit finite difference scheme. Grid points in the radial direction are spaced logarithmically, while grid points in the angular direction are equally spaced, but more refined near the equatorial plane, typically $\Delta\zeta = \pi/150$ for $\pi/2 \geq \zeta \geq \pi/6$ and $\Delta\zeta = \pi/300$ for $\pi/6 \geq \zeta \geq 0$.

2.3   Model Parameters

We consider a Schwarzschild black hole with mass $M_* = 10 M_\odot$ and take the inner boundary radius R_in of the computational domain as $2r_\mathrm{g}$. The model parameters used are listed in Table 1, where R_max is the outer boundary radius. As to the radial mesh size used in each model, $\Delta r/r$ is taken to be 0.035 and 0.1 in model 3 and in other models, respectively. Although the radial mesh sizes do not have a fine resolution to examine detailed disk structure, it is sufficient to examine the overall behaviors of the disk, the luminosity, and the mass outflow.

2.4   Initial and Boundary conditions

The initial conditions consist of a cold, dense, and optically thick disk and a hot, rarefied, and optically thin atmosphere around the disk. The initial disk at $r/r_\mathrm{g} \geq 3$ is approximated by the Shakura-Sunyaev standard model but, at $2 \leq r/r_\mathrm{g} \leq 3$, it is taken to be sub-Keplerian with an inner boundary of freely falling gas at $r/r_\mathrm{g} = 2$.

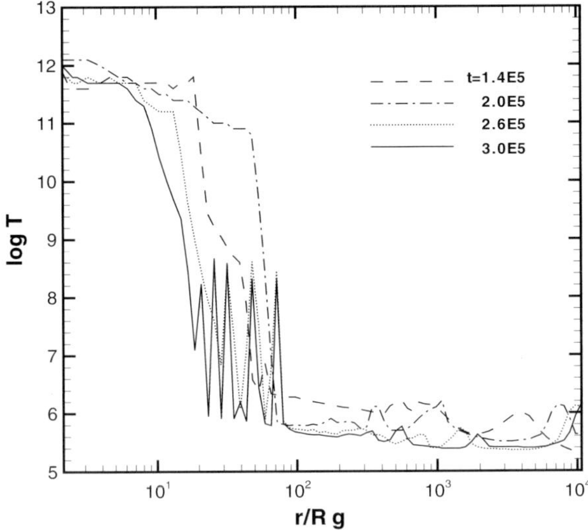

Figure 4.   Temperature profiles on the equatorial plane at $t = 1.4 \times 10^5, 2.0 \times 10^5, 2.6 \times 10^5$, and $3.0 \times 10^5 r_\mathrm{g}/c$ for model 2.

Physical variables at the inner boundary, except for the velocities, are given by extrapolation of the variables near the boundary. However, we impose limited conditions that the radial velocities are given by a free-fall velocity and the angular velocities are zero. On the rotational axis and the equatorial plane, the meridional tangential velocity w is zero and all scalar variables must be symmetric relative to these axes. The outer boundary at $r = R_\mathrm{max}$ is divided into two parts. One is the disk boundary through which matter is entering from the outer disk. At the outer-disk boundary we assume a continuous inflow of matter with a constant accretion rate $\dot{M}$. The other is the outer boundary region above the accretion disk. We impose free-floating conditions on this outer boundary and allow for outflow of matter, whereas any inflow is prohibited here. We also assume the outer boundary region above the disk to be in the optically-thin limit, $|\boldsymbol{F_0}| \rightarrow cE_0$.

3.   Subcritical Cases

3.1   Case of $\dot{m} = 0.06$ and $\alpha = 0.1$

This is a typical case at the subcritical luminosity which has been examined by one-dimensional time-dependent calculations of vertically-integrated models of the transonic accretion disks around black holes (Honma et al., 1991; Szuszkiewicz & Miller, 1998; Watarai & Mineshige, 2003) and they showed limit-cycle behaviors of the disk luminosity connected with thermal

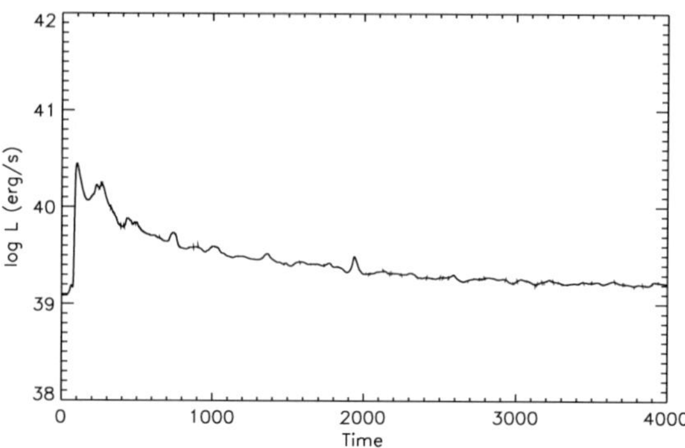

Figure 5.   Time evolution of luminosity L for model 3, where time is given in units of the Keplerian orbital period at the inner boundary $2r_{\mathrm{g}}$.

instability, depending on the viscosity description. However, these results have not been confirmed yet by two-dimensional simulations.

Figure 1 shows the time evolutions of the total luminosity L emitted from the outer boundary radius, the disk luminosity L_{d} emitted from the disk surface, and the mass-inflow rate $\dot{M}_{\mathrm{in}}$ swallowed into the black hole through the inner edge of the disk. L and L_{d} are given by calculating $\int \mathbf{F_0}d\mathbf{S}$ where the surface integral is taken over the outer boundary surface and the disk surface, respectively. Although it is difficult for us to specify correctly the disk surface, particularly in the inner disk, we defined it tentatively as a location where the density decreases to a tenth of the central density on the equatorial plane. If the atmosphere above the disk is optically thin, L_{d} must agree with L. L_{d} has an initial value of $\sim 2 \times 10^{39}$ erg s^{-1} from the initial model, while L is initially small and becomes comparable to L_{d} at the phase t $= 1.2 \times 10^{4} r_{\mathrm{g}}/c \sim R_{\mathrm{max}}/c$ which is the radiation transit time from the disk surface to the outer boundary radius. As far as the evolutionary time of $\sim 2 \times 10^{5} r_{\mathrm{g}}/c$ (~ 20 sec) is concerned, we never find remarkable variabilities of L and L_{d} corresponding to the limit-cycle behavior. This may mean that the evolutionary times are not sufficiently long to denote the limit-cycle of the luminosity, because one-dimensional calculations with the same viscosity prescription show a much longer limit-cycle period (~ 780 sec) than $\sim 2 \times 10^{5} r_{\mathrm{g}}/c$ (10 sec). Instead, we find another quasi-periodic phenomena with a period of ~ 0.25 sec of the disk luminosity and the mass-inflow rate which vary regularly by a factor 2 and 6, respectively. The temperature profiles on the equatorial plane at $t \sim 1.6 \times 10^{5} r_{\mathrm{g}}/c$ are shown in Figure 2, where the numbers of $1-6$ in legend denote six successive time number during one cycle in the QPO-like phenomena of the disk luminosity and a small temperature peak beyond $r/r_{\mathrm{g}} \sim 10^{3}$ shows the outgoing heating wave generated initially near the inner edge of the disk. From examinations of the

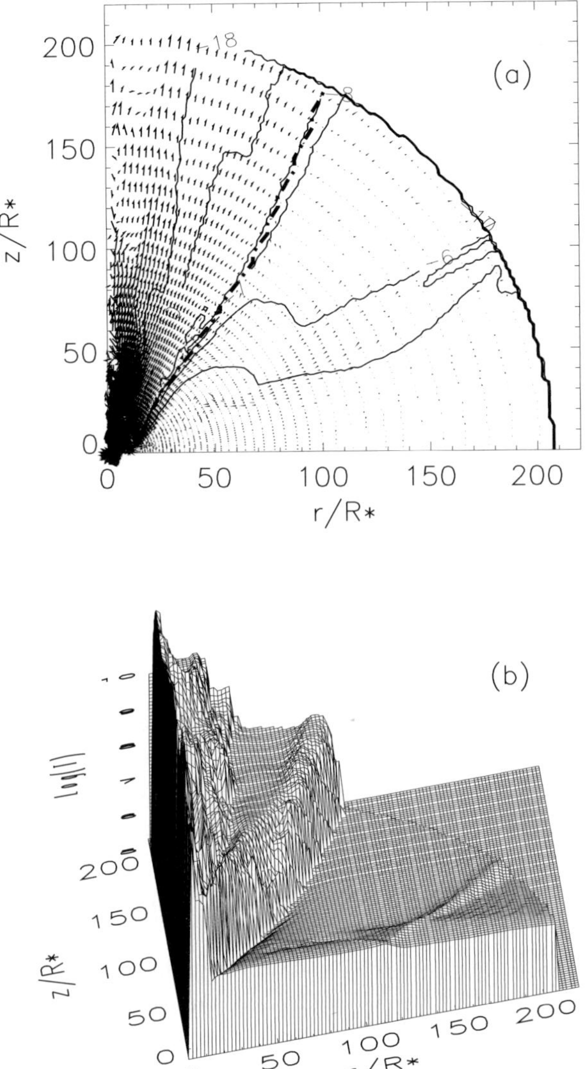

Figure 6.   Velocity vectors and contours of the density, ρ (g cm^{-3}), in logarithmic scale (a) and a bird's-eye view of the gas temperature T (K) (b), on the meridional plane at the evolutionary time $t = 1.4 \times 10^4 r_\mathrm{g}/c$ for model 3. The density contours are denoted by the labels log ρ = -5, -6, -7, -8, -10, and -18, and the velocity vectors show the maximum velocity $0.4c$ at $\zeta \sim 80$ and $r/r_\mathrm{g} \sim 400$. The thick dot-dashed line in (a) shows the disk boundary between the disk and the high-velocity region.

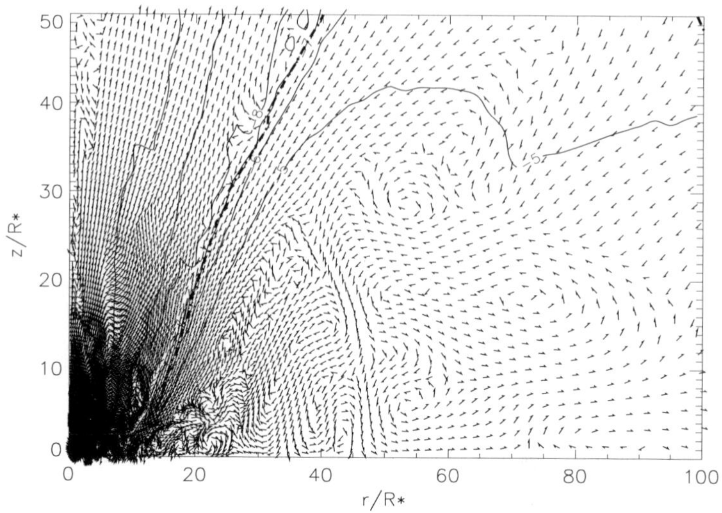

Figure 7. Unit-velocity vectors and the density contours of log ρ = -5, -6, -7, -8, -10, and -18 in the inner disk and the surrounding high-velocity jets region for model 3, where the dot-dashed line denotes the disk boundary between the disk and the high-velocity jets region. A transition between the inner advection-dominated zone and the outer convection-dominated zone occurs at $r \sim 10R_* = 20r_{\rm g}$.

temperatures and Mach numbers of the flow on the equatorial plane, we find that the QPO-like phenomena are attributed to existence of heating and cooling waves which are formed at $10 \leq r/r_{\rm g} \leq 30$ and oscillate at $2 \leq r/r_{\rm g} \leq 30$. The unstable waves strongly influence on the mass-inflow rate and $\dot{M}_{\rm in}$ also oscillates. The periodic phenomena of $L_{\rm d}$ and $\dot{M}_{\rm in}$ are very interesting from a QPO viewpoint, as is observed in the black hole candidate GRS 1915+105. GRS 1915+105 exhibits various types of QPOs with frequencies ranging from $\sim$0.001 – 10 Hz to 67 Hz which are classified as (1) the low-frequency QPO ($\sim$0.001 – 0.01 HZ), (2) the intermediate frequency QPO ($\sim$1 – 10 Hz), and (3) the high frequency QPO ($\sim$67 Hz) (Chakrabarti & Manickam, 2000). The QPO with $\sim$0.25 sec in model 1 corresponds to the intermediate frequency QPO ($\sim$4 Hz).

3.2   Case of $\dot{m}$ = 0.2 and α = 0.8

This model of $\dot{m}$ = 0.2 and α = 0.8 belongs to a case which suggests a limit-cycle behavior (Gu & Lu, 2000) but it has not been confirmed by 1D and 2D time-dependent simulations. Figure 3 shows the time evolution of L and $L_{\rm d}$, which exhibit a limit-cycle behavior with period of $1.5 \times 10^5 r_{\rm g}/c$ ($\sim$15 sec). This is the first confirmation of a limit-cycle behavior of the slim disk by a two-dimensional time-dependent hydrodynamical calculation. Due to the large α, the period of the limit cycle is expected to be small as $\sim$15 sec, compared with the period $\sim$780 sec calculated in model 1 (Szuszkiewicz & Miller, 1998).

Figure 4 denotes the temperature profiles on the equatorial plane at $t = 1.4 \times 10^5, 2.0 \times 10^5, 2.6 \times 10^5$, and $3.0 \times 10^5 r_{\mathrm{g}}/c$ for model 2. This shows the temperature evolutions during one cycle. The relation between the time evolutions of the temperature profiles and the luminosity curve is similar to that obtained in 1D-simulations by Honma et al. (1991) and Szuszkiewicz & Miller (1998). In model 2, the high temperature region with $\sim 10^8 - 10^{12}$ K, where is radiation-pressure dominant and optically thin, reaches about $\sim 100 r_{\mathrm{g}}$ and then backs towards the inner edge of the disk. Another heating waves with low temperatures with $\sim 10^6$ K propagate outward beyond $100 r_{\mathrm{g}}$ where is optically thick.

4.   Supercritical Cases

4.1   Case of $\dot{m} = 3$ and $\alpha = 10^{-3}$

Model 3 is the same as BH-1 in the previous disk model (Okuda, 2002). Figure 5 shows the time evolution of the total luminosity L for model 3. After an initial sharp rise of the luminosity, the luminosity curve descends gradually toward a steady state value. The luminosity L at the final phase is 1.6×10^{39} erg s^{-1} and L/L_{E} is $\sim$1. The disk is convectively unstable.

In model 3, a geometrically and optically thick convection-dominated disk with a large opening angle of $\sim 60°$ to the equatorial plane and rarefied, very hot, and optically thin high-velocity jets region around the disk are formed (figure 6). The thick accretion flow near to the equatorial plane consists of two different zones: an inner advection-dominated zone, in which the net mass-inflow rate, $\dot{M}_{\mathrm{in}}$, is very small, and an outer convection-dominated zone, in which $\dot{M}_{\mathrm{in}}$ increases with increasing radii. A transition between the inner advection-dominated zone and the outer convection-dominated zone occurs at $r \sim 10 R_*(= 20 r_{\mathrm{g}})$. The accreting matter, which is carried to the transition region by convection, partly diverts into the high-velocity jets and partly flows into the inner advection-dominated zone. Figure 7 shows the flow features in the inner disk and the surrounding high-velocity region, where the velocities are indicated by unit vectors and the thick dot-dashed line shows the disk boundary. One of the remarkable features of luminous accretion disks is convective phenomena in the inner region of the disk. The convective motions are clearly found in this figure, and there appear more than a dozen of convective cells. The high-velocity region along the rotating axis is divided into two characteristic regions by the funnel wall, a barrier where the effective potential due to the gravitational potential and the centrifugal one vanishes.

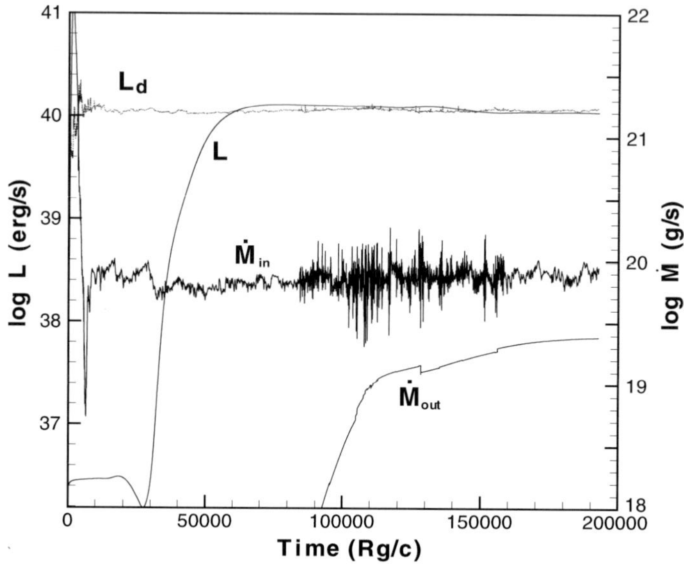

Figure 8. Time evolution of luminosity L, mass outflow-rate $\dot{M}_\mathrm{out}$ from the system, and mass inflow-rate $\dot{M}_\mathrm{in}$ swallowed into the black hole for model 4, where time is shown in units of r_g/c.

4.2 Case of $\dot{m} = 10$ and $\alpha = 10^{-3}$

Figure 8 show the time evolutions of the total luminosity L emitted from the outer boundary, the disk luminosity L_d emitted from the disk surface, the mass outflow rate $\dot{M}_\mathrm{out}$ from the outer boundary, and the mass inflow rate $\dot{M}_\mathrm{in}$ swallowed into the black hole from the inner boundary for model 4. Since the disk luminosity L_d and the mass-inflow rate $\dot{M}_\mathrm{in}$ are strongly dependent on the initial structure of the inner disk, they fluctuate largely at the initial evolutionary phases but settle to their steady state values in a time of $\sim 10^4 r_\mathrm{g}/c$. Here $L_\mathrm{d} = 1.2 \times 10^{40}$ erg s^{-1} and $\dot{M}_\mathrm{in} \sim 8 \times 10^{19}$ g s^{-1}. The total luminosity L attains to the disk luminosity L_d near a phase $t = R_\mathrm{max}/c \sim 5 \times 10^4 r_\mathrm{g}/c$ when radiation from the disk arrives at the outer boundary. On the other hand, after $t = R_\mathrm{max}/0.4c \sim 10^5 r_\mathrm{g}/c$ when the jet with $0.4c$ arrives at the outer boundary, the mass outflow begins and gradually settles down to a steady state value of 2.5×10^{19} g s^{-1}.

The evolutionary features of the high velocity jet are shown in figure 9. The high velocity jet propagates vertically to the equatorial plane. The jet expands gradually far from the rotational axis with increasing times. After the phase in figure 9-b, the jet arrives at the outer boundary in the polar direction but the anisotropic nature of the jet seems to be remained even after it passed through the outer boundary. Figure 10 shows velocity vectors and contours of

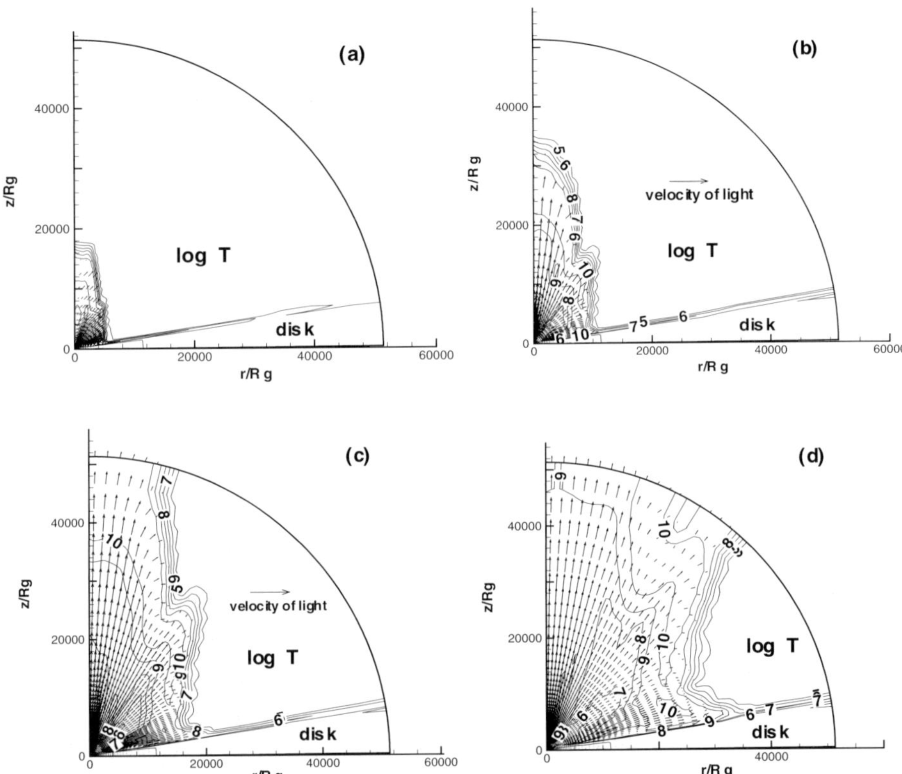

Figure 9.   Velocity vectors and temperature contours in logarithmic scale on the meridional plane at $t = 3 \times 10^4 (a), 5.7 \times 10^4 (b), 1.0 \times 10^5 (c), 1.6 \times 10^5 (d)$ for model 4. The contour lines with labels of $\log T = 5, 6, 7, 8, 9, 10$, and 10.5 are shown. The reference vector of light is shown by a long arrow. The initially anisotropic high velocity jet (a) generated in the inner region of the disk evolves into the isotropic outflow (d) in far distant region.

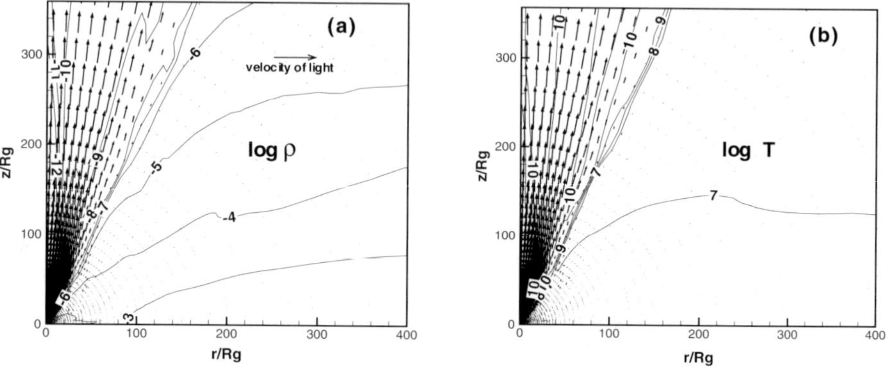

Figure 10.   Velocity vectors and contours of the density (a) and the temperature (b) in the inner region of the disk and the outflow for model 4.

the density and the temperature in the inner region of the disk at $t = 10^5 r_{\mathrm{g}}/c$ for model 4. Here we find characteristic features of the inner disk and the high velocity jet region as is found in model 3.

5.   Summary and Discussion

We have examined subcritical and supercritical accretion disks around the stellar black holes. These luminous disks have been discussed in terms of the thermal instability of the slim disk. In model 1 with $\dot{m} = 0.06$ and $\alpha = 0.1$ at the subcritical luminosity, the transonic solutions of the disk and the 1D time-dependent studies show that the model exhibits a limit-cycle behavior of the luminosity. However, our 2D-simulations of the model do not show such periodic behavior as is found in 1D simulations. It is suggested that the present result might be a transient behavior in the course of a much longer time evolution, that is, the evolutionary times performed here may be not sufficiently long to examine the long evolution. Instead of the limit-cycle behavior, the model exhibits another periodic phenomena with a short period of $\sim$0.15 sec of the disk luminosity and the mass-inflow rate. This may be related to the QPO phenomena with the intermediate frequency as is found in the microquasar GRS 1915+105.

In model 2 with $\dot{m} = 0.2$ and $\alpha = 0.8$ at the subcritical luminosity, a limit-cycle behavior with a period $\sim$15 sec of the disk luminosity is obtained. This is the first confirmation by a 2D-simulation for the limit-cycle behavior predicted from the slim disk model (Gu & Lu, 2000), although we do not show many cycles of the periodic luminosity because it need too many computational times in the 2D-simulations.

The disk and the jet in model 3 seem to attain almost to their steady states. However, Teresi et al. (2004) suggested that these results in model 3 might be a transient behavior during a long disk evolution. Actually from the result of the transonic solutions by Abramowicz et al. (1988), the disk with $\dot{m} = 3$ at $r/r_{\mathrm{g}} = 5$ belongs marginally to the middle branch in the $\dot{m}$ - Σ curve and should be thermally unstable. So we need to examine the time evolution of this model over a long time-scale. On the other hand, the inner disk with $\dot{m} = 10$ in model 4 corresponds to the stable upper branch. Therefore it is reasonable that the steady states of the disk and jet in model 4 are obtained.

References

Abramowicz, M. A., Czerny, B., Lasota, J. P., and Szuszkiewicz, E. Slim Accretion Disks. *The Astrophsical Journal*, 332, 646-658, 1988.

Abramowicz, M. A., Chen, X., Kato, S., Lasota, J.-P., and Regev, O. Thermal Equilibria of Accretion Disks. *The Astrophsical Journal*, 438, L37-L39, 1995.

Chakrabarti, S. K. and Manickam, S. G. Correlation among Quasi-Periodic Oscillation Frequencies and Quiescent-State Duration in a Black Hole Candidate GRS 1915+105. *The Astrophsical Journal*, 531, L41-L44, 2000.

Gu, W.-M. and Lu, J.-F. Bimodal Accretion Disks: Shakura-Sunyaev Disk-Advection-dominated Accretion Flow Transitions. *The Astrophsical Journal*, 540, L33-L36, 2000.

Honma, F., Matsumoto, R., and Kato, S. Nonlinear Oscillations of Thermally Unstable Slim Accretion Disks around a Neutron Star or a Black Hole. *Publications of the Astronomical Society of Japan*, 43,147-168, 1991.

Kato, S., Fukue, J., and Mineshige, S. Black Hole Accretion Disks. Kyoto University Press, Kyoto, 1998.

Kley, W. Radiation Hydrodynamics of the Boundary Layer in Accretion Disks. I - Numerical Methods. *Astronomy and Astrophysics*, 208, 98-110, 1989.

Levermore, C. D. and Pomraning, G. C. A Flux-Limited Diffusion Theory . *The Astrophysical Journal*, 248, 321-334, 1981.

Okuda, T., Fujita, M., and Sakashita, S. Two-Dimensional Accretion Disk Models: Inner Accretion Disks of FU Orionis Objects. *Publications of the Astronomical Society of Japan*, 49, 679-697, 1997.

Okuda, T. Super-Eddington Black-Hole Models for SS 433. *Publications of the Astronomical Society of Japan*, 54, 253-266, 2002.

Paczyńsky, B. and Wiita, P. J. Thick Accretion Disks and Supercritical Luminosities. *Astronomy and Astrophysics*, 88, 23-31, 1980.

Papaloizou, J. C. B. and Stanley, G. Q. G. The Structure and Stability of the Accretion disc Boundary Layer. *Monthly Notices of the Royal Astronomical Society*, 220, 593-610, 1986.

Pringle, J. E. and Rees, M. J. Accretion disc Models for Compact X-Ray Sources. *Astronomy and Astrophysics*, 21, 1-9, 1972.

Shakura, N. I. and Sunyaev, R. A. Black Holes in Binary Systems: Observational Appearance. *Astronomy and Astrophysics*, 24, 337-355, 1973.

Shapiro, S. L., Lightman, A. P., and Eardley, D. M. A Two-Temperature Accretion Disk Model fo Cygnus X-1 - Structure and Spectrum. *The Astrophysical Journal*, 204, 187-199, 1976.

Szuszkiewicz, E. and Miller, J. C. Limit-Cycle Behaviour of Thermally-Unstable Accretion Flows onto Black Holes. *Monthly Notices of The Royal Astronomical Society*, 298, 888-896, 1998.

Teresi, V., Molteni, D., and Toscano, E. SPH simulations of Shakura-Sunyaev instability at intermediate accretion rates. *Monthly Notices of The Royal Astronomical Society*, 348, 361, 2004.

Watarai, K. and Mineshige, S. Model for Relaxation Oscillations of a Luminous Accretion Disk in GRS 1915+105: Variable Inner Edge. *The Astrophysical Journal*, 596, 421-428, 2003.

SEPARATE CIRCUMNUCLEAR STELLAR AND GASEOUS DISKS IN DISK GALAXIES

O. K. Sil'chenko
Sternberg Astronomical Institute of Moscow State University,
Guest Investigator of the UK Astronomical Data Centre
olga@sai.msu.su

Abstract   A careful examination of the very central parts of nearby disk galaxies reveals often a presence of compact elongated structures which seem to be separated from the outer large-scale galactic components. By using a complex photometric and kinematical analysis, we can prove that they may be circumnuclear stellar disks, sometimes inclined with respect to the main symmetry plane of galaxies. Similarly inclined circumnuclear gaseous disks are also often present; among those, a spectacular phenomenon of inner polar rings in normal spiral galaxies is discovered by us several years ago. We argue that a hypothesis of external origin of the inner polar rings, say, by gas accretion from a satellite galaxy, meets substantial difficulties when a large-scale gas distribution in the main symmetry plane also is observed at the same time. We discuss some qualitative intrinsic mechanisms to put gas onto polar orbits and stress a need for 3D hydrodynamic simulations of gaseous disk evolution.

Keywords: galaxies: structure, evolution, kinematics and dynamics

1.   Introduction

A rapid progress in observational technique results in continuing discovery of quite new structures even in nearby galaxies.

A classical point of view starting from Hubble (1926) is that galaxies have two main global components: a large-scale stellar spheroid (bulge) and a large-scale stellar disk. Bulges are thought to be three-dimensional stellar bodies which projected surface brightness obeys to a de Vaucouleurs' law. Stellar disks are rather thin so often are considered as two-dimensional structures; their surface brightness falls with a radius as an exponential. Different morphological types of galaxies are characterized by varying relation between a spheroid and a disk: an elliptical galaxy, according to Hubble, represents a

Alexei M. Fridman, et al. (eds.), Astrophysical Disks; Collective and Stochastic Phenomena 275–290
(c) 2006 Springer. Printed in the Netherlands

naked spheroid, whereas early-type disk galaxies have bulges more luminous than the disks, and late-type spirals are dominated by disks.

However, when accuracy of photometric measurements approached 1% which has become possible with modern CCD detectors, even homogeneous elliptical galaxies were found to possess subtle structures. As azimuthal Fourier decomposition has shown, bright elliptical galaxies have mostly boxy isophotes, with a negative fourth cosine term contribution, and moderate-luminosity ellipticals, on the contrary, demonstrate pointed, or 'disky', shape of the isophotes. Perhaps it means that a large part of elliptical galaxies have compact stellar disks, with scalelengths of a half of kiloparsec, which are embedded into their large spheroids (Nieto et al., 1991).

And what about early-type *disk* galaxies? Has something been found within their large bulges? By photometric methods, separate inner stellar disks were detected in lenticular and early-type spiral galaxies more than once. I can mention here well-studied cases of the Sa NGC 4594 (Burkhead, 1991; Hes & Peletier, 1993) and of the very nearby S0 NGC 3115 (Scorza & Bender, 1995). Seifert & Scorza (1996) analyzed a sample of 15 edge-on lenticulars and have found separate inner stellar disks in 50% of them. After subtracting model bulges with exactly elliptical isophotes, they obtained residual brightness profiles consisting of two exponential disks with a gap between them. The scalelengths of the inner and outer disks were very different, with typical scalelengths of the inner disks of a few hundred parsecs. Our early long-slit kinematical study of a sample of spiral galaxies (Afanasiev et al., 1988a, Afanasiev et al., 1988b, Afanasiev et al., 1991, Afanasiev et al., 1992) has revealed prominent circumnuclear circular rotation-velocity peaks in 30% of all cases. A proximity of the peaks to the nuclei signified the compactness of the central gravitating structures. We then suggested that these structures were stellar disks inside the large-scale bulges (Afanasiev, Sil'chenko, & Zasov, 1989).

The best way to look for kinematically decoupled subsystems and to clarify their structure is 2D spectroscopy. When you see something elongated in the center of a galaxy after subtracting a bulge, it may be a highly inclined disk or a bar. To make a choice between these alternatives, one needs a 2D velocity field. A behaviour of isovelocities differs strongly in these two cases: within a bar the isovelocities align with the bar, and within a disk the zero-velocity line is always orthogonal to the isophote major axis. Since the integral-field spectroscopy provides us with two-dimensional line-of-sight velocity fields, we are now able to analyse both character of rotation and central structure of the galaxies. If we have an axisymmetric mass distribution and rotation on circular orbits, the direction of maximum central line-of-sight velocity gradient (we shall call it 'kinematical major axis') should coincide with the line of nodes as well as the photometric major axis should; whereas in a case of triaxial potential the isovelocities align with the principal axis of the ellipsoid, and generally the kinematical and photometric major axes diverge showing turns in opposite senses with respect to the line of nodes (Monnet et al., 1992; Moiseev & Mustsevoy, 2000). In a simple case of cylindric (disk-like) rotation we have a convenient analytical expression for the azimuthal dependence of central line-of-sight velocity gradients within the area of solid-body rotation:

$$dv_r/dr = \omega \sin i \cos (PA - PA_0),$$

where ω is the deprojected central angular rotation velocity, i is the inclination of the rotation plane, and PA_0 is the orientation of the line of nodes, coinciding in the case of an axisymmetric ellipsoid (or a thin disk) with the photometric major axis. So by fitting azimuthal variations of the central line-of-sight velocity gradients with a cosine curve, we can determine the orientation of the kinematical major axis from its phase and the central angular rotation velocity from its amplitude. It is our main tool of kinematical analysis. A finite spatial resolution (atmospheric seeing conditions) may affect angular rotation velocity estimate making it smaller if the size of the solid-body rotation area is comparable to it, but in any case the orientation of the kinematical major axis is measured correctly.

2. Observational approaches

The spectral data which we analyse in this work are obtained mostly by means of 2D spectroscopy. Integral-field spectroscopy is a rather new approach which was firstly proposed by Prof. G. Courtes some 15 years ago – for a description of the instrumenthal idea see e.g. Bacon et al. (1995). It allows to obtain simultaneously a set of spectra in a wide spectral range from an extended area on the sky, for example, from a central part of a galaxy. A 2D array of microlenses provides a set of micropupils which are put onto the entry of a spectrograph. Having reduced the full set of spectra corresponding to the individual spatial elements, we obtain a list of fluxes in continuum and in emission lines, of line-of-sight velocities, both for stars and ionized gas, and of absorption-line equivalent widths which are usually expressed as indices in the well-formulated Lick system (Worthey et al., 1994). This list can be transformed into two-dimensional maps of the above mentioned characteristics for the central part of a galaxy which is studied. Besides the panoramic view benefits, such an approach gives an unique opportunity to overlay various 2D distributions over each other without any difficulties with positioning.

The 2D spectroscopic observations before August of 1998 were made with the old variant of the Multi-Pupil Field Spectrograph (MPFS) of the 6m telescope (Afanasiev et al., 1990). The panoramic view was provided by a 8×12 microlenses array, and the spectrum sets were registered with a CCD detector of 520×580. The spatial scale was $1.3''$ per microlens. The reciprocal dispersion was 1.6 Å per pixel with the spectral resolution of 4–6 Å slightly varying over the frame.

Later in 1998 a new variant of the MPFS became operational in the prime focus of the 6m telescope.[1] Compared with the previous variant, in the new MPFS the field of view is increased and the common spectral range is larger due to the usage of fibers: they transmit light from 16×15 square elements of the galaxy image to the slit of the spectrograph together with additional 16 fibers that transmit the sky background light taken $\sim 4'$ away from the galaxy, so the separate sky exposures are not necessary now. The size of one spatial element is approximately $1'' \times 1''$; a CCD TK 1024×1024 detector has been used until 2002, now a CCD $2k \times 2k$ is used. The reciprocal dispersion was 1.35 Å per pixel for the small detector and 0.75 Å for the big one, with a spectral resolution of 5 Å rather stable over the field of view.

We obtain the MPFS data mostly in two spectral ranges, the green one, including 4300–5600 Å wavelength interval, and the red one, including 5800–7000 Å interval. The green spectra are used to calculate the Lick indices $H\beta$, Mgb, Fe5270, and Fe5335 which are suitable to determine metallicity, age, and Mg/Fe ratio of old stellar populations (Worthey, 1994). They are used also for cross-correlation with a spectrum of a template star, usually of G8III–K3III spectral type, to obtain in such a way a line-of-sight velocity field for the stellar component and a map of stellar velocity dispersion. The red spectral range contains strong emission lines $H\alpha$ and [NII]λ6583 and is used to derive line-of-sight velocity fields for the ionized gas.

Another 2D spectrograph which data we use in this work is a new instrument, SAURON, operated at the 4.2m William Herschel Telescope (WHT) on La Palma – for its detailed description see Bacon et al. (2001). Briefly, the field of view of this instrument is $41'' \times 33''$ with the spatial element size of $0.94''$. The sky background taken $2'$ from the center of the galaxy is exposed simultaneously with the target. The fixed spectral range is 4800-5400 Å, the reciprocal dispersion is 1.11 Å-1.21 Å varying from the left to the right edge of the frame, and the spectral resolution is about of 4 Å. The comparison spectrum is of neon, and the linearization is made by a polynomial of the 2nd order with a mean accuracy of 0.07 Å.

To study the structure of the central regions of the galaxies under consideration and to compare some characterisctic orientations with the kinematical data, we involve photometric observational data into our analysis. For almost all galaxies we have retrieved the WFPC2/HST data from the HST Archive. Some ground-based photometry is also used.

---

[1] For a description of the spectrograph, see http://www.sao.ru/hq/lsfvo/devices/mpfs/

All the data, spectral and photometric, except the data obtained with the new MPFS, have been reduced with the software produced by Dr. V.V. Vlasyuk in the Special Astrophysical Observatory (Vlasyuk, 1993). Primary reduction of the data obtained with the MPFS was done in IDL with a software created by Prof. V.L. Afanasiev.

3. Circumnuclear Stellar Disks

When having two-dimensional spectral data, we can diagnose a presence of the compact circumnuclear stellar disk in a galaxy by a simple visual inspection of the kinematical maps of the galaxies. Fig. 1 shows a characteristic example of such a situation: the left plot presents a qualitative model, and the right plot – observational data for the early-type spiral galaxy NGC 3623. The data were obtained with the integral-field spectrograph SAURON at the 4.2m William Herschel Telescope in March 2000 and were retrieved by us from the open ING Archive of the UK Astronomical Data Centre. The circumnuclear stellar disk, being a dynamically colder stellar system than the surrounding bulge, demonstrates decoupled fast rotation which forces a characteristic 'pointed' shape to the central isovelocities. The case of NGC 3623 is a 'pure' case of the decoupled circumnuclear stellar disk that has been formed in some secondary star formation burst in the center of the galaxy, because it is decoupled not only by its kinematics, but also by properties of the stellar population. The next Fig. 2 presents two maps which look like a mirror reflection of each other: the map of the stellar velocity dispersion and the map of the Lick magnesium index Mgb (in practice, of the equivalent width of this strong metal absorption line). The stellar velocity dispersion is increased in a ring zone which is obviously relates to the compact bulge of the galaxy; in the very center one can see an elongated area of low stellar velocity dispersion which belongs to the cold dynamical stellar system, namely, to the circumnuclear disk. The magnesium map, on the contrary, reveals a metal-rich elongated central structure and a surrounding zone of the low magnesium-line strength which is obviously the more metal-poor bulge. The sizes of all these features are consistent with each other and imply the presence of the compact metal-rich circumnuclear stellar disk with the radius of 300 pc within the kiloparsec-sized metal-poor bulge.

The circumnuclear stellar disk in NGC 3623 lies strictly in the main symmetry plane of the galaxy: its line of nodes coincides with the line of nodes of the global disk within a few degrees, despite the presence (and influence?) of the large-scale bar in this galaxy. However, such regularity of the circumnuclear stellar disk orientation is not very frequent. By analysing the analogous SAURON data for the lenticular galaxy NGC 3384 (Sil'chenko et al., 2003), we have found kinematical evidences for the presence of the circumnuclear stellar disk in this galaxy too: fast stellar rotation near the center, minimum of

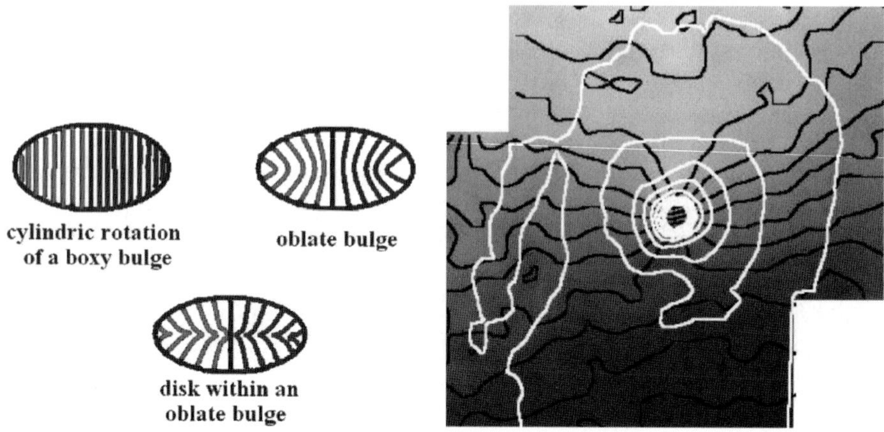

Figure 1.   Stellar velocity field signatures typical for compact inner disks embedded into the bulges: (*left*) The qualitative model presentation (*right*) The stellar line-of-sight velocity field of NGC 3623 obtained with the SAURON, with the isovelocities traced by black; the continuum brightness distribution is shown by white isophotes.

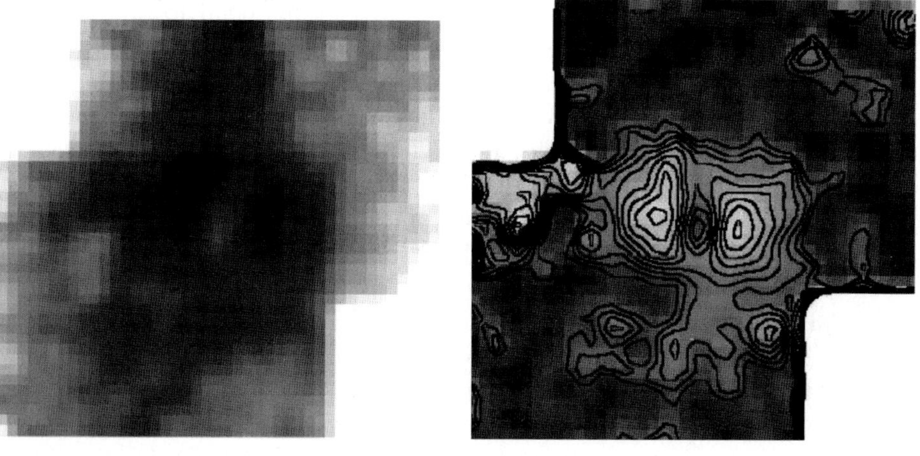

Figure 2.   Other signatures of the inner stellar disk in NGC 3623: (*left*) The stellar velocity dispersion map (*right*) The magnesium-index map shown by gray scale overlaid by isolines.

the stellar velocity dispersion near the major axis, and the coincidence of the kinematical and photometric major axes within the central $10''$. But this time the plane of the circumnuclear disk is inclined to the outer disk plane by some $5° - 7°$. And it is not the only case of inclined circumnuclear stellar disks in our sample galaxies: among 41 galaxies studied by us in detail with the MPFS, the inclined circumnuclear disks have been found also in NGC 488 (Sil'chenko, 1999), NGC 615 (Silchenko et al., 2001), NGC 4036 (Sil'chenko & Vlasyuk, 2001), NGC 7013 (Sil'chenko & Afanasiev, 2002), NGC 7280 (Afanasiev & Sil'chenko, 2000), NGC 7331 (Sil'chenko, 1999), and NGC 7457 (Sil'chenko et al., 2002).

As we have already mentioned in the Introduction, the photometric study of lenticular galaxies by Seifert & Scorza (1996) implies that at least 50% of all objects considered have separate circumnuclear stellar disks. More or less reliable statistics of the kinematical evidences for the circumnuclear disks is still absent. However very recently a survey of the stellar kinematics in the centers of 48 nearby early-type galaxies by Emsellem et al. (2004) has been published; all the observational data are obtained with the SAURON. The visual analysis of their V_{los}-maps based on the qualitative models of Fig. 1(*left*) allows us to evaluate the frequency of inner stellar disks in their sample. The whole impression is striking: among 24 lenticulars, 16 galaxies demonstrate obvious kinematical signatures of the central cold fast-rotating stellar subsystems, another two – central minibars, and there are three more where the mean central velocity gradients look too flat for the disk presence diagnosis, but which have counterrotating cores, according to the Emsellem et al. (2004). So, in general, even more than 50% of early-type disk galaxies possess compact circumnuclear stellar disks. This conclusion has important implications for many common approaches to the study of the central parts of galaxies, in particular, for central black hole mass determination. For this purpose the model approach is widely used which assumes spherical (isotropic) stellar distribution and constant stellar mass-to-light ratio along the radius. Taking in mind all said above we can conclude that it is quite wrong. The masses of the compact inner stellar disks are comparable to the masses of the black holes; it means that one must suppose a two-component, very anisotropic stellar distribution and strongly variable M/L because the inner disks and their surrounding bulges have different mean stellar ages and metallicities.

4. Inner Polar Gaseous Rings

4.1 Observational findings

Another topic of this contribution is a rather new phenomenon which is only coming into evidence, namely, that of inner polar GAS rings in disk galaxies. In contrast to large-scale polar rings, very spectacular but rare features, the

Table 1. The inner polar rings found in regular disk galaxies upto date

NGC	Type (NED)	Reference
2655	SAB(s)0/a	Sil'chenko & Afanasiev 2004, AJ **127**, 2641
2732	S0	Sil'chenko & Afanasiev 2004, AJ **127**, 2641
2768	S0/E6	Sil'chenko & Afanasiev 2004, AJ **127**, 2641
2787	SB(r)0+	Sil'chenko & Afanasiev 2004, AJ **127**, 2641
2911	SA(s)0:pec	Sil'chenko & Afanasiev 2004, AJ **127**, 2641
3414	S0pec	Sil'chenko & Afanasiev 2004, AJ **127**, 2641
4111	SA(r)0+	Sil'chenko & Afanasiev 2004, AJ **127**, 2641
4233	S0	Sil'chenko & Afanasiev 2004, AJ **127**, 2641
7280	SAB(r)0+	Afanasiev & Sil'chenko 2000, AJ **119**, 126
2841	SA(r)b	Sil'chenko et al., 1997, A&A **326**, 941
3368	SAB(rs)ab	Sil'chenko et al., 2003, ApJ **591**, 185
4548	SB(rs)b	Sil'chenko 2002, Pis'ma v AZh **28**, 243
6340	SA(s)0/a	Sil'chenko 2000, AJ **120**, 741
7217	SA(r)ab	Sil'chenko & Afanasiev 2000, A&A **364**, 479
2855	(R)SA(rs)0/a	Corsini et al., 2002, A&A **382**, 488
7049	SA(s)0^0	Corsini et al., 2003, A&A **408**, 873

inner gas polar rings are confined to central bulge-dominated areas of the otherwise morphologically regular galaxies. The sizes of the inner polar rings are usually a few hundred parsecs, so they cannot be surely detected 'by eye' but only from a kinematical study. Table 1 presents a list of galaxies where the inner polar rings were found during the last years, mainly by our efforts. Two more disk galaxies with the inner polar rings are also reported by the Padova group of astronomers.

Lenticular galaxies are the most convenient targets because the gas that we see in their central parts is often their main gaseous content, so the whole picture is not complicated by outer interstellar medium on the line of sight. So among lenticulars we can pre-detect inner polar ring candidates by searching for circumnuclear dust lanes orthogonal to the photometric major axes. Fig. 3(*left*) presents a mosaic of some images to give an impression how it looks like. All the images are obtained with the Hubble Space Telescope Wide-Field Planetary Camera-2 through the green filters and are retrieved from the Hubble Space Telescope Archive. The dust lanes orthogonal to the major axes of the central isophotes are clearly seen. The Fig. 3(*right*) presents two-dimensional line-of-sight velocity distributions for the stars and for the ionized gas obtained with the MPFS of the 6m telescope of the Special Astrophysical Observatory for the same lenticular galaxies as the images above. One can see that the orientations of the rotation axes of the stars and of the gas are decoupled, moreover, they are often almost orthogonal (the more detailed presentation of the same data can be found in (Sil'chenko & Afanasiev, 2004)). When the isophotes of the ionized-gas emission brightness are elongated, their orientations coincide

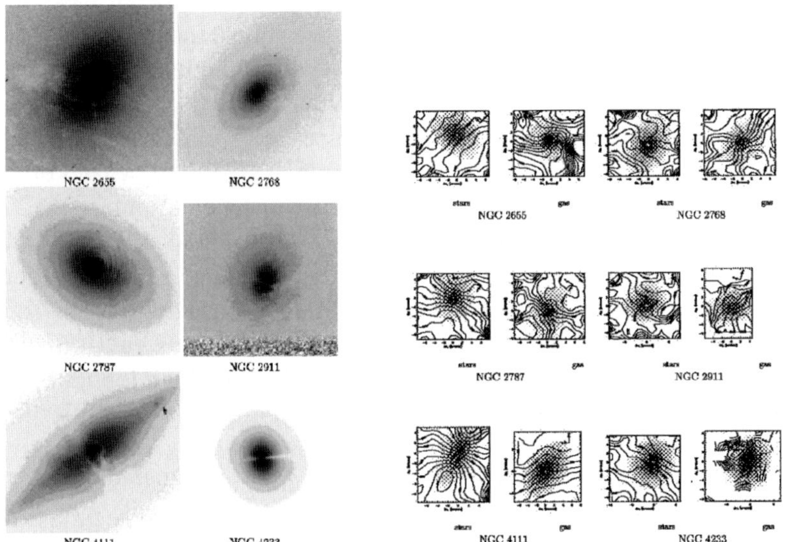

Figure 3.   Some lenticular galaxies with the inner polar rings. (*left*) The HST/WFPC2 view of the inner dust lanes (*right*) The stellar and ionized-gas velocity fields obtained with the MPFS.

with the gas kinematical major axes. Taking into account also the presence of similarly elongated dust lanes, we conclude that we see circumnuclear gas disks or rings, regularly rotating, and they are nearly polar ones.

4.2   Discussion on the inner polar rings origin

With 16 galaxies known to possess inner polar rings (see Table 1), it is time to probe some statistics which may help us to reveal an origin of the new phenomenon. In the Table 2 we compare orientations of kinematical and photometric axes for some galaxies for which we have undertaken a detailed analysis. The first three rows contain the orientations of the line of nodes of the main symmetry planes, the orientations of the inner isophotes and the directions of kinematical stellar major axes. Except the particular case of NGC 3414 (where recently Emsellem et al., 2004 have shown the presence of STELLAR counterrotating subsystems), all three position angles agree rather well. It means that the stars in the central parts of the galaxies with inner polar rings are distributed and rotate axisymmetrically, irrespective of the presence or absence of a large-scale bar. We know, for example, that NGC 2787 is a bona-fide barred galaxy, in several other galaxies bars are suspected, but NGC 2911 is surely unbarred (Jungwiert et al., 1997), both in the optical and in the infrared light. By inspecting the global properties of the lenticular galaxies under consideration we also deduce that dense environments often accompany the phenomenon of the inner polar rings: five galaxies are in groups, two – in clusters; but again, there is an exception: NGC 2787 is a quite isolated lenticular dwarf. The most interesting thing is that all but one lenticular galaxies with the inner polar rings possess a detectable amount of neutral hydrogen that is not very common for lenticulars. For example, over all normal S0 galaxies the HI detection rate is 28% (Wardle & Knapp, 1986), and over early-type galaxies with large-scale polar rings it is 66% (Huchtmeier, 1997). The presence of extended large-scale HI disks in galaxies with otherwise old stellar populations seems to confirm a hypothesis of recent gas accretion from a late-type donor. However for two galaxies, NGC 2655 and NGC 2787, we have not only detections – we have HI maps, and the orientations of the whole neutral hydrogen distributions differ strongly from the orientations of the inner ionized gas. For NGC 2655, we can refer to the HI isophote map in the photometric atlas of Erwin & Sparke (2003): the HI isophotes are strongly elongated, and if the intrinsic HI distribution is a roundish disk, its line of nodes is $PA_{HI} \approx 120°$ – let us compare this position angle to the inner dust lane orientation of 12° and to the circumnuclear ($R = 4'' - 6''$) gas kinematic major axis of 20°. For NGC 2787, there are several detailed studies of HI, and for example, Shostak (1987) having mapped the distribution and kinematics of the neutral hydrogen in this galaxy had reported an outer ring which rotates substantially with the kinematical

Table 2. Orientation parameters of the galaxies

NGC	2655	2732	2768	2787	2911	3414	4111	4233
PA_0	$85°$	$67°$	$95°$	$109°$	$140°$	$179°$	$150°$	$176°$
$PA_{*,inn}$	$89° \pm 2°$	$63° \pm 2°$	$96° \pm 2°$	$110°$	$132° \pm 3°$	$178° \pm 2°$	$147°$	$178°$
$PA_{*,kin}$	$88° \pm 1.7°$	$63° \pm 1.8°$	$88.5° \pm 0.5°$	$105° \pm 1.8°$	$317°(137°) \pm 2°$	$324°(144°) \pm 2°$	$150° \pm 3°$	$162° \pm 3°$
PA_{dust}	$12°,{}^1$	$-$	$\sim175°,{}^2$	$\sim40°,{}^3$	$43°,{}^4$	$93°,{}^5$	$Polar^6$	$77°,{}^5$
$PA_{gas,kin}$	$45° \rightarrow 20°(R > 4'')$	$-9°$	$-13° \pm 2.9°$	$72° \pm 3.4°$	$63° \pm 2.6°$	$68° \pm 3°$	$198° \pm 7°$	$76° \pm 3.2°$
$PA_{[NII]}$	$174°$	$168°$	round	$61°$	$\sim80°$	round	round	$84°$

[1] Erwin & Sparke (2003)

[2] Bertola et al. (1992)

[3] Erwin et al. (2003)

[4] Sil'chenko & Afanasiev (2004)

[5] Tran et al. (2001)

[6] Barth et al. (1998)

major axis of $140° \pm 5°$ – let us compare it to the inner gas kinematical major axis of $72° \pm 3.5°$ and to the dust lanes orientation of $\sim 40°$ (Erwin et al., 2003). We have no HI data to compare inner and outer gas distributions and rotations in other 6 galaxies, but at least these two cases permit us to test some possible mechanisms for inner gas polar ring formation. If the inner polar-ring gas and the outer neutral hydrogen in these two galaxies have common origin, we must admit that when drifting to the centers, the gas changes its spin orientation by almost $90°$. Can it be at all, and if yes, under what conditions? Otherwise, we can admit several independent accretion events, then the polar orientation of the inner gas with respect to the outer one is a pure chance. In NGC 2655 where the circumnuclear stellar population is very young, $T \approx 2$ Gyr, there are *three* gaseous subsystems with different spin orientations because the very central gas is confined to the main disk plane of the galaxy. In principle, the simultaneous existence of rather young nuclear stellar population and the cir-cumnuclear gas in the main plane is consistent with the idea of Friedli & Benz (1993) that nuclear star formation bursts are effectively fueled only by the gas corotating with the stars in the main disk planes – only this gas can reach the center. Then in NGC 2655 we identify at least three independent, asynchro-nous gas accretion events: we suggest that the oldest of them has formed a polar ring in the radius range of $4'' - 6''$, the next one which gas spiraled to the center in the main disk plane has fueled the recent nuclear star formation burst, and the last, the most recent event has provided an extended HI disk with the plane orientation intermediate between two preceding ones. However, it is surprising that even after at least three accretion (minor merger?) events NGC 2655 remains to be a disk galaxy, though of rather early type, however inside its dense environment this galaxy has successfully avoided major merger events.

The problem of the inner polar ring origin becomes even more obscure when we try to consider their appearance in spiral galaxies. Let us discuss two char-acteristics examples: NGC 2841, a highly inclined Sb galaxy, and NGC 7217, an almost face-on Sab galaxy. The inner polar ring in NGC 2841 was the first one found by us with the Multi-Pupil Spectrograph of the 6m telescope (Silchenko et al., 1997). Later, the orthogonality of the gas and star rotations in the center of the galaxy has been confirmed by the data obtained with an-other integral-field spectrograph, 2D-FIS (for the spectrograph and observation descriptions – see Peletier et al., 1999, for their analysis – Sil'chenko, 2002b). The warp of the gaseous disk by $90°$ in the center of NGC 7217 detected with the last modification of the Multi-Pupil Spectrograph of the 6m telescope is also confirmed by our analysis of the long-slit data retrieved from the Isaac Newton Group Archive (of the La Palma Observatory) which is the part of the UK Astronomical Data Centre: within $3''$ from the center we measure a zero gas velocity gradient along the major axis and strong gas rotation close to the minor axis (Sil'chenko & Afanasiev, 2000). Both NGC 2841 and NGC 7217 are quite isolated. And both galaxies have large-scale massive neutral-hydrogen disks lying at the symmetry planes of the stellar bodies and rotating in these planes quite regularly so with the global gas spins polar with respect to the inner gas ones (for the HI mapping of NGC 2841 see Rots, 1980, Bosma, 1981; for NGC 7217 – see Buta et al., 1995). In these cases the hypothesis of the external origin of the inner gas becomes quite improbable: if accreted, the polar-spin gas cloud must collide many times with the own gas of the galaxies, suffering shock consequences, loosing its vertical momentum component and settling finally into the main planes of the galaxies under consideration. We must look for another, intrinsic mechanism to put the inner gas onto a polar orbit. In particular, all the spiral galaxies where we have found the inner polar rings possess some triaxiality: it may be a bar, or a triaxial bulge, or an oval inner lense. This general property of the spiral galaxies with the inner polar rings stimulates us to appeal to dynamical models of gas behaviour within a triaxial potential.

Unfortunately, the majority of dynamical models treat the gas as an infinitely thin disk, so calculations are usually restricted to a two-dimensional case. But I have found some three-dimensional simulations made by Friedli with co-authors, and indeed, strongly inclined gaseous disks form within a barred potential from the gas with the initial spin orthogonal to the main symmetry plane of the galaxy without any external gas supply, if some retrograde gas is present in the model disk (Friedli & Benz, 1993). We can try to test a hypothesis of secular evolution of the retrograde gas within a barred potential, if it is applicable, for example, to the origin of the inner polar ring in SB0-galaxy NGC 2787. If we suppose that the inner polar ring in NGC 2787 may be formed by outer gas drifting to the center through the bar-dominated region, then the mechanism of Friedli & Benz would work if the outer gas counterrotates with respect to the global stellar disk. I have checked the sense of the HI rotation by using the HI observations of Shostak (1987) and stellar kinematics from Erwin et al. (2003). Unfortunately, the outer gas demonstrates prograde rotation, so this mechanism does not work in this particular case. Another attracting possibility to use the mechanism of Friedli & Benz is to suppose an *intrinsic* origin of the counter-rotating gas. Indeed, counterrotating stars are always present inside a triaxial potential on banana-like orbits, so as they lost gas during their evolution (and stellar wind is a common phenomenon for old red giants) it would be partially counterrotating. Then nothing prevents this gas to drift to the center within a triaxial potential and to pass to highly inclined orbits. When Bertola with co-workers (Bertola et al., 1995; Bertola et al., 1984) discussed the problem of the origin of dynamically hot ionized gas in the centers of lenticular galaxies, they demonstrated some simple estimates showing that the mass loss of old stars of a typical bulge could provide up to $10^7\ M_\odot$ of gas. On the other hand, dynamical N-body simulations of bars reveal up to 30% of all stellar mass on retrograde orbits within the corotation radius (Wozniak & Pfenniger, 1997). So we may conclude that a triaxial bulge can easily supply $10^5 - 10^6\ M_\odot$ of counterrotating gas necessary to explain the observed inner polar rings.

One more remark about the stability of the inner polar rings. If the gas accumulates in one of two main planes of a triaxial structure – that perpendicular to the long axis or that perpendicular to the short axis of the ellipsoid, – its rotation orbits would be stable, and the polar rings must be long-lived. However some weak accretion is still observed in the centers of all our galaxies: though powerful active nuclei are absent in the sample, almost all the galaxies have point-like non-thermal sources in the nuclei, either radio or X-ray. So these sources are evidently fueled, and there must be weak radial gas inflow. The interplay of rotational and radial motions in the areas of the gas inner polar rings may explain some deviations of $PA_{gas,kin}$ from PA_{dust} or $PA_{[NII]}$ in the Table 2.

However, the question of the inner gas polar ring origin remains still open and needs thorough discussion among observers as well as among theoreticians.

Acknowledgments

I thank my collaborators V. L. Afanasiev, A. N. Burenkov, A. V. Moiseev, and V. V. Vlasyuk, of the Special Astrophysical Observatory of RAS. The 6m telescope is operated under the financial support of Science Ministry of Russia (registration number 01-43). During the data analysis we have also used the Lyon-Meudon Extragalactic Database (LEDA) supplied by the LEDA team at the CRAL-Observatoire de Lyon (France) and the NASA/IPAC Extragalactic Database (NED) which is operated by the Jet Propulsion Laboratory, California Institute of Technology, under contract with the National Aeronautics and Space Administration. The research is partly based on the data taken from the ING Archive of the UK Astronomical Data Centre and on observations made with the NASA/ESA Hubble Space Telescope, obtained from the data archive at the Space Telescope Science Institute, which is operated by the Association of Universities for Research in Astronomy, Inc., under NASA contract NAS 5-26555. The study of the young nuclei in lenticular galaxies has been supported by the grant of the Russian Foundation for Basic Researches 01-02-16767 and the study of the evolution of galactic centers – by the Federal Scientific-Technical Program – contract of the Science Ministry of Russia no.40.022.1.1.1101.

References

Afanasiev V.L., Burenkov A.N., Zasov A.V., and Sil'chenko O.K.: 1988, *Astrofizika* **28**, 243
Afanasiev V.L., Burenkov A.N., Zasov A.V., and Sil'chenko O.K.: 1988, *Astrofizika* **29**, 155

Afanasiev V.L., Sil'chenko O.K., and Zasov A.V.: 1989, *A&A* **213**, L9

Afanasiev V.L., Vlasyuk V.V., Dodonov S.N., and Sil'chenko O.K.: 1990, *Preprint SAO RAS*, N54 (Nizhnij Arkhyz: Special Astrophys. Obs.)

Afanasiev V.L., Burenkov A.N., Zasov A.V., and Sil'chenko O.K.: 1991, *Astron. Zh.* **68**, 1134

Afanasiev V.L., Burenkov A.N., Zasov A.V., and Sil'chenko O.K.: 1992, *Astron. Zh.* **69**, 19

Afanasiev V.L. and Sil'chenko O.K.: 2000, *AJ* **119**, 126

Bacon R., Adam G., Baranne A., Courtes G., Bubet D., et al.: 1995, *A&AS* **113**, 347

Bacon R., Copin Y., Monnet G., Miller B.W., Allington-Smith J.R., et al.: 2001, *MNRAS* **326**, 23

Barth A.J., Ho L.C., Filippenko A.V., and Sargent W.L.W.: 1998, *ApJ* **496**, 133

Bertola F., Bettoni D., Rusconi L., and Sedmak G.: 1984, *AJ* **89**, 356

Bertola F., Buson L.M., and Zeilinger W.W.: 1992, *ApJ* **401**, L79

Bertola F., Cinzano P., Corsini E.M., Rix H.-W., and Zeilinger W.W.: 1995, *ApJ* **448**, L13

Bosma A.: 1981, *AJ* **86**, 1791

Burkhead M.S.: 1991, *AJ* **102**, 893

Buta R., van Driel W., Braine J., Combes F., Wakamatsu K., Sofue Y., and Tomita A.: 1995, *ApJ* **450**, 593

Corsini E.M., Pizzella A., and Bertola F.: 2002, *A&A* **382**, 488

Corsini E.M., Pizzella A., Coccato L., and Bertola F.: 2003, *A&A* **408**, 872

Emsellem E. et al.: 2004, *MNRAS* **352**, 721

Erwin P. and Sparke L.S.: 2003, *ApJS* **146**, 299

Erwin P., Vega Beltran J.C., Graham, A.W., and Beckman, J.E.: 2003, *ApJ* **597**, 929

Friedli D. and Benz W.: 1993, *A&A* **268**, 65

Hes R. and Peletier R.F.: 1993, *A&A* **268**, 539

Hubble E.: 1926, *ApJ* **64**, 321

Huchtmeier W.K.: 1997, *A&A* **319**, 401

Jungwiert B., Combes F., and Axon D.J.: 1997, *A&AS* **125**, 479

Moiseev A.V. and Mustsevoy V.V.: 2000, *Astronomy Letters* **26**, 565

Monnet G., Bacon R., and Emsellem E.: 1992, *A&A* **253**, 366

Nieto J.-L., Bender R., Arnaud J., and Surma P.: 1991, *A&A* **244**, L25

Peletier R.F., Vazdekis A., Arribas S., del Burgo C., Garcia-Lorenzo B., Gutièrrez C., Mediavilla E., and Prada F.: 1999, *MNRAS* **310**, 863

Rots A.H.: 1980, *A&AS* **41**, 189

Scorza C. and Bender R.: 1995, *A&A* **293**, 20

Seifert W. and Scorza C.: 1996, *A&A* **310**, 75

Shostak G.S.: 1987, *A&A* **175**, 4

Sil'chenko O.K.: 1999, *Astronomy Letters* **25**, 140

Sil'chenko O.K.: 1999, *AJ* **118**, 186

Sil'chenko O.K.: 2000, *AJ* **120**, 741

Sil'chenko O.K.: 2002, *Astronomy Letters* **28**, 207

Sil'chenko O.K.: 2002, In: *"Galaxies: The Third Dimension"/ Eds. Rosado M., Binette L., Arias L. ASP Conf. Ser.* **282**, 121

Sil'chenko O.K. and Afanasiev V.L.: 2000, *A&A* **364**, 479

Sil'chenko O.K. and Vlasyuk V.V.: 2001, *Astronomy Letters* **27**, 15

Sil'chenko O.K. and Afanasiev V.L.: 2002, *A&A* **385**, 1

Sil'chenko O.K. and Afanasiev V.L.: 2004, *AJ* **127**, 2641

Sil'chenko O.K., Vlasyuk V.V., and Burenkov A.N.: 1997, *A&A* **326**, 941

Sil'chenko O.K., Vlasyuk V.V., and Alvarado F.: 2001, *AJ* **121**, 2499

Sil'chenko O.K., Afanasiev V.L., Chavushyan V.H., and Valdes J.R.: 2002, *ApJ* **577**, 668

Sil'chenko O.K., Moiseev A.V., Afanasiev V.L., Chavushyan V.H., and Valdes J.R.: 2003, *ApJ* **591**, 185

Tran H.D., Tsvetanov Z., Ford H.C., Davies J., Jaffe W., et al.: 2001, *AJ* **121**, 2928

Vlasyuk V. V.: 1993, *Astrofiz. issled. (Izv. SAO RAS)* **36**, 107

Wardle M. and Knapp G.R.: 1986, *AJ* **91**, 23

Worthey G.: 1994, *ApJS* **95**, 107

Worthey G., Faber S.M., González J.J., and Burstein D.: 1994, *ApJS* **94**, 687

Wozniak H. and Pfenniger D.: 1997, *A&A* **317**, 14

BENDING INSTABILITY IN GALAXIES: THE STELLAR DISK THICKNESS AND THE MASS OF SPHEROIDAL COMPONENT

N. V. Tyurina,[1] A. V. Khoperskov,[2] and D. V. Bizyaev,[3]

[1]*Moscow State University, Russia*
tiurina@sai.msu.ru

[2]*Volgograd State University, Russia*
khopersk@sai.msu.ru

[3]*Moscow State University, Russia*
dmbiz@sai.msu.ru

Abstract   We present results of numerical N-body simulations of a galactic stellar disk embedded into a spherical dark halo. The non-linear dynamics of bending instabilities developing in the disk is studied. The bending modes, axisymmetric and not, are considered as main factors increasing the disk thickness. The relation between the disk vertical scale height and the halo+bulge-to-disk mass ratio is inferred. The method of estimation of the spherical-to-disk mass ratio for edge-on spiral galaxies based on this relation is studied and applied to constrain the spherical subsystem mass and the mass of dark halos (M_h/M_d) in seven edge-on galaxies. The values of M_h/M_d are of order 1 for our galaxies.

Keywords: edge-on galaxies, bending instability, halo, N-body simulations, galactic structure

1.   Introduction

The luminous matter represents only a part of overall mass in galaxies. One of arguments for massive dark matter component comes from numerical simulations of galaxies. Thus, numerical models evaluated for disk galaxies with light spheroidal subsystem show that their stellar disks heat themselves significantly during the evolution achieving the equilibrium state with too high

Alexei M. Fridman, et al. (eds.), Astrophysical Disks; Collective and Stochastic Phenomena 291–306
© 2006 *Springer. Printed in the Netherlands*

velocity dispersion (Ostriker & Peebles, 1973; Carlberg & Sellwood, 1985; Athanasoula & Sellwood, 1986; Bottema & Gerritsen, 1997; Fuchs & von Linden, 1998, Khoperskov et al., 2003). To explain the low ratio of velocity dispersion to the gas rotational velocity c^{obs}/V_c in outer parts which is often observed in many galaxies, one should consider a spherical subsystem as massive as the galactic disk in its optical limits.

The vertical scale height of stellar disk depends on the local disk surface density and the spheroidal subsystem mass M_s. Then, for the case of edge-on galaxies we are enabled to include the observed disk thickness into the numerical simulations. The disk thickness is available from observations for many edge-on spiral galaxies. On the other hand, if the velocity dispersion is close to the required for the marginal stability of the stellar disk, its thickness is tied up with the mass ratio of disk to the spherical component (Zasov et al., 1991; Mikhailova et al., 2001).

An important mechanism of the vertical velocity dispersion c_z growth is the bending instability. The bending instability in galactic disks has been considered with the help of N-body simulations. Important results were obtained when the tidal interactions were taken into account (Hernquist et al., 1993; Weinberg, 1998; Velazquez & White, 1999; Mayer et al., 2001; Bailin & Steinmetz, 2003; Reshetnikov & Combes, 2002). The conditions for emerging the bending instabilities were investigated by (Raha et al., 1991; Sellwood, 1996; Sellwood & Merritt, 1994; Patsis et al., 2002; Griv et al., 2002; Binney et al., 1998), (Sotnikova & Rodionov, 2003).

In this paper we point our attention to the radial distribution of c_z/c_r and to the thickness of the stellar disk needed to provide its stability against different kinds of bending perturbations. We consider the late type galaxies without a prominent bulge because the rotation curve in very central regions is not surely defined. The presence of bulge would complicate the analysis of bending instability. The bulge plays a stabilizing role for bending instability when all other conditions being similar. If no bulge presents, one can restore the internal part of rotation curve assuming it is defined by the disk component in central ($r \lesssim 2L$) regions (Zasov & Khoperskov, 2003). Here L denotes the exponential scale length of the stellar disk.

We performed numerical N-body simulations for seven edge-on galaxies with known photometric scales, both radial and vertical.

Our basic interest throughout the paper is to follow up the evolution of the disk thickness and the behaviour of c_z/c_r ratio that holds the stellar disk in stability against the bending perturbations.

2.   Dynamical modeling of edge-on galaxies

We evaluated free parameters of the model: $\mu = M_s/M_d$, the radial scale of the halo a_h, and the disk central surface density σ_0, looking for the optimal agreement between the calculated and observed values of the disk thickness. Here M_d is the galactic disk mass, $M_s = M_h + M_b$ is the mass of the spherical component which comprises of halo and bulge in general case. Note that since almost all our galaxies have no visible bulge, their spherical component means the galactic dark halo.

Our dynamical modeling is based on the numerical integration of the motion equations for N gravitationally interacting particles. This system of collision-less particles forms a disk embedded into dark halo and bulge. The steady state distribution of mass in the bulge $\varrho^{(b)}$ and halo $\varrho^{(h)}$ is defined as:

$$\varrho^{(h,b)}(\xi) = \frac{\varrho_0^{(h,b)}}{(1 + \xi^2/a_{(h,b)}^2)^k}, \tag{1}$$

where $\xi = \sqrt{r^2 + z^2}$, and r, z are the radial and vertical coordinates, k equals to 3/2 for the bulge and to 1 for the halo. The dimensional spatial scales for the bulge and halo are denoted as a_b and a_h, respectively. We supposed that the bulge is encompassed by a sphere with radius of $\xi \leq r_b^{\max}$.

The initial vertical equilibrium of the disk is defined by the Poisson equation

$$\frac{\partial}{r\partial r}\left(r\frac{\partial \Phi}{\partial r}\right) + \frac{\partial^2 \Phi}{\partial z^2} = 4\pi G\left(\varrho + \varrho^h + \varrho^b\right) \tag{2}$$

and by balance of forces in the vertical direction in first approximation

$$\frac{\partial\left(\varrho c_z^2\right)}{\partial z} + \frac{\partial\left(r\varrho\alpha_{rz}\right)}{r\,\partial r} = -\frac{\partial \Phi}{\partial z}\varrho. \tag{3}$$

Here Φ is the gravitational potential, ϱ is the disk volume density, c_z is the vertical component of velocity dispersion, and $\alpha_{rz} = \langle uw \rangle$ is the chaotic part of the radial u and vertical w velocity components. The radial component of the Jeans equation defines the rotation velocity in the stellar disk (Valluri, 1994):

$$V^2 = V_c^2 + c_r^2\left\{1 - \frac{c_\varphi^2}{c_r^2} + \frac{r}{\varrho c_r^2}\frac{\partial(\varrho c_r^2)}{\partial r} + \frac{r}{c_r^2}\frac{\partial\alpha_{rz}}{\partial z}\right\}, \tag{4}$$

We distinguish between the circular velocity $V_c = \sqrt{r\left(\partial\Phi/\partial r\right)_{|z=0}}$ and the stellar rotation curve $V(r) = r\Omega$ because of the velocity dispersion.

The system of equations (2)–(3) can be reduced to the equation for the dimensionless disk density $f(z;r) = \varrho(z;r)/\varrho(z=0;r)$:

$$\frac{d}{dz}\left(c_z^2\frac{df}{dz}\right) - \frac{c_z^2}{f}\left(\frac{df}{dz}\right)^2 + f\frac{d^2c_z^2}{dz^2} + 4\pi Gf\cdot\left(\varrho + \varrho^h + \varrho^b + E\right) + f\frac{d}{dz}\frac{E_\alpha}{\varrho} = 0, \tag{5}$$

$$E = -\frac{1}{4\pi Gr}\frac{\partial V_c^2}{\partial r}, \qquad E_\alpha = \frac{\partial(r\varrho\alpha_{rz})}{r\,\partial r},$$

where the surface density is equal to $\sigma = 2\varrho(z = 0) \cdot z_0$, and the vertical scale of the disk is defined as $z_0 = \int_0^\infty f dz$. In the case $c_z = \text{const}$ and $\alpha_{rz} = 0$ the equation (5) coincides with result of Bahcall (1984). The circular velocity is defined in the galactic plane $z = 0$. For the case of $c_z = \text{const}$, $\varrho_s = 0$, $V_c = \text{const}$ and $E_\alpha = 0$, the disk volume density profile in the vertical direction is:

$$\varrho(z) = \varrho(z = 0) \cdot \text{sech}^2(z/z_0), \qquad (6)$$

where we designated the vertical disk scale as $z_0 = c_z^2/\pi G\sigma$. We suppose that the density distribution in a galaxy follows the brightness distribution. It corresponds to the constancy of the mass-to-luminosity ratio. We consider two possible laws for the vertical density distribution $\varrho(z)$: $\exp(-z/h_z)$ and $\text{sech}^2(z/z_0)$. The values of h_z and z_0 are computed as functions of time and location in disk. The radial density profile in the disk is assumed to be exponential $\varrho(r) = \exp(-r/L)$. The mass of the dark halo M_h is calculated inside the maximum disk radius $R_{\max}$. As a rule, $R_{\max} \approx 4\,L$ (van der Kruit & Searle, 1981; van der Kruit & Searle, 1982; Pohlen et al., 2002; Holley-Bockelmann & Mihos, 2001).

We assume this kind of axysimmetric models in the equilibrium state as initial templates for our simulations. The initial radial distributions of the radial (c_r) and azimuthal ($c_\varphi = c_r \alpha/2\Omega$) velocity dispersions keeps the disk in gravitationally stable state.

Considering edge-on galaxies, we encounter a well known problem that their observed structural parameters are averaged along the line of sight. Once we compare the model parameters with observations, this effect has to be taken into account when we reproduce the model rotation curves as well as the radial and vertical photometric profiles. Deriving the surface density, we integrated the volume density of the model disk along the line of sight. The same integration was performed for the rotation curves as well, see also (Zasov & Khoperskov, 2003).

3.   The bending instabilities in stellar disks

The disk scale height depends on the vertical velocity dispersion $c_z(r)$, and the latter is dependent of the radial velocity dispersion $c_r(r)$. A discussion about the relation between c_r and c_z was opened firstly by Toomre (Toomre, 1966). Considering a simplified model of an infinitely thin uniform self-gravitating layer, (Poliachenko & Shuhman, 1977) found that an essential condition for the system stability against the small-scale bending perturbations is $c_z/c_r \geq 0.37$.

The dynamics of the bending instabilities taking a nonuniform volume density distribution in z-direction into account was also considered by (Araki,

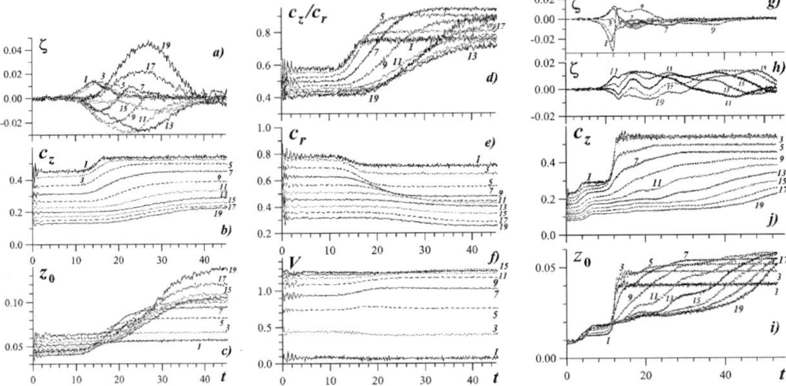

Figure 1. Evolution of the stellar disk parameters when the axisymmetric bending mode being developed in the model with $\mu = 1$, $a = L$: a) the vertical coordinate of the disk's local barycenter ζ, b) vertical component of the velocity dispersion c_z, c) disk's scale height z_0, d) the vertical to radial velocity dispersion ratio c_z/c_r, e) c_r, f) rotational velocity of particles in the disk V. g)–i) The same for the model with $\mu = 4$. The vertical coordinate of barycenter ζ for the internal and external regions is shown in separated plots g) and h) respectively. Different curves are drawn for a set of distances to the center $r_j = 4L \cdot (0.05j - 0.025)$, where j is shown in the plots. All the parameters are averaged by the azimuthal coordinate.

1986) where a lower value of the critical ratio $c_z/c_r \gtrsim 0.3$ was inferred, see also discussion in (Merrit & Sellwood, 1994). The linear analysis of stability against global bending perturbations was discussed by (Poliachenko & Shuhman, 1979) and generalized by (Vandervoort, 1991).

3.1 The bending instability

Let's consider the dynamics of global bending perturbations. The stellar disks at the initial moment ($t = 0$) were: 1) axisymmetric, 2) in an equilibrium state along the radial and vertical directions, 3) gravitationally stable in the disk's plane (it was made by assuming a high value for the radial velocity dispersion $c_r \gtrsim (1.5 \div 4) \cdot c_T$, the ratio $Q_T = c_r/c_T$ is a function of r (Khoperskov et al., 2003). The disks could be either stable or unstable against the bending perturbations depending on $c_z(r)$ distribution. The scale height z_0 and velocity dispersion depend on each other (Bahcall, 1984), hence lesser value of c_z corresponds to a thinner disk.

The evolution of systems with a small initial value of c_z/c_r (i.e. the system that is dynamically cold in the vertical direction) reveals a gradual growth of the global bending instability which heats the disk up in the vertical direction, and rises its thickness.

The bending mode $m = 0$. The developing of unstable axisymmetric bending mode ($m = 0$) leads to the significant disk heating in the vertical direction (Sellwood, 1996). Let's consider the stellar disk where the axisymmetric mode is developing. The evolution of the vertical coordinate of the disk's local barycenter $\zeta(r, t)$, the vertical velocity dispersion $c_z(r, t)$, the disk scale height $z_0(r, t)$, the dispersion ratio $\alpha_z \equiv c_z(r, t)/c_r(r, t)$, the radial velocity dispersion $c_r(r, t)$, and the rotational velocity $V(r, t)$ is shown in Fig. 1. We fix the units by the assumption $M_d = 1$, $G = 1$, $4L = 1$.

As it is seen in Fig. 1, the parameters of the disk stay unchanged for the first 2.5 turns ($t \approx 10$ in our units). This time the modes are formed at the linear stage of the bending instability evolution. After $t \gtrsim 10$ it changes to the nonlinear development of the bending instability and the barycenter $\zeta(r)$ oscillates with larger amplitude in z-direction (see Fig. 1a). In the central regions of the disk the amplitude $\zeta(r)$ rises up faster to its maximum value and falls then down to the initial value (see curves 1-5 in Fig. 1a), whereas the growth of $\zeta(r)$ occurs much slower at the outer parts of the disk (curves 11-19). The rapid growth of the velocity dispersion c_z and of the vertical scale height z_0 begins after a time delay from $\zeta(r)$, see Fig. 1b and 1c. The flare emerges in the disk's central region and then propagates toward its periphery. The value of c_z/c_r rises up mostly because of c_z growth, and due to c_r decrease as well (Fig. 1 d, e). The azimuthal component of velocity dispersion c_φ follows c_r and the relation $c_\varphi \simeq c_r \mathit{æ}/2\Omega$ is valid almost everywhere. The equation can be failed only either in a very thick disk or in regions where $z_0/L \gtrsim 0.4$. Note that $c_r/c_\varphi < 2\Omega/\mathit{æ}$ that means the anisotropy is lesser in those regions because of the system spherization. The decrease of c_r and c_φ is a result of conversion of the kinetic energy of random motion in the disk plane to the kinetic energy of vertical motions. Note that the development of this bending mode takes place in the disk which is axisymmetric in all parameters.

The value of ζ has different signs in disk's center and periphery (see Fig. 1a), what causes a "mexican hat-like" structure formation as a result of this instability (Fig.2). The isolines of ζ have a concentric shape. Up to the moment $t = 25.4 = 6\tau$ the internal ring and the periphery are shifted off the disk plane toward opposite directions, whereas the central region ($r \lesssim 0.4$) has returned to its initial state. The disk vertical structure driven by the bending mode $m = 0$ is shown in Fig.2b. One can see a box-like shape of the disk which is typical for all models where the axisymmetric bending mode dominates and this feature is not relevant to a bar.

A gradual change of the velocity dispersion c_z and c_r, and rotational velocity V begins when the amplitude of the vertical oscillations ζ grows significantly. The vertical disk heating is accompanied by increase of c_z/c_r but starting with certain values of c_z/c_r the favourable conditions for developing of the bending instability disappear. As a result, new stable and thick disk forms. The

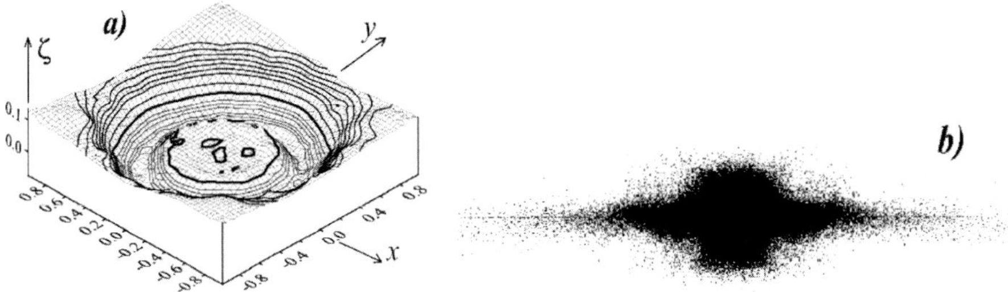

Figure 2. The shape of the stellar disk under the development of the bending instability at the moment $t = 25.4$ for the model shown in Fig. 1 a–f. Distribution of the vertical coordinate of the barycenter $\zeta(x, y)$ are shown in the $x - y$ plane (a). The bold solid line represents $\zeta = 0$. b) The edge-on view to the disk on the stage of the global axisymmetric mode development. The vertical/plane aspect ratio is increased 4 times.

characteristic time for this process depends significantly on the model parameters and is of order ten turns of the disks' outer regions. If the initial distribution of $\alpha_z = c_z/c_r$ is subcritical, the linear stage may take longer time due to low increment of instability, and heating at the nonlinear stage appears feeble. Hence, the amplitude of the bending mode and even the ability of its emerging mostly depend on the initial radial distribution of c_z/c_r.

The important consequence is that the final distribution of $\alpha_z(r)$, as a result of bending perturbations, depends on its initial value $\alpha_z(r, t = 0)$. Thin disks which have a low value of the scale height after the action of the global bending instability get the larger ratio c_z/c_r than it would required to keep the disk stable. The explanation lays in the essentially nonlinear nature of the disk heating. The parameter c_z/c_r passes through the "zone of stability" and stays off it and c_z/c_r jumps over its marginal threshold.

If the disk is hot enough to provide the gravitational stability against the bar mode and the mass of halo is low ($\mu \lesssim 1$), the main cause of heating the previously thin cold disk is the axisymmetric bending mode (m = 0) whereas the modes m = 1,2 do not emerge at all. Note that the mode m = 0 might start forming far from the center ($r \gtrsim L$) not penetrating into the central part of disk. It can happen by two ways: 1) the initial disk has $c_z/c_r \geq \alpha_z^{crit}$ and is stable in the central regions whereas the peryphery of the disk is thin and unstable (Fig. 3 a, b); 2) when the galaxy has a concentrated and massive bulge (Fig. 3 c). In such models the bending modes in the center have lesser amplitude and, in general, a bulge, as well as a halo, plays a stabilizing role. Hence, having all other equal conditions, the disks of galaxies owing bulges are thinner. Note that we don't consider dynamical models for the bulge and as a correct approach a non-stationary model for the bulge has to be evaluated,

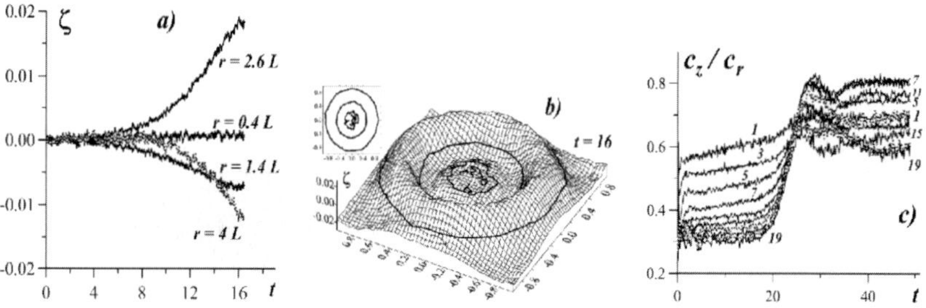

Figure 3.   The model with $\mu = 4$. a) Initial moments of the vertical displacement evolution at different distances from the center of the disk. b) The shape of the stellar disk (i.e. distribution of z-coordinate of its local barycenter through the disk). The thick line corresponds to $\zeta = 0$. c) The flare of the disk without a halo but in presence of a bulge ($M_b = 0.25\,M_d$, $b = 0.2\,L$) after the development of the mode $m = 0$. In the regions of bulge the disk thickening is not significant. The notation is kept the same as in Fig. 1.

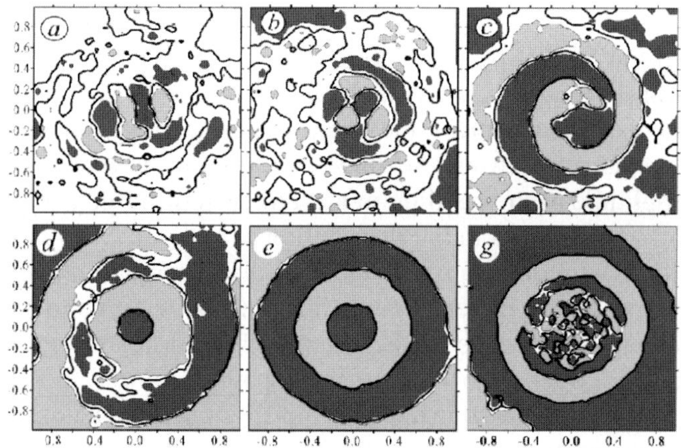

Figure 4.   The structure of the vertical displacement $\zeta(x, y)$ in the disk plane at different moments: a) $t = 2.7$ – the mode $m = 2$ developes itself in the central region, b) $t = 3.1$ – $m = 2 \rightarrow m = 1$ reorganization, c) $t = 4.2$ – a well-developed one-arm mode $m = 1$, d) $t = 10.1 - m = 1 \rightarrow m = 0$ reorganization, e) $t = 16.4$ – a well-developed symmetric mode $m = 0$, g) $t = 51$ – all perturbations in the center $r \lesssim L = 0.25$ have almost gone out. In the middle regions one can see the axisymmetric perturbations and the mode $m = 2$ takes place at the outer parts of the disk.

hence this our result require further investigation. Meanwhile we constrain our study in § 4 by the case of galaxies without large bulges.

The bending modes $m = 1, 2$.   In the models with a massive halo ($\mu \gtrsim 2$) the axisymmetric bending modes $m = 1$ and $m = 2$ may develop themselves and heat up the disk in the vertical direction. The surface density of the disk remains axisymmetric. The evolution of initially thin disk for $\mu = 4$ is shown in Fig. 1 g-i. In the latter case the bending mode $m = 2$ of saddle-type is being developed firstly in central disk regions ($r \lesssim 2L$). Fig. 4 shows the distribution of ζ in the disk plane for different moments of $t = 0 \div 40$. The vertical heating and disk's flare due to the mode $m = 2$ is very modest in this case. However, after $t \gtrsim 3$ the non-linear stage of one-arm asymmetric mode $m = 1$ develops itself and the vertical heating gets more significant. The third stage of the heating begins at $t \gtrsim 10$ when the mode $m = 0$ begins to dominate in the disk. The growth rate of the vertical dispersion c_z is especially high at this time; the disk scale height increases by a factor of 2–3. The example considered shows the process of transformation between the bending modes and transition from the asymmetric mode $m = 2$ to the axisymmetric one.

As well as for the case of the spherical subsystem of low mass, the favourable conditions for the emerging of the global bending instability get worse and, starting with some $\alpha_z(r)$, the disk warp does not develop itself for at least 20 rotation turns if the initial $\alpha_z = c_z/c_r$ (and $z_0(r)$ respectively) is high enough.

A stellar disk developing the bending instability with $m = 1, 2$ is not in a quasy steady-state in the vertical direction. The distribution of c_z remains axisymmetric, in a contrary to z_0. Hence, the condition $c_z^2/z_0 = const$ is failed and this instability evolves rapidly. For the axisymmetric mode the condition $c_z^2 \propto z_0$ is fulfilled much better except the very central region of the disk in stages when the thickness rises significantly.

Apparently, an essential part of real stellar disks did not undergo the heating due to the axisymmetric bending mode $m = 0$ because this mode would produce too thick disks. As an example, in the experiment shown in Fig. 1 ($\mu = 1$) the disk scales ratio $\langle z_0 \rangle / L \simeq 0.4$ (here $\langle z_0 \rangle$ denotes the average thickness over the disk). The relation c_z/c_r exceeds its critical value for the final state of disk. The relative thickness is less ($\langle z_0 \rangle / L \simeq 0.16$) for the case of massive halo ($\mu = 4$) after the relaxation of the axisymmetric mode. On the other hand, there are galaxies which have the ratio $\langle z_0 \rangle / L \leq 0.15$ According to our assumptions either the very massive halos are required for this case, or the axisymmetric bending mode $m = 0$ was never developed in those galactic disks.

Bendings of bar. Let's consider the experiments with rather light halo ($\mu \lesssim 1.5$). If the initial state of disk is gravitationally unstable, a bar can be developed. The bar formation is accompanied by its warps (Raha et al., 1991). The bendings of a bar can emerge as a result of global instability of the bar-mode during its initial stage when the bar forms in initially thin cold disk. Fig. 5 shows the bar bending which increases the vertical velocity dispersion with time. The amplitude of the bar warps falls essentially with increasing the bar thickness. Let's stress that once the bar has been formed, it stops the further possible developing of the global bending modes, thus it destroys the axisymmetric mode $m = 0$ first of all.

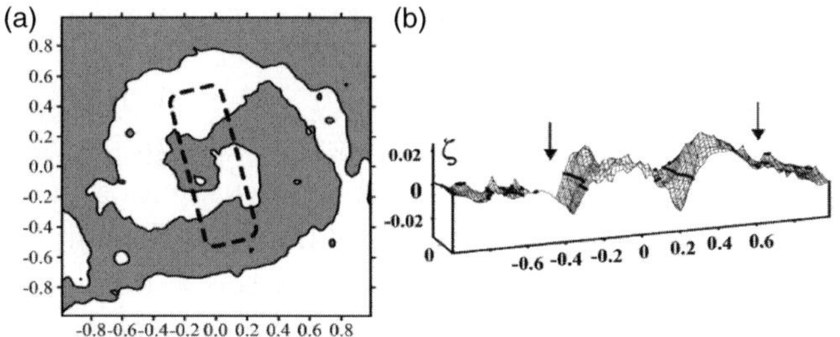

Figure 5.   a) The distribution of ζ in the disk plane (dash-line shows the bar position, the solid line marks $\zeta = 0$). b) The vertical profile of ζ taken along the bar. The arrows mark the bar edges).

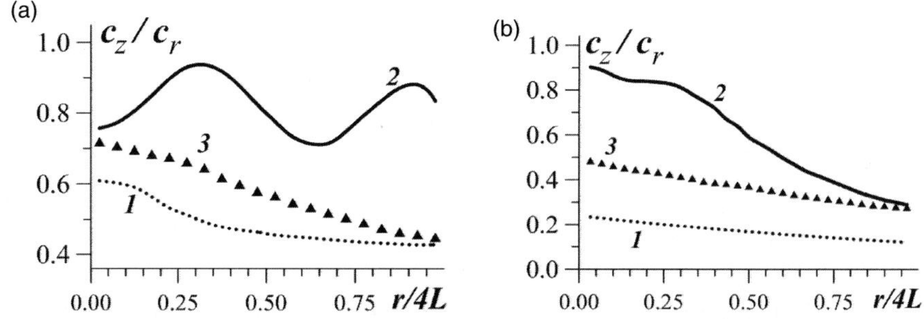

Figure 6.   The radial distribution of $\alpha_z = c_z/c_r$ for: a) $\mu = 1$ (see Fig. 1a–f), b) $\mu = 4$ (see Fig. 1g–i). 1 – the initial distribution, 2 – the final distribution, and 3 – the critical level of α_z.

The bar formation occurs faster in the case of cold initial disk (for small values of the Toomre stability parameter Q_T) and as a result, the bar bending amplitude rises (Fig. 5). If the initial disk was in a marginally subcritical state (i.e. c_r was just below the critical value which provides stability against the global bar-mode), the bar forms very slowly and it is stable against the bending perturbations.

The ratio c_z/c_r. With the similar values of parameters in our model, the key parameter responsible for the stellar disks stability is the ratio $\alpha_z = c_z/c_r$. To stabilize the global bending perturbations in the case of low relative mass of the spherical subsystem, the values of c_z/c_r have to be larger than $0.3 - 0.37$ what was figured out from linear analysis in simple models (Poliachenko & Shuhman, 1977; Araki, 1986; Merrit & Sellwood, 1994). Let's consider bulgeless model with a moderate halo $\mu = 1$ (see Fig. 6). The initial (curve 1) and final (curve 2) distributions of $\alpha_z(r)$ are shown in Fig. 6a. The vertical heating of disk originated from the axisymmetric bending mode development is as strong that the averaged throughout the disk ratio $\langle c_z/c_r \rangle = 0.82$. If the initial ratio $\alpha_z(r)$ follows the curve 3, then the global bending modes remain stable.

Once the relative mass of halo rises gradually, the ratio c_z/c_r needed to marginally stabilize the bending perturbations (we will denote it as the critical α_z ratio) is getting less. The initial (curve 1) and final (curve 2) distributions of $\alpha_z(r)$ for the case of very massive halo of $\mu = 4$ are shown in Fig. 6b. The curve 3 shows the initial distribution of $\alpha_z(r)$ needed to provide the stability against the global bending modes. At the same time in the disk periphery $0.27 < c_z/c_r < 0.37$.

A distinctive feature of the considered models at the threshold of bending stability is the non-uniformity of c_z/c_r along the radius (see Fig. 6). For the case of moderate halo ($\mu \lesssim 1$) and when the initial distribution of $c_r(r)$ suppresses the bar instability, the critical value of α_z^{crit} is a declining monotonous function of r: its typical values range from $0.7 \div 0.8$ at the central regions to $0.4 \div 0.5$ at the periphery.

In both cases of large and small μ, the ratio of velocity dispersions c_z/c_r falls exponentially with radius and can be approximated as $\alpha_z^{crit} \propto \exp(-r/L_\alpha)$ with scale $L_\alpha \simeq (4-6) \cdot L$.

There are two reasons of decreasing the disk thickness with the rising of the halo mass. At first, the ratio c_z/c_r needed to stabilize the bending instability is lower. On the other hand, the halo suppresses the gravitational instability in the plane of disk, therefore c_r/V falls (Khoperskov et al., 2001; Khoperskov et al., 2003). Note that in real galaxies several more factors such as density waves, scattering on giant molecular clouds, and tidal interactions heat up the disk and increase its scale height. Then α_z^{crit} and the corresponding disk thickness gives us the lower estimate for the halo mass.

4. The results of dynamical modeling of edge-on galaxies

In order to compare our model predictions with observations, we consider seven spiral edge-on galaxies. Four of them, NGC 4738, UGC 6080, UGC 9442, and UGC 9556 have no bulges and their structural parameters were derived by (Bizyaev & Kajsin, 2002). The radial distribution of their stellar disks' thickness is available. The galaxy UGC 7321 is interesting for us because of its superthin disk and low surface brightness nature (Matthews, 2000). For two large and nearby galaxies, NGC 891 and NGC 5170, the data on the radial component of stellar velocity dispersion are available from published data (Bottema et al., 1987; van der Kruit & Searle, 1981; Bahcall & Kylafis, 1985; Shaw & Gilmore, 1989; Morrison et al., 1997; Xilouris et al., 1999; Sancisi & Allen, 1979; Begeman et al., 1991; Bottema et al., 1991). It enables us to incorporate those data together with the structural parameters of disks. The name of galaxy, the adopted distance D, the radial scale length L, the averaged scale heights

Table 1. Parameters of the edge-on galaxies

Name	D	L	$\langle z_0 \rangle$	$\langle h_z \rangle$	R_{max}	μ
	Mpc	kpc	kpc	kpc	kpc	
UGC 6080	32.3	2.9	0.69	0.48	7.32	1
NGC 4738	63.6	4.7	1.3	0.7	19.2	0.7
UGC 9556	30.6	1.5 (3.6)	0.51	–	9	1.1
UGC 9422	45.6	3.5	0.80	0.51	14.6	0.8
NGC 5170	20	6.8	0.82	–	26.2	1.9
UGC 7321	10	2.1	0.17^a	0.14^b	8.2	3
NGC 891	9.5	4.9	0.98	0.49	21	1

[a] the scale is shown for the disk's periphery in the case of $sech(z/z_{ch})$.
[b] the scale is shown for the disk's central region.
Here D is the distance to the galaxy ($H_0 = 75$ km s^{-1} Mpc^{-1}), L is the exponential disk scale length, $\langle z_0 \rangle$ is the averaged over the disk value of the vertical scale height for $sech^2$ vertical profile, $\langle h_z \rangle$ is the scale height averaged over the disk for the exponential vertical profile, R_{max} is the maximum radius of the stellar disk.

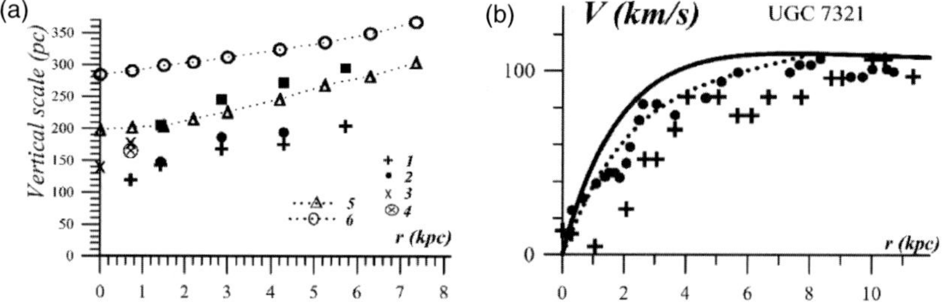

Figure 7. a) The radial distribution of the scale height in UGC 7321. The observational data: 1 and 2 – for the case of $sech(z/h_{ch})$ density distribution in the vertical direction, 3 and 4 – for the exponential law $\exp(z/h_z)$. b) The rotation curve of UGC 7321. The symbols $+$ and $\bullet$ represent the observational data for both sides from the center.

$\langle z_0 \rangle$ and $\langle h_z \rangle$ (corresponding to $sech^2$ and exp distributions in the vertical direction respectively), and the stellar disk maximum radius R_{max} are shown in Table 1.

As an example, we consider the modeling of UGC 7321. The galaxy UGC 7321 reveals itself as a superthin (Goad & Roberts, 1981) and bulgeless (Matthews et al., 1999) disk galaxy. The disk thickness was derived by (Matthews, 2000). We choose the systematic radial velocity of 394 km/s as that the maximum rotational velocity at the periphery has the similar values ($V \simeq 100$ km/s) in both sides of the galaxy. The curves 5 and 6 in Fig. 7 are calculated for the models with $\mu = 2.3$ and 1.5, respectively, inside of $r = R_{\mathrm{max}}$. As it is seen, the model with $\mu > 3$ gives a better agreement with the observed radial distribution of the vertical thickness. The best fit model with $\mu = 2.3$ has $M_d = 0.62 \cdot 10^{10} \, M_\odot$ and $M_h = 1.4 \cdot 10^{10} \, M_\odot$.

The evaluated best-fit parameters μ (spherical-to-disk mass ratios) are shown in Table 1. Since five our galaxies have no bulges, and the bulges of another two are relatively small, the values of μ represent the relative dark halo masses. As it can be seen from the Table 1, the dark and luminous matter contributions are comparable inside the stellar disk limits in our galaxies. The exception is the superthin galaxy, UGC 7321, which has a massive dark halo and a low surface brightness disk.

5.   Conclusions

1 The development of the bending instabilities in stellar galactic disks is studied with the help of N-body numerical simulations. The axisymmetric bending mode ($m = 0$) is found to be the strongest factor which may heat up stellar disks in the vertical direction. If a bar formation was suppressed, the role of ($m = 1$) and ($m = 2$) bending modes in the disk thickening would be low. The most significant thickening of the stellar disk occurs at the initial non-linear stage of the bendings formation. Once the bendings are destroyed, the vertical heating becomes less effective. The lifetime for the bending mode $m = 0$ increases with the growth of the relative mass of the spherical subsystem μ. We show that the initially thin disks increase their thickness much more rapidly than those started from a marginally subcritical state. It is important to notice that the final disk thickness and c_z depend on its initial state and the definition of stability boundary requires a special approach.

2 The critical values of the ratio $\alpha_z^{crit} = c_z/c_r$ are considered as a function of the spherical subsystem parameters. At the threshold of stability, the value of $\alpha_z^{crit}(r)$ falls with the distance to the center. The value of α_z^{crit} can be twice less at the periphery in comparison with its central value.

3 The averaged relative disk scale height $\langle z_0 \rangle / L$ falls when the relative halo mass increases. We use this kind of relation to estimate the mass of the spherical subsystem for edge-on galaxies.

4 We conduct N-body modeling for seven edge-on galaxies based on published rotation curves and surface photometry data (for two galaxies we incorporate the data on the observed stellar radial velocity dispersion in addition). The relative mass of the spherical subsystem (dark halo in most cases) is inferred for all the galaxies. The evaluated mass of the dark halo in our galaxies is of order of their disk's mass. As an exception, the superthin LSB galaxy UGC 7321 own the dark halo which contains more than 2/3 of overall galaxy mass.

Acknowledgments

This work is supported by the Russian Foundation for Basic Research through the grants RFBR 04-02-16518, 04-02-96500 and by the Technology Program "Research and Development in Priority Fields of Science and Technology" (contract 40.022.1.1.1101).

References

Araki, S. PhD Thesis 1986, Massachusetts Institute of Technology

Athanasoula, E., Sellwood, J.A. 1986, MNRAS, 221, 213

Bahcall, J. 1984, ApJ, 276, 156

Bahcall, J.N., Kylafis, N.D. 1985, ApJ, 288, 252

Bailin, J., Steinmetz, M. 2003, Ap. & Space Sci., 284, 701

Binney J., Jiang, I.-G., Dutta, S. 1998, MNRAS, 297, 1237

Bizyaev, D., Kajsin, S. 2002, AAS, 201, 46.05

Bottema, R., Gerritsen, J. 1997, MNRAS, 290, 585

Bottema, R., van der Kruit, P., Freeman, 1987, A&A, 178, 77

Bottema, R., van der Kruit, P.C., Valentijn, E.A. 1991, A&A, 247, 357

Begeman, K.G., Broeils, A.H., Sanders, R.H. 1991, MNRAS, 249, 523

Carlberg, R.G., Sellwood, J.A. 1985, ApJ, 292, 79

Fuchs, B., von Linden, S. 1998, MNRAS, 294, 513

Goad, J., Roberts, M. 1981, ApJ, 250, 79

Griv, E., Yuan, C., Gedalin, M. 2002, ApJ, 2002, 580, 27L

Hernquist, L., Heyl, J., Spergel, D. 1993, ApJL, 416, 9

Holley-Bockelmann, J., Mihos, J., 2001, AAS, 198, 08.15

Khoperskov, A., Zasov, A., Tyurina, N. 2001, Astron. Rep., 45, 180

Khoperskov, A., Zasov, A., Tyurina, N. 2003, Astron. Rep., 47, 357

Mayer, L., Governato, F., Colpi, M., et al. 2001, ApJ, 559, 754

Matthews, L.D., Gallagher, J.S., van Driel, W. 1999, AJ, 118, 2751

Matthews, L.D. 2000, AJ, 120, 1764

Merritt, D., Sellwood, J. 1994, ApJ, 425, 551

Mikhailova, E., Khoperskov, A., Sharpak, S. 2001, in "Stellar dynamics: from classic to modern", ed. L.P. Ossipkov, I.I. Nikiforov, (Saint Petersburg), 147

Morrison, H., Miller, E., Harding, P., Stinebring, D., et al. 1997, AJ, 113, 2061

Ostriker, J., Peebles, P. 1973, ApJ, 186, 467

Patsis, P.A., Athanassoula, E., Grosbol, P., Skokos, Ch. 2002, MNRAS, 335, 1049

Pohlen, M., Dettmar, R.-J., Lutticke, R., Aronica, G. 2002, A&A, 392, 807

Poliachenko, V., Shuhman, I. 1977, Sov. Astron. Lett., 3, 134

Poliachenko, V.L., Shukhman, I.G. 1979, Sov. Astr. Rep., 23, 407

Raha, N., Sellwood, J., James, R., Kahn, F. 1991, Nature, 352, 411

Reshetnikov, V., Combes, F. 2002, A&A, 382, 513

Sancisi, R., Allen, R.J. 1979, A&A, 74, 73

Sellwood, J. 1996, ApJ, 473, 733

Sellwood, J.A., Merritt, D. 1994, ApJ, 425, 530

Shaw, M.A., Gilmore, G. 1989, MNRAS, 237, 903

Sotnikova, N., Rodionov, S. 2003, Astronomy Letters, 29, 321

Toomre, A. 1966, Geophys. Fluid Dyn., No. 66–46, 111

Valluri, M. 1994, ApJ, 430, 101

van der Kruit, P., Searle, L., 1981, A&A, 95, 116

van der Kruit, P., Searle 1982, A&A, 110, 61

Vandervoort, P. 1991, ApJ, 377, 49

Velazquez, H., White, S. 1999, MNRAS, 304, 254

Weinberg, M. 1998, MNRAS, 299, 499

Xilouris, E.M., Byun, Y.I., Kylafis, N.D., et. al. 1999, A&A, 344, 868

Zasov, A., Makarov, D., Mikhajlova, E. 1991, Sov. Astron. Lett., 17, 374

Zasov, A.V., Khoperskov, A.V. 2003, Astron. Letters, 29, 437

DISK-TO-HALO MASS RATIO EVALUATIONS BASED ON THE NUMERICAL MODELS OF COLLISIONLESS DISKS

Anatoly V. Zasov,[1] Alexander V. Khoperskov, [2] and Nataly V. Tyurina, [3]

[1] *Sternberg Astronomical Institute of Moscow State University, 119992, University prospect, 13, Moscow, Russia*

zasov@sai.msu.ru

[2] *Volgograd State University, Volgograd, Russia*

khopersk@sai.msu.ru

[3] *Sternberg Astronomical Institute of Moscow State University, 119992, University prospect, 13, Moscow, Russia*

tiurina@sai.msu.ru

Abstract   We propose that the lower bound of the stellar radial velocity dispersion c_r of an equilibrium stellar disk is determined by the gravitational stability condition. We compared the estimates of stellar velocity dispersion at radii $r > (1.5 - 2) \cdot R_0$ (where R_0 is the photometric radial scalelength of a disk), found in the literature, with the minimal values of c_r necessary for the disk to be in a stable state, using the results of numerical simulations of 3D collisional disks. This approach enables to estimate an upper limit of the local surface density and (if R_0 is known) a total masses of a disk and a dark halo. We argue that the old stellar disks of spiral galaxies with active star formation usually have the velocity dispersions which are close to the expected marginal values. A rough values of disk-to-total mass ratios (within the fixed radius) are found for about twenty spiral galaxies. Unlike spirals, the disks of the "red" Sa–S0 galaxies are evidently "overheated": their radial dispersion of velocities at $r \simeq 2R_0$ exceeds significantly the marginal values for gravitational stability.

Keywords: galaxies, numeric simulations, dynamics

1.   Introduction

It is usually accepted that substantial amounts of dark matter are needed to explain flat rotation curves in the outer regions of disk galaxies. However the fraction of a dark matter belonging to a disk and to a halo is still an open

Alexei M. Fridman, et al. (eds.), Astrophysical Disks; Collective and Stochastic Phenomena 307–318

question. Several different methods were proposed how to get a separate estimate of a mass or a local density of a disk. One may try, for example,

- To decompose a rotation curve of a galaxy into spherical and disk –
  related components;

- To assume that the mass-to-luminosity M_d/L ratio of stellar disk is
  known from the models of stellar population and to convert L into M_d;

- To model either the formation or the kinematic properties of such structural features as a bar or spiral arms;

- To compare the observed scaleheight of gaseous layer with the expected
  one for different M_d, assuming that the velocity dispersion of gas is
  known;

- To use the available information about the velocity dispersion of an old
  stellar disk population (in addition to the rotation curve).

The first two approaches suffer from the well known ambiguities and usually
cannot get the unique solution. The third method needs the development of
some theoretical background which will allow to connect the observational
data with theoretical expectations. The fourth method may be applied (with
some inevitable assumptions) to edge-on galaxies only. The applicability and
the effectivety of the last method is still an open question. In this work we
tried to verify to what extent the observed velocity dispersion of an old disk
population may be used to estimate the local density or total mass of a disk.

Galactic disks, consisting mostly of old stars, may be considered as collisionless systems in quasi-stationary equilibrium with a very slow evolution
(if to exclude the cases of a strong interactions or mergings with neighbor systems). The disk in equilibrium is characterized by certain radial distribution
of stellar velocity dispersions in the plane of a disk (c_r, c_φ), that ensures its
stability to gravitational perturbations. If the stability criteria were known, it
would allow to develop a self-consistent model for the disk of real galaxy and
to estimate its density when both the rotational velocities and stellar velocity
dispersions are measured. It is convenient to describe a local radial velocity
dispersion c_r in terms of dimensionless Toomre parameter (which equals to
unit for the uniformly rotating thin disk, marginally stable to radial perturbations):

$$Q_T = c_r \cdot \mathfrak{x}/3.36G\sigma , \qquad (1)$$

where $\mathfrak{x}$ is the epicyclic frequency, and σ is a non-perturbed local density of
a disk. If the local value of c_r is known from observations, one may estimate
the local density from (1). To follow this way, one may either a) to assume
that Toomre stability parameter Q_T is known from some theoretical stability
criterium (see Zasov, 1985, Bottema, 1993 and more recent papers, cited in

Khoperskov et al., 2003), or to apply N-body models to galaxies to verify that the observed velocity dispersion satisfies the condition of stable equilibrium (e.g. Zasov, 1985, Khoperskov et al., 2001, Zasov et al., 2002). This method is rather cumbersome, but it may be used in a more simple way if we know a minimum value of Q_T of marginally stable disks which may be applied to models with a wide range of parameters.

2. The Choice of q_t for Marginally Stable Disks

To find the minimum radial velocity dispersion sufficient to ensure stability of the disk against perturbations of arbitrary shapes is highly important if, as some authors have suggested, real galaxies may be in a state of threshold (marginal) stability. Note that in the general case, the old stellar population of a galactic disk can have an excess velocity dispersion in the presence of other factors heating the disk (which may act both from inside and from outside), which are not related directly to the gravitational instability. However, even in this case, the conditions for marginal stability still provide a valuable information by yielding an upper limit for the mass of the disk that enables it to be stable.

There are two different approaches used to find the threshold value of velocity dispersion: either to seek for analytical solution of the problem or to use the numerical models to reproduce the observed properties of real galaxies assuming different mass distribution in the disk, bulge and halo. Together with certain advantages over numerical simulations (the mathematical rigorousness of the solutions in the framework of the problem formulated), the analytical approach to the dynamics of perturbations in a disk and the conditions for its stability has the drawback that it can be implemented only for very simple models (usually 2D disks and the local perturbations are considered analytically, based on the analysis of dispersion equation in the epicyclical approximation) and can yield only coarse estimates for the parameters of the disk when applied to real objects. Numerical simulations of collisionless systems are more flexible in terms of the choice of model. They make it possible to go beyond simple two-dimensional models and directly follow the development of perturbations in a disk that is initially in equilibrium. However, this approach has drawbacks of its own. The most serious problems of N-body simulations include (1) certain inevitable mathematical simplifications, and (2) the dependence of the final state of the system (after it reaches quasi-equilibrium) on the initial parameters, which are poorly known for real galaxies.

Both numerical and analytical estimates lead to conclusion that the marginal stability of a disk corresponds to $Q_T > 1$, rising to $Q_T \geq 3$ in the outer parts, although different approaches may give different values for the same disk parameters. We refer to our paper (Khoperskov et al., 2003) for the more detailed discussion. We analyzed there the conditions for the gravitational stability of a three-dimensional collisionless disk with exponential density profile, embedding in the gravitational field of two rigid spherical components – a bulge and a halo, whose central concentrations and radial scales were varied over wide ranges. The initially weakly unstable disks in our models started their evolution from the subcritical equilibrium state. The results of dynamical simulations allowed to determine the disk parameters at the stability limit (when the velocity dispersion ceases to change, and the disk reaches a quasi-steady state after 5-20 rotations of the outer edge of the disk). The stability of the solutions against the choice of computational method was verified by comparing results for several models obtained by two very different methods of computing the gravitational force: the TREE-code method and direct "particle-to-particle" (PP) integration, in which each particle interacts with each of the others, for $N = (20 - 80) \times 10^3$. A comparison of the two results revealed no significant differences between the final disk states.

We constructed different numerical models (the number N of equal mass particles was up to 500×10^3), where the ratios of halo-to disk masses inside of a disk radius varied between 0.5 and 3. In the case of a low-mass or nonexistent halo, the evolution of the disk is determined by the bar mode, and the disk is heated due to the formation of an non-axisymmetric bar and associated two-armed spiral. Models with sufficiently compact bulge or massive halos do not show any enhancement of the bar mode, however they develop a complex transient system of small-scale spiral waves. The decrease of the amplitudes of these waves with time is accompanied by a transfer of rotational kinetic energy to the chaotic component of the velocity, resulting in heating of the disk. In turn, the increase of the radial-velocity dispersion c_r slows down with decreasing wave amplitude. The heating virtually ceases after the decay of the transient spiral waves. If the disk is initially cool ($Q_T \leq 1$), its heating is very efficient, and its dynamical evolution clearly demonstrates that the wave-decay process has a certain inertia: the velocity dispersion is already high enough to maintain the stability of the disk, however the spiral waves have not yet decayed (as is confirmed by Fourier analysis of the density perturbations in the disk) and continue to heat the disk. Therefore, to obtain the minimum velocity dispersion required for disk stability, we used an iterative algorithm, seeking for a subcritical starting point to make the initial velocity dispersion approach the stability limit.

The other essential initial parameter is the disk thickness or vertical-to-radial velocity dispersion ratio. The thicker is the disk initially, the lower is the minimum radial velocity dispersion c_r, which determines its stability. This circumstance also shows that the minimum critical dispersions in both coordinates z and r are reached if the disk begins to evolve from a subcritical state for both radial and bending perturbations, slowly increasing radial and vertical velocity dispersion at the initial state of evolution. As expected, the minimum

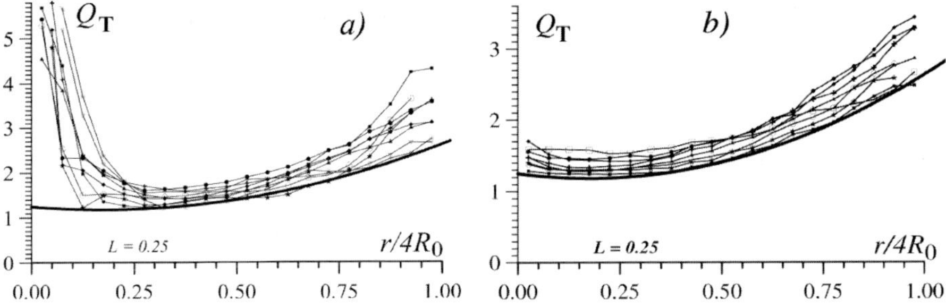

Figure 1. Parameter Toomre Q_T for marginally stable disks for different models with bulges (a) and without bulges (b). Radius of disk corresponds to $r = 4\,R_0$.

radial-velocity dispersion at the end of the simulations (expressed in units of the circular velocity) is higher in the models where the relative mass of the halo is lower, the initial disk thickness is less, and the degree of differential disk rotation is higher.

The radial dependences of Q_T, calculated for our models, are different, being determined primarily by the relative mass and degree of concentration of the spherical components. Yet it is essential that in all cases we considered the run of $Q_T(r)$ along the radius passes through a minimum $Q_T \simeq 1.2 - 1.6$ just beyond the region controlled by a bulge, at a galactocentric distance of $(1-2) \cdot R_0$, where R_0 is the radial scale of a disk (see Figs. 1a and b), and this behavior depends only slightly on the choice of model. If a bulge is of low mass or absent the low value of Q_T keeps down to the very centre (Fig. 1 b) These properties can be used for a rough estimate of the density (and, consequently, the mass) of a galactic disk (or to put limits on these quantities) from the observed c_r at these radial distances without the use of numerical simulations or analytical stability criteria.

3.   The Application of the Method

The measurements of stellar velocity dispersion obtained by absorption line spectroscopy are usually restricted by the central parts of galaxies due to steep brightness gradient of stellar disks and the difficuties of analysing the low intensity absorption line spectra. There are not so many galaxies where the line-of-sight dispersion is measured at $r \geq 2R_0$. At this distance the brightness of a bulge is usually negligible or at least is not overwhelming (some exceptions may exist however among the early type disky galaxies *Sa–S0*). For our purpose we took the data obtained for spiral and lenticular galaxies from the papers listed below:

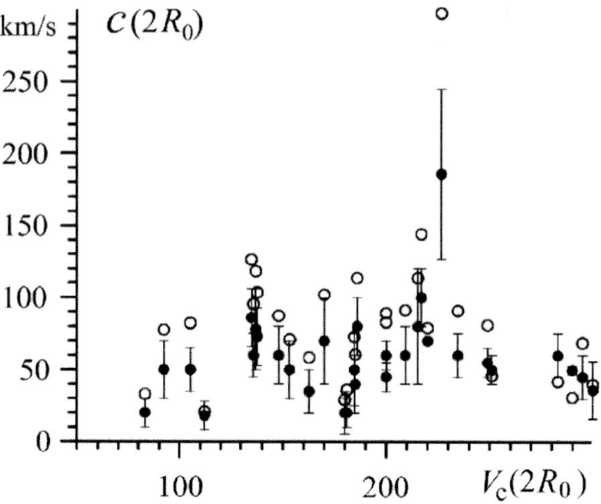

Figure 2.   A comparison of velocity dispersion at $r \simeq 2R_0$ with the velocity of rotation of galaxy. Filled circles are for the observed velocity dispersion c_{obs}, open circles – for the estimates of radial component c_r. Error bars are given for c_{obs} only.

- Shapiro et al. (2003): NGC 1068, NGC 2460, NGC 2775, NGC 4030;

- Bershady et al. (2002): NGC 3982;

- Bottema (1993): NGC 1566, NGC 2613, NGC 3198, NGC 5247, NGC 6340, NGC 6503;

- Heraudeau et al. (1999): IC 750;

- Beltran et al. (2001): NGC 470, NGC 4419, NGC 7782;

- Simien, Prugniel (2000, 2002): NGC 2962, NGC 3630, NGC 4143, NGC 4203, NGC 4578, NGC 5273;

- Neistein et al. (1999): NGC 584, NGC 2549, NGC 2768, NGC 3489, NGC 4251, NGC 4649; NGC 4753, NGC 5866.

The radial scalelengths R_0 (reduced to $H_0 = 75$ km/s/Mpc, if necessary) were taken from the cited papers or, if they are absent there, from (Baggett, 1998) or (Grosbol, 1985). For our Galaxy, which was added to the list above, we came from $c_r = 38$ km/s in the solar vicinity (Dehnen, Binney, 1998), at radius $r_\odot \simeq 2.7R_0$. In most cases the estimations of the line-of-sight velocity dispersion c_{obs} are related to the major axis of a galaxy. The radial velocity dispersion c_r was calculated from c_{obs} using the equation

$$c_r = c_{obs} \cdot \left[(c_\varphi/c_r)^2 \sin^2 i + (c_z/c_r)^2 \cos^2 i \right]^{-1/2} . \tag{2}$$

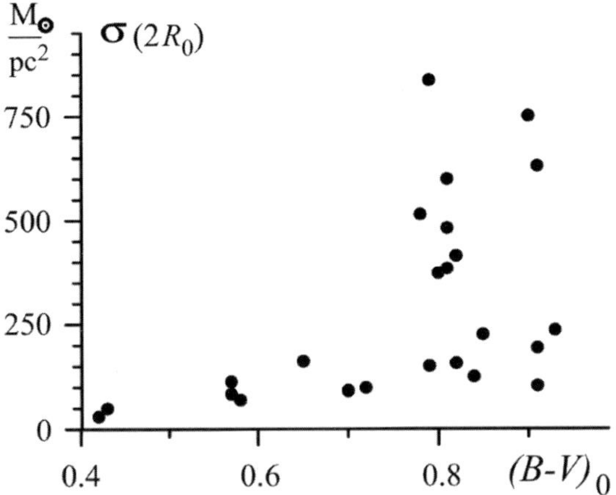

Figure 3. Threshold values of local surface density of galactic disks at $r \simeq 2R_0$, corresponding to the observed velocity dispersion, over the corrected total color indices of galaxies.

It was accepted that $c_\varphi/c_r = \mathit{æ}/2\Omega$ (epicyclic approximation), and $c_z/c_r \simeq 1/2$. To get $\mathit{æ}$ and Ω we used the rotation curves obtained from gas emission lines. These curves were taken either from the original papers (see the papers cited above) or from the references presented in the Catalogue of kinematically resolved data (see HYPERLEDA database). If only the "stellar" rotation curve is available, we applied to it the approximate correction for the asymptotic drift at $r = 2R_0$ (in the original paper by Neistein et al., 2002 the corrected velocities are already given).

Fig. 2 reproduces the estimates of the observed velocity dispersion at $r \simeq 2R_0$ (filled symbols) and the corresponding values of c_r (open symbols) in comparison with the maximal velocity of rotation of galaxies. The correlation between the velocity of rotation and the velocity dispersion is practically absent. Error bars are given only for the observed values. These bars are rather of illustrative nature, being taken by eye from $c_{obs}(r)$ diagrams given in the original papers. They demonstrate rather low accuracy of the estimates, especially for galaxies with low velocity dispersion, for which the errors of c_{obs} and, hence the local density estimates $\sigma(2R_0)$, in some cases may exceed a factor of two. It means that the results we obtain from these data for individual galaxies may be considered only as rather preliminary ones. Nevertheless it makes a sense to compare them with those expected for the marginally stable disks.

In Fig. 3 we show the local surface densities of the marginally stable disks of chosen galaxies at $r = 2R_0$, calculated for the adopted parameter $Q_T = 1.5$. They are plotted against the corrected values of color indices of galaxies, taken from HYPERLEDA database. It is worth to remember that if the disks are overstable, that is if the velocity dispersion of stars exceeds the threshold value for the gravitational stability, these estimates may be considered as an upper bound of the density at a given radius. The candidates to galaxies with the overheated disks are several "red" galaxies (most of them are lenticulars) which stand out in Fig. 3 by their incredibly high marginal surface densities exceeding 400 $M_\odot/\text{pc}^2$, whereas most of the other galaxies have $\sigma(2R_0) \simeq (50 - 200)\ M_\odot/\text{pc}^2$. For comparison, in solar neighborhood the column density of the galactic disk does not exceed 60 $M_\odot/\text{pc}^2$ (Khoperskov, Tyurina, 2002).

A similar conclusion about the disk "overheating" of some galaxies follows from the estimates of the ratio of disk masses to total masses of galaxies, which we describe as the indicative mass inside of the sphere with radius $r = 4R_0$:

$$\frac{M_d}{M_t} = \frac{\int\limits_0^{4R_0} 2\pi r \sigma(r)\,dr}{4V^2 R_0/G}\,. \tag{3}$$

The galaxies of the "red group" mentioned above have an unphysical estimates of the ratio $M_d/M_t > 1$, which means that their radial velocity dispersions definitely exceed the marginal values even if to admit that the whole mass of a galaxy contains in a disk. Hence either such galaxies have very "overstable" stellar disks, or the measurements of dispersion were influenced by the light of the bulge, which caused the overestimations of c_{obs}. Both versions are possible and in principle the situation has to be analyzed separately for every single galaxy. It is worth noting however that for three of the presumably overheated galaxies (NGC 4251, NGC 4578 and NGC 5273) the existing estimates of the velocity dispersion extend to $r > 2R_0$, reaching the distances r/R_0 = 3.6, 3.3 and 3.1 correspondingly, that is their dispersion is obtained at radii where the influence of a bulge is much lower than at $r = 2R_0$. Although the uncertainty of Q_T becomes more severe there, we may admit that its value still does not exceed 3, as numerical models demonstrate, which allows to obtain the upper limit of the disk mass using c_{obs} at large radial distances. However even in these cases the ratio M_d/M_t remains unphysically high (M_d/M_t = 1.3, 2.0 and 1.2 correspondingly for the galaxies in question), that is the conclusion about the "overheating" of their disks is confirmed.

It's essential that for all galaxies (but two early type lenticulars: NGC 2768 and NGC 4203), which do not belong to a group of red galaxies with a high marginal disk density at Fig. 3, the ratio $M_d/M_t < 1$ (Fig. 4). It means that their radial velocity dispersions at $r = 2R_0$, obtained from the observational data, cannot be explained without the presence of massive haloes – independently on whether the disks are marginally stable or overheated. Thence a dark matter, if exists in a galaxy, cannot be concentrated in a disk. A fraction of mass of dark matter in galaxies seems to change significantly from one galaxy to another.

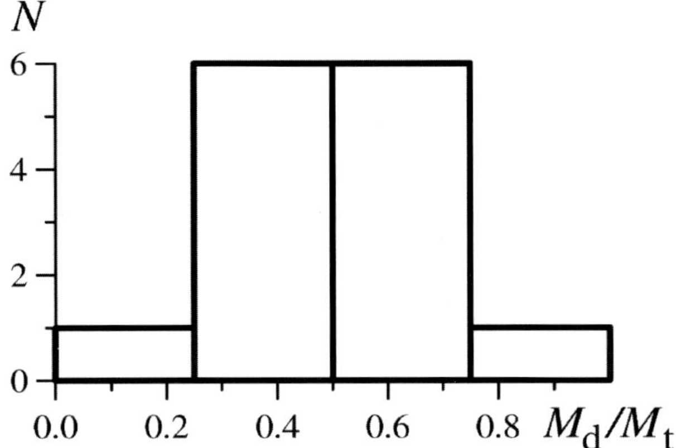

Figure 4. A histogram of M_d over M_t ratio, where M_d is the mass of a marginally stable disk, M_t is the indicative total mass of a galaxy within the radius $r = 4R_0$.

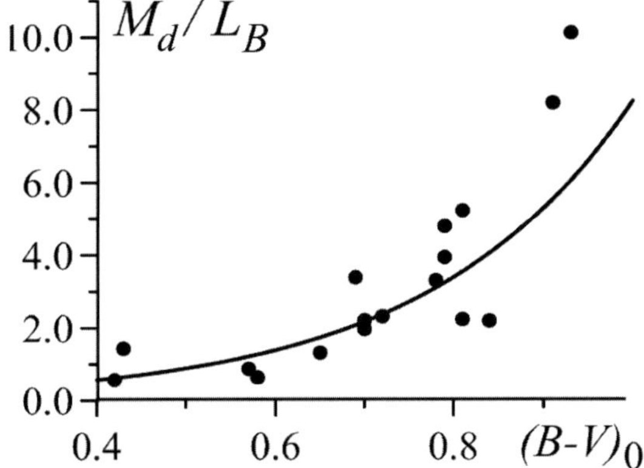

Figure 5. A relationship between mass-to-luminosity ratio M_d/L_B and the corrected color index for spiral galaxies. The curve describes the relationship found from the evolution modeling of stellar system (from Bell, deJong, 2001, closed box model).

However we should have in mind that M_d is no more than the crude estimate of the upper limit of the disk mass. It is worth trying to verify whether the real disk mass is close to M_d. If this proposition is correct, one can expect that the mass-to luminosity ratio for a disk is lower in galaxies with the less evolved stellar population. These galaxies should possess lower color indices due to the presence of young stars.

In Fig. 5 we compare the ratio M_d/L_B with the corrected color index $(B - V)$ for spiral galaxies of our sample, where the disk gives the main input to a total luminosity (lenticular galaxies were omitted). As the evolutionary models of stellar population show, the ratio M_d/L_B increases along the sequence of color indices, weakly depending (for a given color) on the history of star formation (Bell, de Jong, 2001). A curve drawn in Fig. 5 is model relationship, taken (without any normalization) from Bell, de Jong 2001, which was obtained for a modified Salpeter Initial Mass Function and presented in an analytical form (see their Table 3 for a "closed box" model). Two reddest galaxies are IC 750 (Sab) and NGC 2962 (S0a), the other galaxies belong to later types. In spite of the significant spread of points (which is not surprising due to the crudeness of the estimates), a general agreement between the model and the observed dependencies is evident: masses of the disks estimated under proposition of their marginal stability are close to those expected for galaxies with the observed luminosity and colour. In enables to propose that in most spiral galaxies a radial velocity dispersions of old stars in the disks (at least for $r \simeq 2R_0$) is really close to the minimum values necessary for the disks to be stable. If this is the case, one may conclude that the mechanisms of dynamical "heating" were not too efficient for the late-type galaxies we considered after they reached a stable state.

This conclusion evidently cannot be applied to the overstable disks of early type galaxies. It is possible that high color indices (low star formation rate) and a low gas content, typical for these galaxies are caused by the same events as the dynamical heating of their stellar disks (like a merging or a capture of small galaxies followed by the fast gas consumption).

4. Conclusion

- Numerical modeling of 3D stellar disks evolving from the sub-critical (to gravitational instability) to stationary state allows to find the minimal value of local parameter Toomre $Q_T \simeq 1.2 - 1.6$ which reaches at the radial distance $r \simeq (1 - 2) \cdot R_0$ for a wide ranges of masses and radial scales of bulges, disks and halos of model galaxies.

- The observed stellar velocity dispersion in the disks of spiral galaxies well agrees with the proposition that the radial velocity dispersion (at least at $r \simeq 2R_0$) is close to the expected one for the marginally stable disks.

- The observed velocity dispersion in the disks of most of spiral galaxies we considered makes the presence of a dark halo unavoidable for the disks to be gravitationally stable. Dark matter in a galaxiy, if exists, cannot be entirely concentrated in a disk.

- Galaxies with the high color index $(B - V)$ (most of them are lenticular galaxies) may possess strongly "overstable" disks, with the radial velocity dispersion exceeding the threshold level for gravitational instability. Hence this instability cannot be responsible for the observed velocity dispersion.

Acknowledgments

This work is supported by the Russian Foundation for Basic Research through the grants RFBR 04-02-16518, 04-02-96500 and by the Technology Program "Research and Development in Priority Fields of Science and Technology" (contract 40.022.1.1.1101).

References

Bell E., de Jong R.S., 2001, Astrophys. J. **550**, 212

Baggett W.E., Baggett S.M., Anderson K.S.J., 1998, Astron. J., **116**, 1626

Beltran J.C.V., Pizzella A., Corsini E.M. et al., 2001, Astron. Astrophys. **374**, 391

Bershady M., Verheijen M., Anders D.A., 2002, ASP Conf. Proc. **275**, 43

Bottema R., 1993, Astron. Astrophys. **275**, 16

Dehnen W., Binney J.J., 1998, MNRAS **298**, 387

Grosbol P. J., 1985, Astron. Astrophys. Suppl., **60**, 261

Heraudeau Ph., Simien F., Maubon G., Prugniel Ph., 1999, Astron. Astrophys. Suppl. **136**, 509

HYPERLEDA   http://www-obs.univ-lyon1.fr/hypercat/

Khoperskov, A.V., Zasov, A.V., Tyurina, N.V. 2003, ARep, 47, 357

Khoperskov, A.V., Tyurina, N.V. 2003, ARep, 47, 443

Khoperskov, A.V., Zasov, A.V., Tyurina, N.V. 2001, ARep, 45, 180

Neistein E., Maoz D., Rix H,-W., Tonry J.L., 1999, Astron.J. **117**, 2666

Shapiro K.L., Gerssen J., van der Marel R.P., 2003, Astron. J., **126**, 2707

Simien F., Prugniel Ph., 2000, Astron. Astrophys. Suppl. Ser. **145**, 263

Simien F., Prugniel Ph., 2002, Astron. Astrophys. **384**, 371

Zasov A.V., 1985, PAZh, 11, 730

Zasov, A.V., Bizyaev, D.V., Makarov, D.I., Tyurina, N.V. 2002, AstL, 28, 527

III

POSTERS

GLOBAL IRREGULARITIES OF SPIRAL PATTERNS IN GALAXIES: MANIFESTATION OF HYDRODYNAMIC INSTABILITIES?

A. D. Chernin
Sternberg Astronomical Institute, Moscow University
Universitetskij prosp. 13, Moscow 119899, Russia
chernin@sai.msu.ru

V. V. Korolev
Department of Physics, Volgograd State University
Universitetskij prosp. 100, Volgograd 400062, Russia
ppvito@vlink.ru

I. G. Kovalenko
Department of Physics, Volgograd State University
Universitetskij prosp. 100, Volgograd 400062, Russia
igk@vlink.ru

Abstract   We advocate the hydrodynamic origin of polygonal shape of spirals observed in some spiral galaxies building our arguments upon two-dimensional hydrodynamic simulations.

Keywords: spiral galaxies, hydrodynamical instability, shock waves

1.   Introduction

Observations of spiral patterns in flat galaxies show that a considerable portion of them possesses rather piecewise-smooth, faceted, than purely rounded shape. Instead of smooth spirals, one can talk about polygonal structures composed of line fragments, 'rows'. The extended straight segments of arms are distinctly traced both by young star clusters (Waller et al., 1997) and by

Alexei M. Fridman, et al. (eds.), Astrophysical Disks; Collective and Stochastic Phenomena 321–328
© *2006 Springer. Printed in the Netherlands*

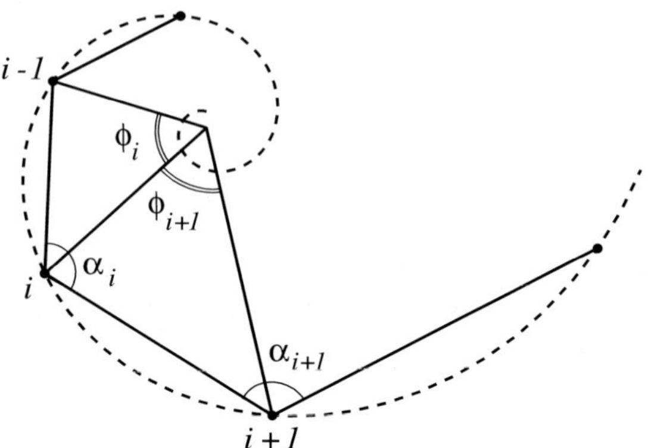

Figure 1.   A roughcast of the polygonal fragmentation of a spiral. The opening angles $\alpha_j \approx 2\pi/3$.

distribution of interstellar gas and dusty matter (Young & Scoville, 1991) which allows us to state with confidence that this is a real physical phenomenon rather than misinterpretation of observations. The number of galaxies with straight segments in a spiral design approaches 200 (Chernin et al., 2001). The fraction of galaxies with rows totals $6 \div 8\%$ of all spiral galaxies with a well-developed spiral design. Most of them are normal galaxies (Chernin et al., 2000).

Chernin (1999) noticed that the polygonal patterns have some geometrical traits in common: "(i) the length of a segment increases with distance from the centre almost linearly; (ii) the segments intersect one another at an angle which is on average $2\pi/3$; (iii) two polygonal patterns made by the segments are similar and comprise a significant part of the length of both arms". Fig. 1 illustrates these properties schematically.

The mechanisms of spiral pattern segmentation are still not completely clear.

As in the whole history of development of theories of the spiral structure origin, one can observe a 'battle' between stellar and gas dynamics concepts.

Contopoulos & Grosbøl (1986) referred the straight line segmentation in the galaxy NGC 5247 to manifestation of resonances in stellar dynamics.

The hydrodynamical version was suggested by Chernin (1999) who hazarded a conjecture that straight segments are manifestation of instability of a galactic shock wave in a curved gravitational well of a spiral arm.

The small size of the present poster paper allows us to examine carefully only one, hydrodynamical, hypothesis.

An idea of Chernin (1999) is based on a simple observational fact that slightly curved concave shock fronts have a tendency to become flat. The most vivid evidence of this phenomenon is the Mach reflection or Mach interaction of two shocks at small angles when two shock surfaces prefer to split into several flat pieces rather than remain indivisible (Landau & Lifshitz, 1986).

The aim of present study is to demonstrate the possibility in principle that the hydrodynamical instability is indeed an apt mechanism for generation of polygonal structures.

2. The Model, Computational Algorithm, Initial and Boundary Conditions

In a present paper we study numerically the behavior of a shock wave inside a curved potential well. We consider the interaction of interstellar gas flow with a sector of a spiral arm in a plane of a galaxy disk. Presuming a local character of instability we thus do not particularize the details of the global distribution of the gravitational potential in a galaxy. Therefore we disregard the effects of a finite disk thickness considering the problem as a two-dimensional one. The contribution of 3D-effects can generally be noticable; their role is discussed in our preceding paper in the present volume (Kovalenko et al., 2005).

We consider interstellar matter as a perfect adiabatic gas with a specific heats ratio $\gamma = 5/3$, ignoring non-adiabatic processes as well as self-gravitation and magnetic forces.

Gas flows into the potential well supersonically and passes through the well losing velocity behind a shock front. The aim of our study is to find possible steady-state, or perhaps quasi-steady stable flow.

An ultimate state is prepared in two steps. Initially the potential is specified as an infinite y-uniform well

$$\psi_{ini}(x) = \left\{ \begin{array}{ll} \psi_0 \cos^2\left(x\pi/2d\right)), & |x| \leq d; \\ 0, & |x| \geq d. \end{array} \right. \tag{1}$$

In this initial state the problem becomes one-dimensional and allows analytical treatment in full. It can be shown that the shock wave has a preferential stable position on a front side of the well at the point x_{cr} where

$$|\psi_{ini}(x_{cr})| = \psi_{cr} \approx \frac{1}{2}\left(\frac{\gamma+1}{\gamma-1}\left(\frac{\gamma-1}{2\gamma}\right)^{2/(\gamma+1)} - 1\right)v_0^2 \tag{2}$$

on the condition that the depth of the well exceeds the critical value ψ_{cr} (Kovalenko & Levy, 1992). Here v_0 is the inflow velocity of gas. The area $|\psi_{ini}(x)| > \psi_{cr}$ in the neighborhood of the well's bottom marks out the so-called 'forbidden zone' where the shock front can never be in a steady state position.

To ensure rapid convergence to a steady state the shock front is mounted on a front side in the area $-d < x_0 < x_{cr}$. After relaxation of a shock front to a predicted steady position, the central segment of the well $-10 < y < 10$ is slowly reshaping into the concave circular arc whereas the outer parts of the well hold their linear shapes (Fig. 2).

The parameters of the flow can be expressed in the dimensionless form in terms of density and sound speed of the unperturbed inflow and the half-thickness of the well

$$\rho_0 = 1 \, , \qquad c_{s0} = 1 \, , \qquad d = 1 \, , \qquad v_{0x} = 1.25 \div 7 \, , \qquad v_{0y} = 0.5 \div 10 \, ,$$

$$-\psi_0 = 0.5 \div 6 \, , \qquad R_c = 10 \div 100 \, , \qquad \tau = 10 \div 20 \gg \frac{2d}{v_{0x}} \, .$$

Parameter τ defines the time of building-up of the well's curvature radius from infinity to the ultimate value R_c. Its value is chosen to be large compared with the typical time of relaxation of a stream to a steady state.

The boundary conditions at the entrance (left in the case of normal inflow or left and bottom in the case of oblique inflow respectively) side(s) are fixed, and 'free-efflux' conditions are prescribed at the exit boundaries.

A second-order TVD scheme is exploited for calculations (Harten, 1983).

3.　　　Basic Results

Numerical simulation shows that the ultimate state strongly depends on the angle of incidence.

The pure normal inflow demonstrates several possible regimens of the flow. At the $1 < v_{0x} < 1.5$ the flow relaxes to a steady state but the shock does not inscribe potential well's shape. The shock front profile resembles that of in Fig. 2 (upper left frame).

When the entry velocity slides into the range $1.5 < v_{0x} < 5$, the shock front loses stability and starts oscillating. The amplitude and the period of oscillation are determined both by the velocity v_{0x} and by the initial location of undisturbed shock front. If the shock front is initially somewhere near the critical point $x_0 \sim x_{cr}$, then only the central part of the front oscillates (Fig. 2) and the amplitude maximizes at $v_{0x} = 2$ (Fig. 3). If, conversely, the shock front takes up a position far from the critical point x_{cr}, oscillations cover the whole front but the amplitude of oscillations is smaller. As v_{0x} grows, the tendency to local straightening of the front is observed. Fig. 2 clearly demonstrates that at some moments straight-line interpolation of the shock can be noted. Subsequent growth of v_{0x} damps oscillations and at $v_{0x} = 5$ the front almost exactly inscribes the well's shape and the flow becomes quiescent. The shock front profile is then geometrically similar to that in a bottom left frame of Fig. 2.

Fig. 4 marks the domains of steady and self-oscillatory regimens on the plane of input parameters.

At the oblique incidence the flow attains a steady state rapidly. At angles greater than $\sim 10°$ the front is all inside the potential well and is not fractured.

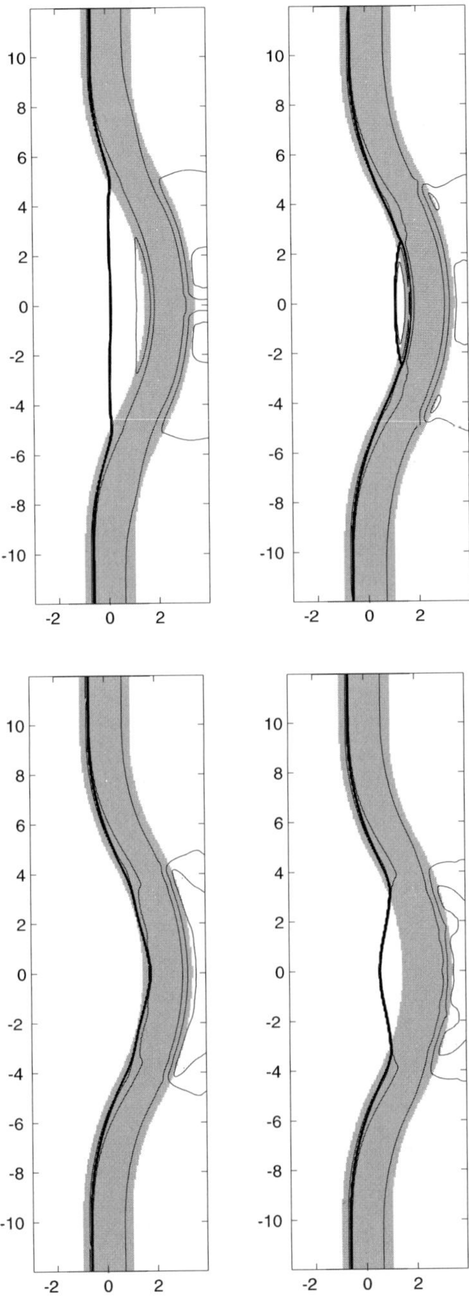

Figure 2. Location of the shock front (bold line) for different moments of time with the interval $\Delta t = 2 \cdot 10^8$ years (from upper left per line). Isolines of modulus of the Mach number are shown. The potential well is depicted by a shaded fringe. Distances are given in half-thicknesses of the well (1 kpc). Inflow velocity is $v_{0x} = 3.0$ (30 km/s), $v_{0y} = 0$.

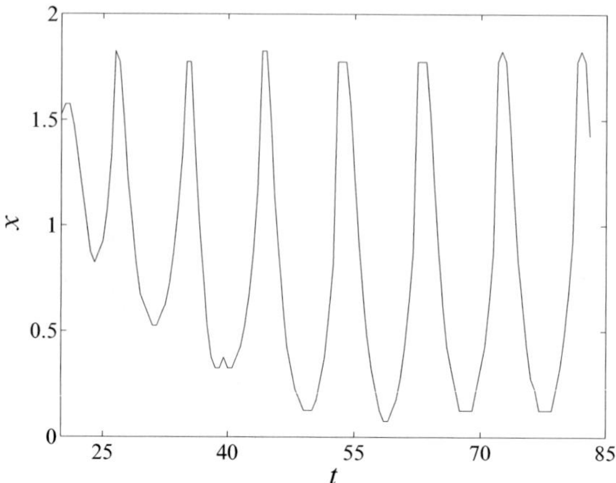

Figure 3.   Displacement-time curve of the equatorial point $y = 0$ of the shock front for $v_{0x} = 3.0$ (30 km/s), $v_{0y} = 0$ and $x_0 = -0.68$. Distance is given in half-thicknesses of the well (1 kpc) and time is in units $9.5 \cdot 10^7$ years.

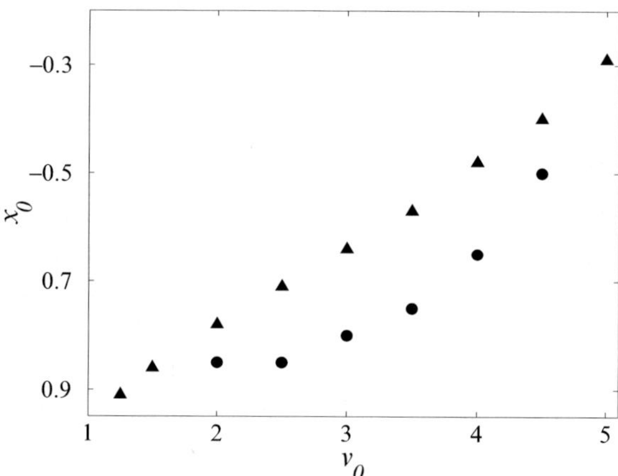

Figure 4.   The plane: inflow velocity v_{0x} - initial position of the shock front x_0. Triangles set lower bound to the 'forbidden zone' (see Sect.2). Below the circles the flow is steady state. Between triangles and circles is the area of oscillatory solutions.

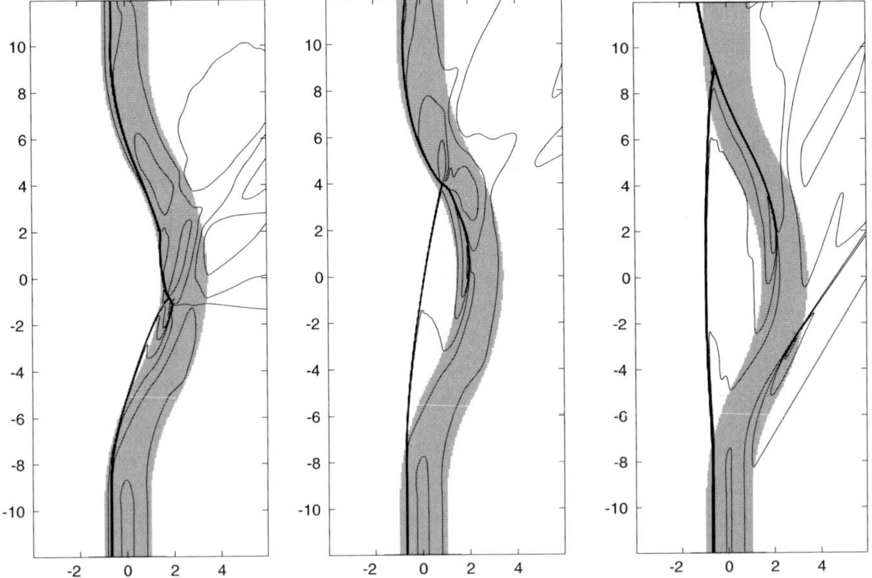

Figure 5. The same as in Fig. 2 but for steady state oblique flow. The incidence angles are (a) $\Theta = 33°$, (b) $45°$, (c) $63°$. Gas flow is north by northeast.

If an angle is between $20° \div 35°$, the front branches into three linear fragments (Fig. 5a). At larger angles two segments are formed one of which pushes out (Fig. 5b, c).

4. Conclusion

Our numerical experiments show that the shock front in a curved potential well does not precisely delineate its shape. There exist steady state configurations when fragmentation of the shock front takes place at the oblique incidence. There may also exist temporal polygonal structures when the shock front oscillates about an equilibrium position. This may serve as the argument in support of hydrodynamical origin of the polygonal segments in the arms of spiral galaxies.

A mechanism of shock wave fragmentation considered here has a local nature. Numerical simulation of disk as a whole will show whether hydrodynamical hypothesis is able to explain other remarkable properties of polygonal spirals such as constancy of opening angles and linear growth of lengths of segments with distance.

Acknowledgments

The work was supported in part by the Russian Foundation for Basic Research (project 04-02-96500).

References

Chernin, A. D.: 1999, *MNRAS* **308**, 321

Chernin, A. D., Zasov, A. V., Arkhipova, V. P. and Kravtsova, A. S.: 2000, *Astron. Lett.* **26**, 285

Chernin, A. D., Kravtsova, A. S., Zasov, A. V. and Arkhipova, V. P.: 2001, *Astron. Rep.* **45**, 841

Contopoulos, G. and Grosbøl, P.: 1986, *A&A* **155**, 11

Harten, A.: 1983, *J. Comput. Phys.* **49**, 151

Kovalenko, I. G. and Levy, V. V.: 1992, *A&A* **264**, 406

Kovalenko, I. G., Eremin, M. A. and Korolev, V. V.: 2006, present volume, p. 231

Landau, L. D. and Lifshitz, E. M.: 1986, *Hydrodynamics*. Nauka, Moscow

Waller, W. H., Bohlin, R. C., Cornett, R. H., et al.: 1997, *ApJ* **481**, 169

Young, J. S. and Scoville, N. Z.: 1991, *ARAA* **29**, 581

THE MORPHOLOGY OF GASEOUS FLOWS IN Z AND IN THE ACTIVE STATE

E. Yu. Kilpio,[1] D. V. Bisikalo,[1] A. A. Boyarchuk,[1] and O. A. Kuznetsov[1,2]

[1]*Institute of Astronomy*
Russian Academy of Sciences
Moscow, Russia

lena@inasan.ru

[2]*Keldysh Institute of Applied Mathematics*
Moscow, Russia

Abstract   The work deals with 2D modeling of gas flow structure in the symbiotic system Z And in the active state. In previous works the quiescent state of the system has been considered using the 2D gasdynamic model and the new possible mechanism of transition from quiescent to active state has been proposed. Here one more step is made in order to understand the outburst process. Namely, the results of previous calculations are used as input data to the modified model where the thermonuclear runaway is simulated as the pressure jump on the accretor's surface. It is shown that for realistic cases the structure of S-shaped shocks forms in the space between components. The changes in the calculated parameters of the system showed agreement with the observed brightness changes.

Keywords: accretion: hydrodynamics, accretion disks

1.   Introduction

Z Andromedae is one of the most intensively observed symbiotic stars with more than 100 years long history of observations. In the course of this period some nova-like outbursts have been detected. The last of them took place not long ago - in 2000-2002 and has been intensively observed in different wavelength ranges. During an outburst the system's brightness increases by 3 magnitudes in tens of days, after that it becomes to decrease and it takes a few hundreds of days for magnitude to reach quiescent values. It is common knowledge now that Z And is the detached binary system consisting of red

Alexei M. Fridman, et al. (eds.), Astrophysical Disks; Collective and Stochastic Phenomena 329–336
© 2006 *Springer. Printed in the Netherlands*

giant and white dwarf surrounded by the nebulosity (Boyarchuk, 1967). The mass exchange in this system is driven by stellar wind from the red giant.

The studies of the energy distribution in the wide spectral range (from UV to IR) during the quiescent period (1978-1982) (Fernandez-Castro et al., 1988) allowed to estimate the characteristics of Z And: the donor (cool M3.5III giant, $M = 2M_\odot$, $R = 77R_\odot$) loses mass at the rate of $2 \times 10^{-7}M_\odot/yr$; the hot compact component (accretor) ($M = 0.65M_\odot$, $R = 0.07R_\odot$) has the temperature $\sim 10^5 K$ and accretes the small part of the gas of the donor's wind (about 2%). The orbital period of the system is 758 days and the separation is $482.53R_\odot$. The gas of the circumbinary envelope has the electron density of $\sim 10^{10} cm^{-3}$ and the temperature $\sim 1.5 \times 10^4 K$. Systematic observations of Z And allowed to estimate its characteristics during the outburst and to find the features that can be explained in terms of the formation and subsequent drop-off of the optically thick shell by the accretor at the rate of about 250-300 km/s.

In order to explain the transition from quiescent to active state various mechanisms were proposed but there is still a lot of indefiniteness. In case of Z And, the nova-like activity is usually explained by changes of the accretion rate on the white dwarf (Mikolajewska & Kenyon, 1992). Previous calculations allowed us to propose a new mechanism that could provide a significant increase of the accretion rate (Bisikalo et al., 2002). The results have shown that the observed value of the giant's wind velocity (~ 25-40km/s) lies near the border dividing different accretion regimes. Namely, if the wind velocity is greater than 35 km/s the steady accretion disk takes place while for greater values the cone shock forms. So we can suppose that in quiescent state the accretion disk takes place in the system but rather minor variation of the giant's wind velocity can result in the accretion regime change. The process of transition between these two cases was studied in details (Bisikalo et al., 2002) and it was found out that the abrupt jump in the accretion rate takes place during the disk destruction process. This jump in the accretion rate could lead to the burning regime change in the accretor's bounder layers and thus to the outburst.

The model used in that calculations allowed to say nothing about the behaviour of the system after the accretion rate jump because it did not take into account the changes on accretor that should occur. Here we assume that the accretion rate jump caused the thermonuclear runaway from the accretor and model it by introducing the pressure jump on the accretor's surface. So we take one more step forward and consider the flow structure after the outburst.

2.   The model

The 2D gasdynamic model has been used to study the gas flow structure in the equatorial plane of the system. The zero point of the coordinate system was placed to the center of the donor, x axis was directed along the line connecting centers of the components, y axis - against the donor's orbital motion. The flow was described by the system of Euler's equations in the coordinate frame rotating with the angular velocity of the binary system Ω.

The force potential used differs from the standard Roche potential by the term corresponding to the additional force responsible for donor's wind acceleration. We use the parametric representation for this force with the parameter value providing the wind velocity change according to the β-law (Lamers law) with $\beta \sim 1$ and the value at infinity that are in agreement with observations.

To complete the system the ideal gas equation of state was used where the ratio of the specific heats γ was accepted to be 5/3.

To solve the system of the 2D gasdynamic equations the TVD-type Roe scheme (Roe P.L., 1986) with the restrictions of fluxes in the Oscher form (Chakravarthy S., Osher S., 1985) was used. The non-uniform rectangular grid consisting of 679×589 nodes was used for our calculations (the maximum grid density was in the vicinity of the accretor). The considered domain has the form of a square $[-A \ldots 2A] \times [-3/2A \ldots 3/2A]$ with excluded circles of radii equal to the ones of components and centers in the centers of components. The free outflow boundary conditions ($u = 0, p = 0$) were accepted on the outer border. The situation before the outburst was calculated assuming free outflow boundary conditions on the accretor. The outburst was modeled by introducing the pressure jump on the accretor's surface.

3.　　Results

In Figure 1 the situation before introducing of the pressure jump is shown. This solution corresponds to the value of the donor's wind velocity $V = 25km/s$. It has been shown in previous works (Bisikalo et al., 2002) that in this case the steady accretion disc exists in the system. This solution was used as the starting point for the outburst modeling. Then the boundary conditions on the accretor have been changed in order to model the pressure jump and further calculations were conducted with these new conditions.

Parameters for the jump modeling (density and temperature values) have been selected in order to bring the modeled energy release of the outburst in correspondence with the observed one. Temperature values in the range from $1.5 \times 10^5 - 1.5 \times 10^6$K have been considered. It should be noted that in our calculations the accretor is considered to be the circle of $3R_\odot$ radius and we should take into account that the real accretor's size is much smaller. So the parameters we consider correspond to the $3R_\odot$ distance from the center of the accretor.

The results of simulations have shown that after the introducing of the pressure jump on the accretor's surface the accretion disc disappears and the structure of two S shaped shocks forms in the space between the components. These changes occur on the timescale of tens of days. The situation for the moment

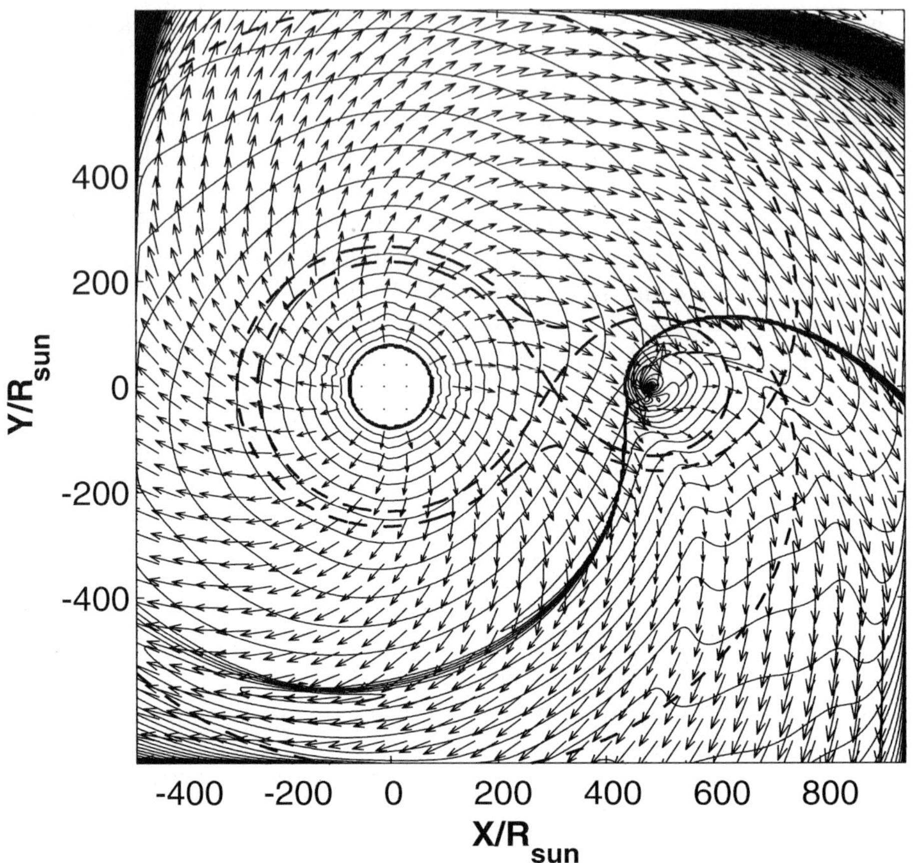

Figure 1.   Density field and velocity vectors for the steady solution with accretion disc. The isolines of the density logarithm are shown. Distances are given in solar radii. Dashed lines show equipotential contours of the standard Roche potential. Donor's surface is shown by the circle with the center in $(0, 0)$ point and the radius R_1. The black point marks the accretor.

of approximately $\sim$100 days after the boundary conditions change is shown in Figure 2. This case corresponds to the $T = 1.2 \times 10^6$K. Shocks are shown as isodences condensations.

In the Figure 3 the area near the shocks is shown in more details.

We can see that after crossing the shocks each of flows (from the donor and from the accretor) change their directions. It should be noted that oppositely directed flows along the tangential discontinuity exist. This fact could be seen in corresponding changes in spectral line profiles.

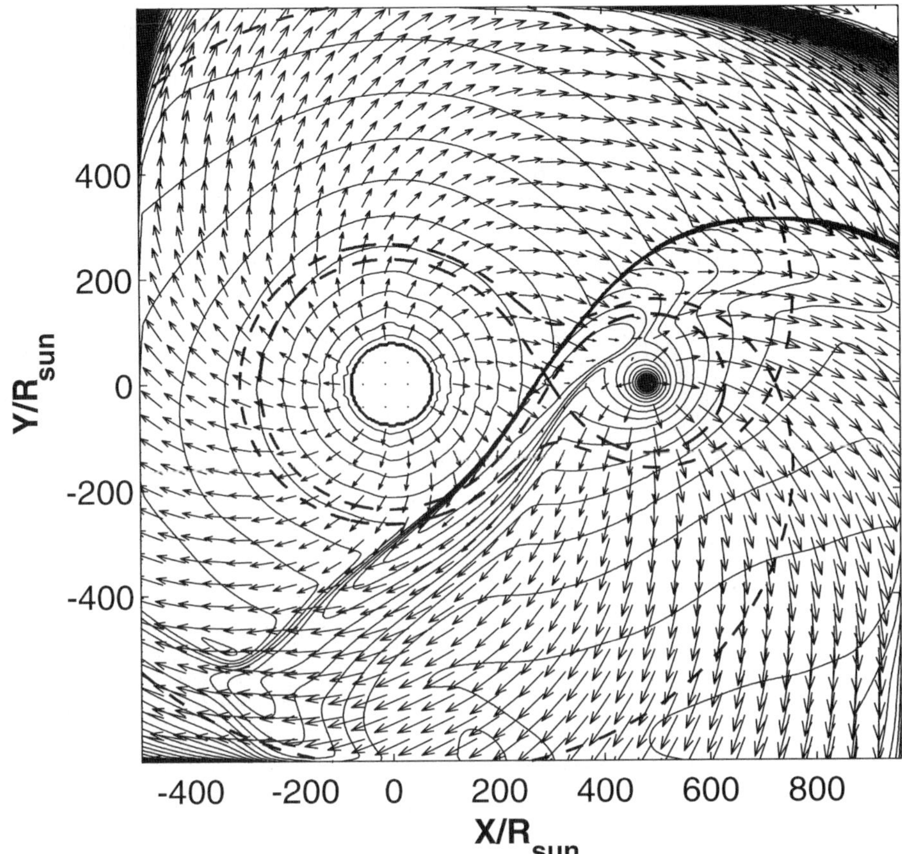

Figure 2.   Density field and velocity vectors for the moment The situation corresponds to the time $\sim$130 days from the outburst beginning. Isolines of the density logarithm are shown. Distances are given in solar radii. Dashed lines show equipotential contours of the standard Roche potential. Donor's surface is shown by the circle with the center in $(0,0)$ point and the radius R_1. The black dot marks the accretor.

It is of interest to consider how the changes in the flow structure can influence the observed brightness of the system. Here some rather rough estimates are presented. We assumed that brightness changes are proportional to changes in energy losses in the system. We used the approximation that the energy loss per unit volume and per unit time is $\rho^2 \Lambda(T)$, where $\Lambda(T)$ is the cooling function (see e.g. Cox et al., 1971). We calculated the integral energy loss in all computational domain and compared it to the one for the pre-outburst state:

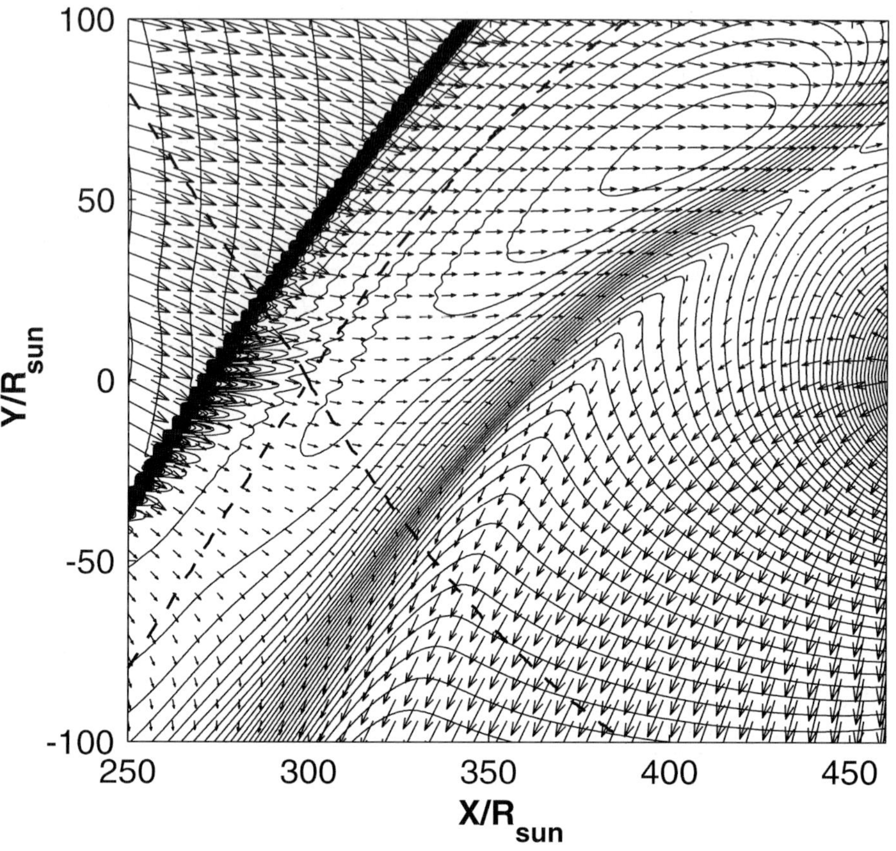

Figure 3.	The part of Fig. 2 near the shocks in more details ($[250R_\odot \ldots 450R_\odot] \times [-100R_\odot \ldots 100R_\odot]$).

$$\frac{\sum \rho^2 \Lambda(T)_{afteroutburst}}{\sum \rho^2 \Lambda(T)_{preoutburst}} \approx 12 \,. \tag{1}$$

The typical amplitude of the outburst for Z And approximately equals $\Delta m \sim 3^m$. This corresponds to the flux ratio ~ 16. So the parameters selected can provide the values of brightness change close to observed ones.

It is also important to consider possible timescales for these processes and to compare them with the real light curve. In order to do this we assumed that the pressure jump on the accretor exists only during the time corresponding to the rise to maximum on the light curve i.e. for ~ 100 days. Then we assume that the accretor returns to the quiescent state and consider the reverse process without a pressure jump. The agreement was also found to be rather good.

Conclusions

The 2D gasdynamic simulations of the flow structure have been carried out for an active state of a classical symbiotic system using the parameters of the prototype of this class - Z Andromedae. We define active state as the situation that takes place in the system after the accretion rate jump that is the result of flow rearrangement during the accretion regime change. We assume that this jump in the accretion rate leads to the change of the burning regime in the accretor's outer layers that is recognized to be the most probable cause of outbursts in systems of this class. The outburst was imitated by introducing the pressure jump on the accretor.

The preliminary results show that for selected values of parameters the structure of two S-shaped shocks located between components takes place. Flows from the donor as well as from the accretor change directions after crossing the shock and split in two oppositely directed flows.

The changes of the modeled parameters caused by formation of shocks show rather good agreement with the observed brightness changes. Namely the relative amplitude of changes of energy losses in the system corresponds to $2^m.7$ that is close to the typical amplitude of Z And outbursts. The timescales of the changes in the modeled characteristics correspond to the observed brightness changes of the system (time of the rise to maximum as well as time required for the reverse process).

Acknowledgments

The work was partially supported by Russian Foundation for Basic Research (projects NN 05-02-16123, 05-02-17070, 05-02-17874, 06-02-16097), by Science Schools Support Program (project N 162.2003.2), by Federal Programme "Astronomy", by Presidium RAS Programs "Mathematical" modelling and intellectual systems", "Nonstationary phenomena in astronomy", and by INTAS (grant N 00-491). O.A.K. thanks Russian Science Support Foundation for the financial support.

References

Boyarchuk, A.A.: 1967, *Izv. CrAO* **38**, 155

Fernandez-Castro, T., Cassatella, A., Gimenez, A., et al.: 1988, *Astrophys. J.* **324**, 1016

Mikolajewska, J., Kenyon, S. J.: 1992, *MNRAS,* **256**, 177

Bisikalo, D., Boyarchuk, A, Kilpio, E., Kuznetsov, O.: 2002, *Astron. Reports* **46**, 1022

Roe, P.L.: 1986, in *Ann. Rev. Fluid Mech.*, **18**, 37

Chakravarthy, S., Osher, S.: 1985, *AIAA Pap* **N 85-0363**

Cox, D.P., Daltabuit, E.: 1971, *Astrophys. J.* **167**, 113

SELF-CONSISTENT GAS AND STELLAR DYNAMICS OF DISK GALAXIES: A PROBLEM OF DARK MASS

Alexander V. Khoperskov, and Sergej S. Khrapov
Volgograd State University, Russia
khopersk@sai.msu.ru

Abstract We present results of numerical modeling made for the galactic stellar and stellar-gas disk embedded in the spherical halo and bulge. The stellar disk is simulated by N-body system, the equations of hydrodynamics are solved by TVD-method. We used TREEcode-algorithm for calculation of a self-gravity in stellar and gaseous components. The possibility of bars birth in a hot stellar disk because of gravitational instability of a cold gas component is investigated. The conditions of occurrence lopsided-galaxies from a axisymmetric disk as a result of gravitational instability are explored. The self-consistent models of double bars are constructed and the dynamical stability of these structures is discussed.

Keywords: galaxies, halo, N-body, gasdynamics

1.   Introduction

Dynamics of many structures in disk galaxies is considerably determined by spherical subsystem properties and, in particular, by characteristics of density distribution of dark halo in stellar disk limits.

The bar formation because of global bar-mode instability is impeded, if halo mass inside of optical radius surpasses disk mass $M_h \lesssim (1 - 1.5) \cdot M_d$ (Polyachenko and Shuhman, 1979); (Bisnovatyi-Kogan, 1984). On the other hand, the observations data and N-body simulation give estimates $M_h/M_d > 1.5$ for some galaxies (Khoperskov et al., 2001b); (Khoperskov, 2002). Gas component is cold because of radiative cooling and can be gravitationally unstable. The unstable modes in massive gas disk are capable to generate the bar even in the hot stellar disk in case of a high dispersion of stars velocities and at presence of the massive halo.

Alexei M. Fridman, et al. (eds.), Astrophysical Disks; Collective and Stochastic Phenomena 337–344
© 2006 *Springer. Printed in the Netherlands*

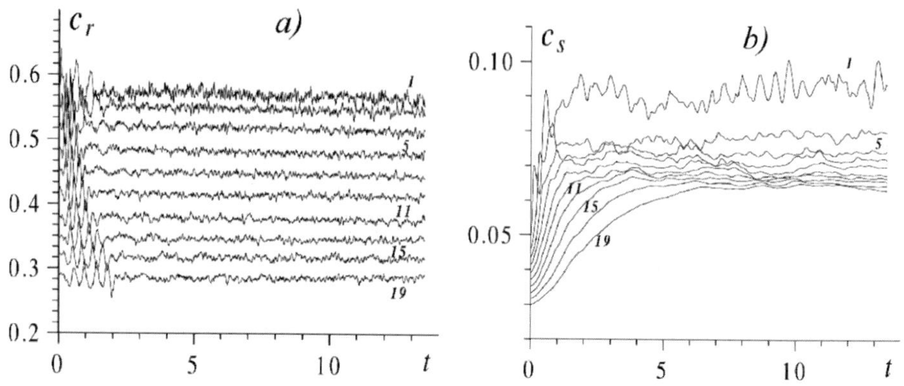

Figure 1.   Time dependences of the dispersion of stars radial velocities c_r (a) and sound speed of gas c_s (b) on various radiuses (curve *1* — disk center, *2* — disk periphery).

A asymmetrical spiral structure (one-arm) and bar displacement concerning disk centre are typical distinctive features of a series SBcd–SBm galaxies (Magellanic type). These properties are observed at LMC, NGC 55, 925, 1313, 1744, 4490, 4618, 4625 etc. (Vaucouleurs and Freeman, 2002); (Odewahn, 1991); (Pisano et al., 2000). The formation mechanism of the displaced bar and other features of lopsided-galaxies can be caused by preferred growth of one-arms modes in gravitationally unstable disk and by subsequent interaction of these perturbations with a bar-mode at a nonlinear stage (Zasov and Khoperskov, 2002). The late type galaxies contain more gas, than early type objects. Therefore question on influence of gas on the bar displacement and asymmetry in disk structure requires of special study.

The small-scale asymmetrical structures at disks center are very important for understanding of a phenomenon of nuclear galactic activity. The double bars can deliver gas to a active nuclei (Shlosman et al., 1989). The photometric data are the basic evidence about presence of the second inner bar at approximately 70 galaxies (Moiseev, 2001). The self-consistent models of double bars were studied by the N-body method (Friedli and Martinet, 1993), (Khoperskov et al., 2001a). Key problem of double bars is the question on dynamic stability of these systems.

2.   Modelling

2.1   The numerical model of stellar-gas disk

The 3D stellar disk simulation is based on N-body model, taking into account an external field of the rigid matter distributions in bulge and halo. The gas disk model is constructed on the non-viscous equations of gasdynamics, and is complemented by gravitational forces on the part of stellar disk, spherical subsystem, and gas self-gravity also.

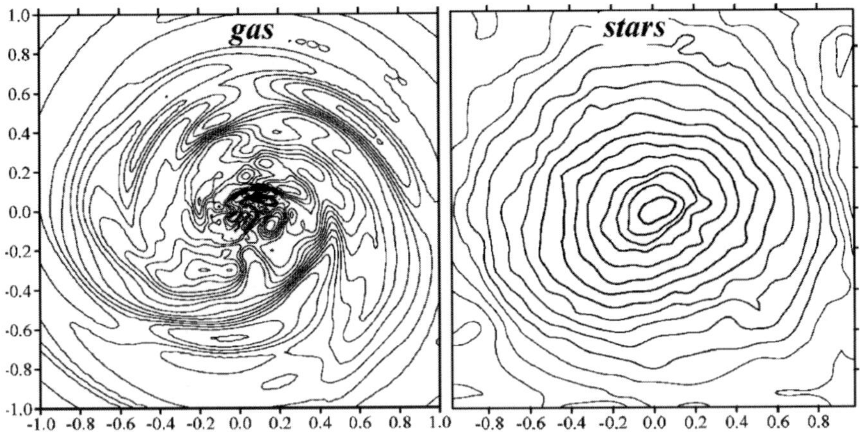

Figure 2. The gravitational instability of gas is the reason of bar formation in central area of the hot stellar disk. The isolines of a surface density are shown.

The gas galactic disks are cold, as the sound velocity c_s is much less than the dispersion of radial velocities in stellar disks c_r. The radiative losses in the equation on energy are defined by quantity Q^-:

$$Q^- = A_c \frac{(c_s^2 - c_{s1}^2)^\alpha}{(c_{s2}^2 - c_s^2)^\beta} \cdot \varrho^2 \text{ for } c_s > c_{s1} \quad (Q^- = 0 \text{ for } c_s < c_{s1}), \quad (1)$$

ϱ — density, parameters A_c, c_{s1}, α, c_{s2}, β are free. The cooling of gas strongly grows in the case $c_s \rightarrow c_{s2}$, therefore restriction $c_s < c_{s2}$ is carried out always. We solved hydrodynamical equations by the method TVD-E. The self-gravity account in gas and stellar disks is based on TREEcode. We simulated disks with an exponential profile of surface density and radial scale L.

If the stellar disk is on threshold of gravitational stability, small mass of cold gas component does not give to an additional heating of the stellar system (fig. 1). The gas mass is equal $M_g = 0.08 \cdot M_d$ inside the stellar disk in this model. And the massive halo ($M_h = 3M_d$) forbids the bar formation, as in stellar component (which besides is hot), and in gas disk.

2.2 The stellar bar formation because of gravitational instability in the gas disk

Let's consider models with halo mass $M_h = (1 \div 2.5) \cdot M_d$. In all cases the initial dispersion of star velocities and the massive halo provide gravitational stability of the stellar disk in absence of gas. The account of gas can qualitatively change evolution of system.

The radiative cooling provides cold, gravitationally unstable state of gas component, it gives in formation in gas of non-axisymmetric structures, which in turn generate disturbances in the stellar disk (fig. 2). There is the prompt

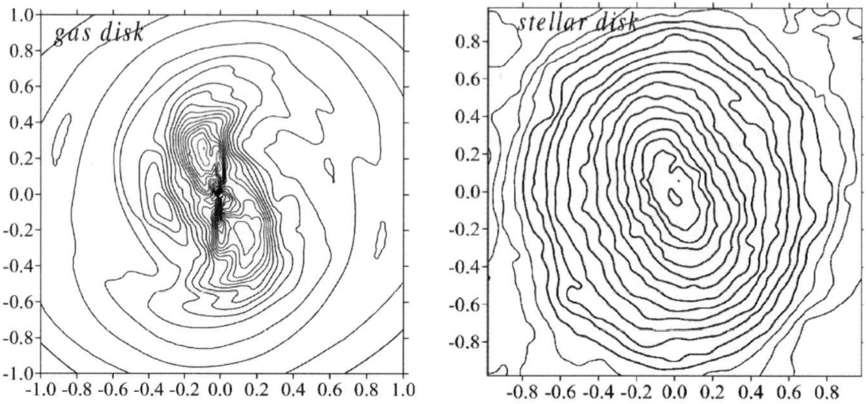

Figure 3.   The isolines of surface density of stellar disk (left panel) and gaseous disk (right panel) after 5 rotation periods of outer part of disk. The standings of shock waves are well visible in region of the bar.

bar formation in the stellar disk because of gravitational gas instability, if the models contain a lot of gas. It is important, that the instability of gas can generate the bar in hot stellar disk (fig. 3), when Toomre's parameter exceeds $Q_T = c_r/c_T \gtrsim 2$ in the region $r \leq L$. Such disks are stable without gas.

2.3   Lopsided-galaxies

Let's consider key influence of the gas component on effects of bars displacement and occurrence of asymmetry in disk structure in the whole. The initial distribution of dispersion of stars velocities and the parameters of spherical subsystem suppose the slow formation of bar without the account of gas component. However the amplitude of one-arm harmonic ($m = 1$) is very small and formation of the lopsided-disk does not occur.

The formation of harmonic $m = 1$ is possible in a massive cold gas subsystem. The nonlinear interaction of a one-arm mode and bar-mode ($m = 2$) in the gas disk is the reason of asymmetry of stellar disk also. The results of simulation in case of $M_{gas} = 0.47 \cdot M_d$ in limits of $r \leq 4L$ are shown in fig. 4. With growth of relative gas mass we have amplifications of the bar displacement concerning centre of disk and power of spiral structure asymmetry. The considered mechanism of formation of lopsided-galaxies is most effective in case of small halo mass and if the halo scale exceeds the exponential disk scale in 2 times and more.

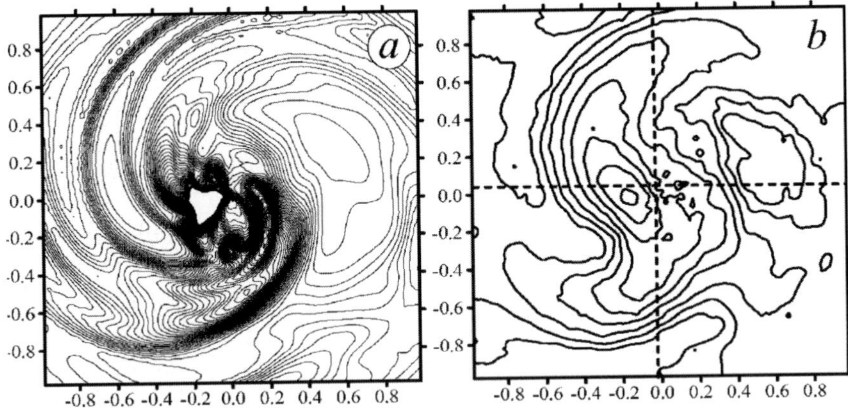

Figure 4. The isolines of surface density of gas (a) and stellar (b) components are shown in the case $M_h/M_d = 1$, $M_{gas}/M_d = 0.47$. The initial Toomre parameter is equal $Q_T \gtrsim 1.3$ for the stellar disk in region $r \leq 2L$.

2.4 The problem of double-bars formation

The bar formation requires not a hot initial disk and the halo mass $M_{halo} + M_{bulge} \lesssim 2M_{disk}$ at halo scale a $(1-4) \cdot L$ (L — exponential disk scale). The birth of inner bar in numerical models occurs at presence of enough massive bulge ($M_{bulge} \gtrsim 0.3M_d$). In fig. 5, we show the distributions of surface density logarithm $lg(\sigma)$ at the different time moments in model with $M_h = M_d$, $M_{bulge} = 0.6M_d$, in which at particular stages there are structures such as double bars.

The features of a kinematics of disk central region ($r < 2L$) at a stage of double bar are shown in a fig.6. The field of velocities in the stellar disk is the important information on existence of inner bar. The radial velocity U demonstrates four-areal structure, both for inner bar, and for primary bar.

The lifetime of double-bar does not exceed 1–2 rotation periods in the most ideal model of the stellar disk. The bending instability of disk and/or bar is the important factor of double-bars decay. The additional account of gas qualitatively changes result. The double-bars are not forming in self-consistent numerical stellar-gas models because of additional nonlinear perturbations. External asymmetrical potential (the tidal influence from the massive companion) gives similar result and such models do not give double-bars also. The conclusion about a dynamically fast phase of existence of "double bar" is agreed with work (Moiseev, 2002), that the secondary bar not is real dynamically allocated structure at observed galaxies, and represent a combination of objects with various morphology.

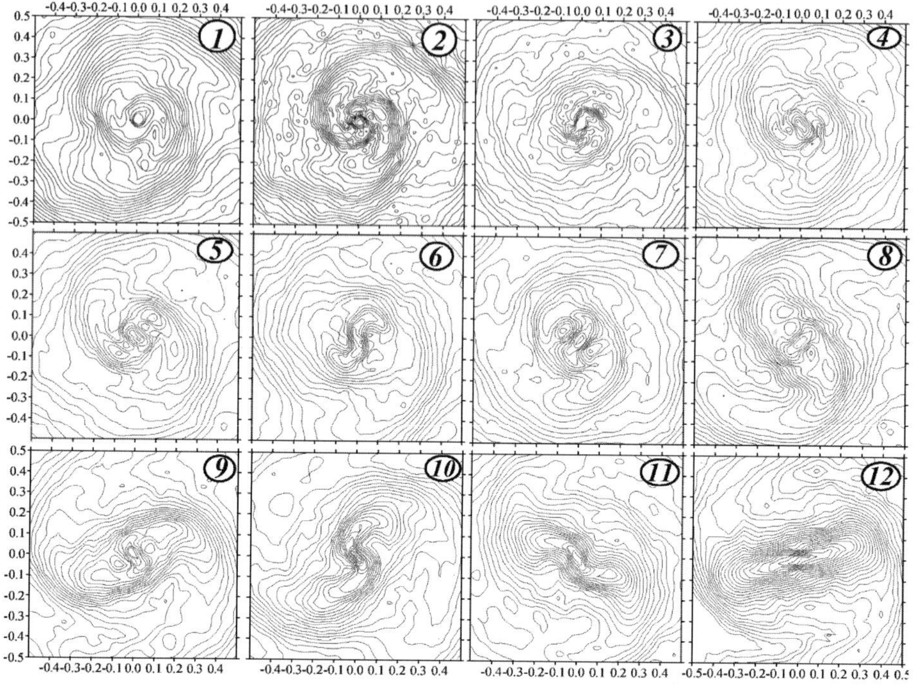

Figure 5.   The surface density distribution in different time.

3.   Conclusions

1 The bar formation in the hot gravitationally stable stellar disk can be generated by the unstable cold gas disk. This mechanism generates the bar even in case of the massive halo $M_h/M_d \simeq 1 - 2$.

2 The account of gas component strengthens the formation of asymmetrical structures in a disk as a result of gravitational instability in case of halo with small mass and large scale in comparison with a disk scale. At Magellanic type galaxies the relation of halo mass to disk mass in limits of optical radius on the average is less, than at systems of early types.

3 The self-consistent models with the double bar are extremely unstable in relation to the various factors (transient spiral waves in a disk plane, bar warps, bending instabilities of a disk, tidal influence, gas component) at initial stages of evolution. The conclusion about a very short phase of existence of systems such as "double bars" is made and similar structures can arise under special conditions at an initial stage of bar-mode development.

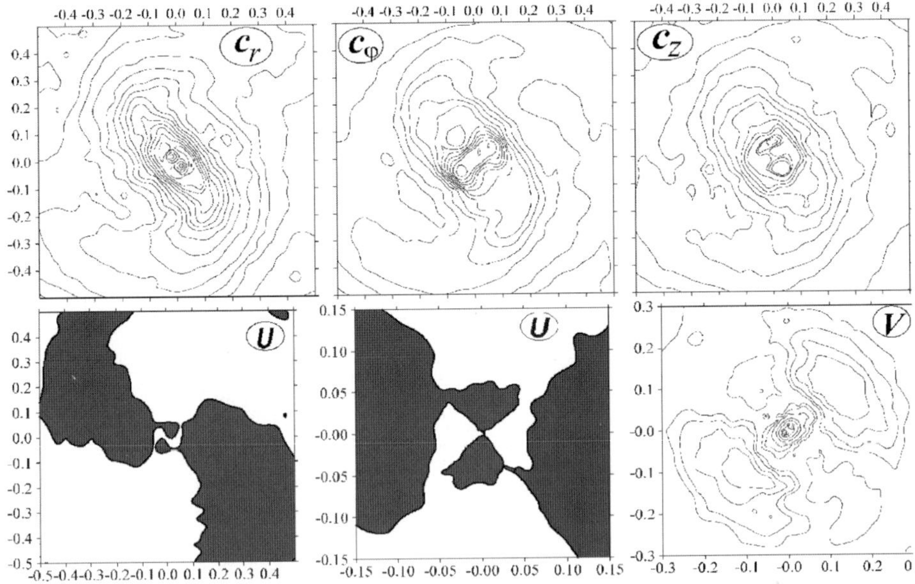

Figure 6. The distributions of the dispersions of velocity components (c_r, c_φ, c_z) and the isolines of azimuthal velocity V and radial velocity U. The areas of positive and negative radial velocity are shown by different shading.

Acknowledgments

This work was supported by the Russian Foundation for Basic Research through the grants RFBR 04-02-16518, 04-02-96500 and Technology Program "Research and Development in Priority Fields of Science and Technology" (contract 40.022.1.1.1101).

References

Bisnovatyi-Kogan, G.S. 1984, Astrofizika, 20, 547

Friedli D., Martinet L. Astron. Astrophys., 1993, 277, 27

Khoperskov, A.V., Moiseev, A.V., Chulanova, E.A. 2001, Bull. SAO, 52, 135

Khoperskov, A., Zasov, A., Tyurina, N. 2001, A. Rep, 45, 180.

Khoperskov, A. 2002, Astr. Lett., 28, 651.

Moiseev, A.V. 2001, Bull. SAO, 51, 140

Moiseev, A.V. AstL, 2002, 28, 755

Odewahn, S.C. 1991, AJ, 101, 829

Pisano D.J., Wilcots E.M., Elmegreen B.G. 2000, AJ, 120, 763

Polyachenko, V.L., Shukhman, I.G. 1979, SvA, 23, 407

Shlosman, I., Frank, J., Begelman, M.C. 1989, Nature, 338, 45

Vaucouleurs, de G., Freeman, K. 1972, Vistas Astron., 14, 163

Zasov, A.V., Khoperskov, A.V. 2002, Astron. Reports, 46, 173

NUMERICAL SIMULATION OF EXPANDING SHOCK WAVES IN THE YOUNG STARS OBJECTS

S.S. Khrapov, and V.V. Mustsevoi
Volgograd State University, Volgograd, 400062, Russia

Abstract   We present the results of nonlinear numerical simulation of the process of expanding shock waves on the early stage of the planetary nebula forming, like the nebula Mz3 (see [Redman et al., 2000, Philips & Mampaso 1988, Quinn et al., 1996]). We suppose that such a planetary nebula forming begins when the pressure and temperature in a young stars object is break growing. So, the portion of matter is thrown to environment out of the young star. This matter transports along the symmetrical axis, because it is collimated by accretion disk. This supersonic eject is forming the strong shock wave in the ambient medium – in the remains of proto – star cloud.

Keywords: hydrodynamics, accretion disk, planetary nebula, jet, shock waves.

1.   Introduction

Our model includes the rotated gaseous quasikeplerian disk around the proto-star nuclear and ambient medium in the gravity field of the nuclear without magnetic field.

We work in spherical coordinates system. The model of matter is ideal gas ($p \equiv (\Gamma - 1)\rho\varepsilon$, where p — is pressure, ρ — is density, ε — is internal energy of the gas, Γ — is adiabatic index). The gravity field is considered to be the spherical symmetric about the origin of the coordinates. So, we ignore the gravity influence of accretion disk. We assume the possibility of gas to be radiative cooling and external heating by radiative of the central star.

The radial function of the basic equilibrium parameters is:

$$p_0(r) \propto r^{-5/2}, \quad \rho_0(r) \propto r^{-3/2}, \quad \varepsilon_0(r) \propto r^{-1}.$$

The cooling function (see [Norman & Stone 1997, MacDonald & Bailey 1981]) is approximated by cubic splines.

We use the TVD–E code in $r - \theta$ plane of the spherical coordinates system.

345

Alexei M. Fridman, et al. (eds.), Astrophysical Disks; Collective and Stochastic Phenomena 345–351
© 2006 *Springer. Printed in the Netherlands*

2.   The discussion of results

The structure of fluxes inside the shell, created by shock wave is very complicated. Together with the wave transmitted into the accretion disk and reflected into the shell by the disk, the structure also includes the hierarchic structure of short–living and long–living vortexes evidence about the tendency to the turbulization of the matter. We should consider here the most important things throwing light on the causes of forming of observation nebulas.

First of all, when about 20 series of experiments are carried out, we can assume that the evolution of fluxes inside the shell is divided into three typical and natural stages.

I. At the first stage the matter of the initial eject, which has the considerably higher pressure, than in the surrounding environment and has already formed the shock wave (shell) starts expanding latitudinal angle, flowing around this dense shell and because of that the matter acquires the corner moment oriented according to z–coordinate (*Figure 1*(a)).

II. At the second stage though the shell has already expanded further along the radius, the matter received the corner moment is forming the toroidal vortex (*Figure 1*(b)). This long–living vortex is formed in the shell under the very inside part of the disk. The inner surface of the vortex forms the Laval nozzle (converge–diverge configuration) and it is here the acceleration of gas to the supersonic speed takes place.

It is important to note that the movement of accelerated gas and the matter rotating in the vortex are co–directional because of the velocity gradient from nozzle to the rotating matter is small, that's why such a configuration is not destroyed by the Kelvin–Helmholtz instability.

In the result of this process in all mentioned above series of experiments the narrow supersonic rotating jet is formed inside the shell which is made from the disk matter drawing into the shell by the toroidal vortex. Just with the forming of such supersonic jets we connect the point sharpening of head–part (later pear–shaped form) of shell – see *Figure 1–3*. The jet overtakes the shell penetrating into its gas, and it creates the inner shock wave and transmits the energy to the head–part of the shell and accelerates it.

III. On the third stage the further the rotating matter of jet under the centrifugal force ejected through the Laval nozzle flows away from the axis of symmetry, the farther it is from the center of symmetry of the system. In the result it makes a windsout–like swirl round the axis of symmetry, and this swirl, in its turn, draws the matter into and throws it away towards and by sideways the shell (the tornado effect). It is important to note that this thrown ejected matter of the accretion disk "overbuilds" the swirl. So the described current is self–sustaining. That's why although the torus–shaped vortex has already disappeared, the simulated emission of the jet has not stopped till termination of simulation in all series of experiments.

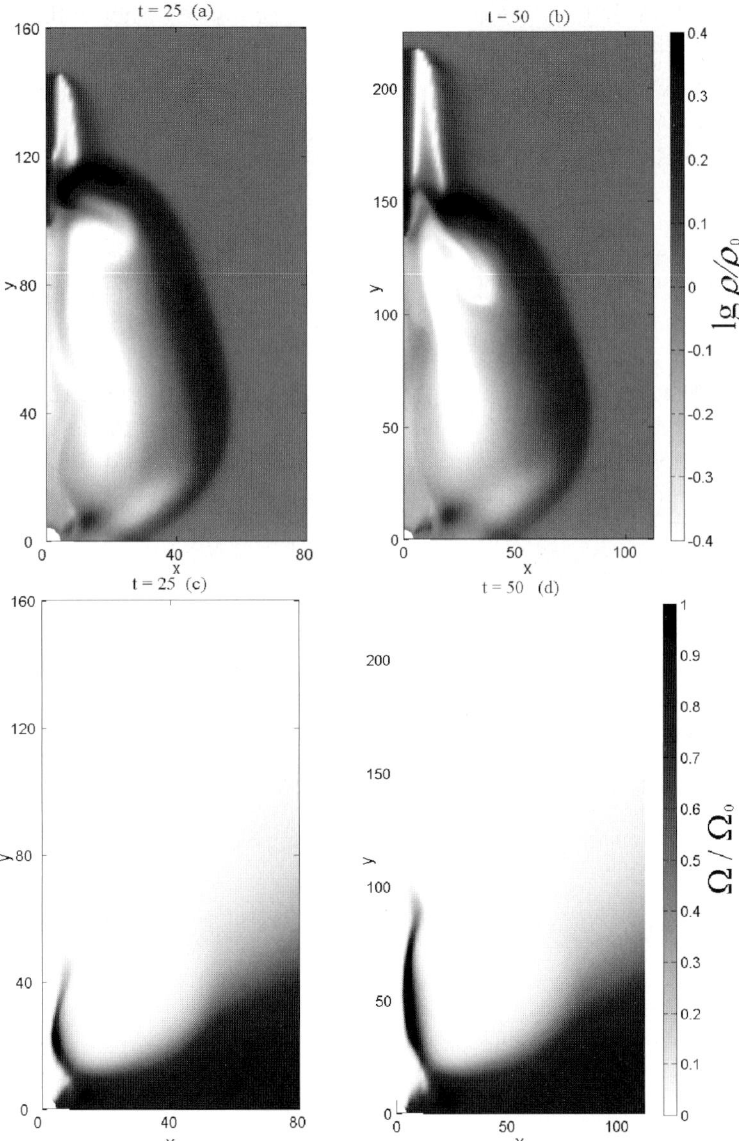

Figure 1.   Contour plots of the relative density logarithm $\ln \rho/\rho_0$ (a), (b) and dimensionless angular velocity of gas (c), (d) for the two different moments of the time for the case of bipolar eject with the initial Mach number $M = 10$ and opening angle $\theta = 20°$. Dimensional unit of the time is 112 yr, unit of the length $- 10^{13}$ m.

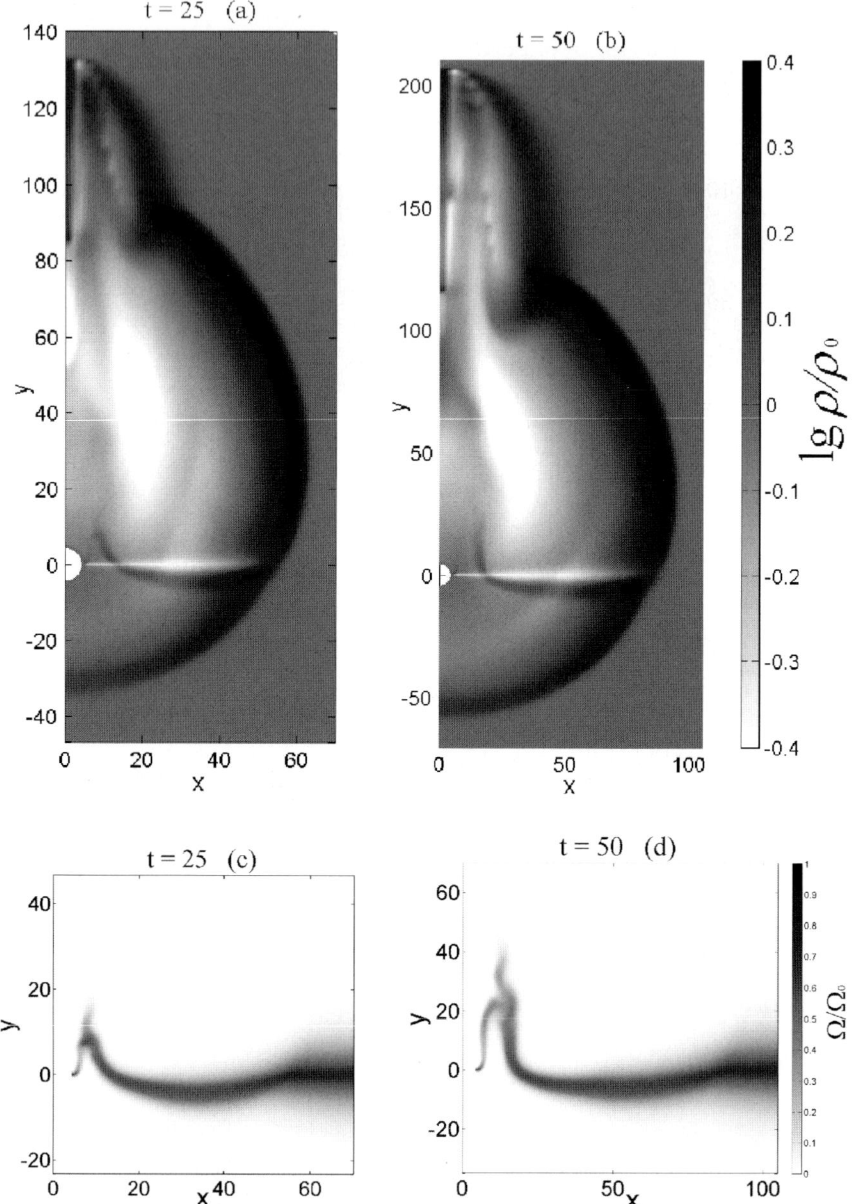

Figure 2.   Contour plots of the relative density logarithm $\ln \rho/\rho_0$ (a), (b) and dimensionless angular velocity of gas (c), (d) for the two different moments of the time for the case of unipolar eject with the initial Mach number $M = 7$ and opening angle $\theta = 18°$. Dimensional unit of the time is 112 yr, unit of the length $- 10^{13}$ m.

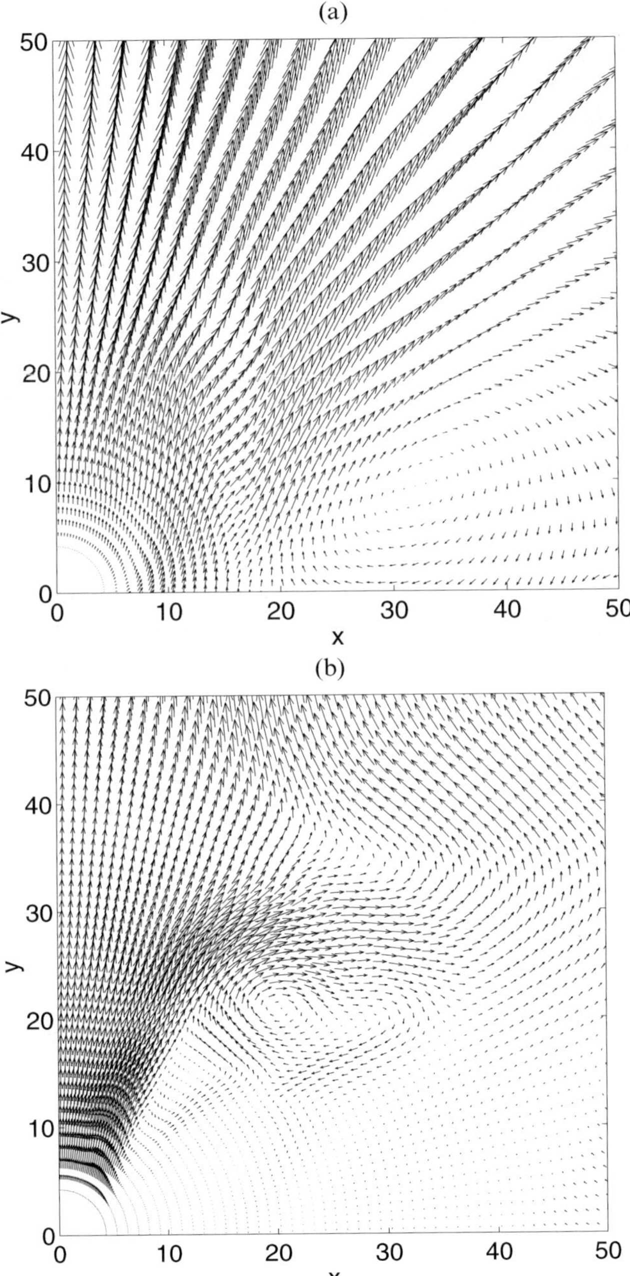

Figure 3.   Examples of the vectors velocity field in the region of long–living torus–like vortex for two different series of experiences.

It is highly important to emphasize that the described hydrodynamic mechanism of jet forming through the development of the sequence "eject–torus–tornado–jet" is based only on the law of conservation of angular momentum and that's why it should operate for all accretion–jet systems. The existence of other additional factors–magnetic fields and so on, can modify but not eliminate it.

There is rather interesting fact that the flexure of accretion disk is observed in the series with one–way eject and there is its movement along the axis of symmetry in the direction opposite to the initial eject – *Figure 3*. This effect is evidently caused by the impulse, transmitted to the matter of the disk by the shock wave. In the inner nearby star region the situation is quite opposite. The main part of the matter is drawn into the shell in the direction of the initial eject, for that the reason the accretion tempo in the real objects should fall sharply. It is possible that the similar effect would take place by the asymmetric eject.

Very gaunt form of the shell observed in all series of experiments during long periods of time – *Figure 2–3*, bears a strong resemblance to morphology of head shock waves in the young stars with jets (as for example HH1/HH2). It is possible that these systems are formed to be the same means at later stages of stars evolution, than those having been under our consideration. The head part of the shell is heat up in the result of this process. Since the pressure in this part and its surroundings is balanced quite quickly, the density in the head shell sinks considerably, and it can be visually perceived as a tendency to the shell break–through. But as our simulation shows the break–through does not appear even at very big times.

As follows from the *Figure 3*, in the case of one–way eject and thin disk the shock wave goes through the accretion disk and it makes another shock wave, i.e. the shell, in the opposite half–sphere relative to the initial eject. The form of this shell is slightly different (practically parabolic). We think that some asymmetric nebulas may be born just like that.

## 3.	Conclusion

The performed analysis resumes in the following considerations:

1 Results of numerical simulations give the evidence that the coming into the jet disk matter obtains the angular momentum.

2 If by some reasons one–side matter eject is realized, instead of bipolar eject the generated shock wave passes disk, producing shock wave behind the disk.

3 Within the high starting velocities of eject, the considerable prolongation of shock wave front becomes appropriate, and that is typical for many young star objects.

4 Numerical simulation shows that jet formation is caused by the availability of long–living torus–like vortex, where gas rotation along the middle line and around it takes place.

5 The formation of density loss at the shock wave front between its head and other parts is right appropriate then.

6 Luminous filaments can be caused by irregular in extinction and absorption of the central star radiation on the gas and dust of the head shock wave due to differential of shock wave front density distribution.

Acknowledgments

We would like to thank Dr. Moiseev A.V. for the helpful discussions and comments. This work was supported by the Russian Foundation for Basic Research through the grant RFBR 04-02-96500.

References

Redman, M.P., O'Connor, J.A., Holloway, A.J., Bryce, M., Meaburn, J., 2000, MNRAS, **312**, L23.
Philips, J.P., Mampaso, A., 1988, A&A, **190**, 237.
Quinn, D.E., Moore, T.J.T., Smith, R.G., Smith, C.H., Fujiyoshi, T., 1996, MNRAS, **283**, 1379.
Norman, M.L., Stone, J.M., 1997, Astrophys. J., **483**, 121.
MacDonald, J., Bailey, M.E., 1981, MNRAS, **197**, 995.

Volume 312: *High-Velocity Clouds*, edited by H. van Woerden, U. Schwarz, B. Wakker
Hardbound ISBN 1-4020-2813-X, September 2004

Volume 311: *The New ROSETTA Targets- Observations, Simulations and Instrument Performances*, edited by L. Colangeli, E. Mazzotta Epifani, P. Palumbo
Hardbound ISBN 1-4020-2572-6, September 2004

Volume 310: *Organizations and Strategies in Astronomy 5,* edited by A. Heck
Hardbound ISBN 1-4020-2570-X, September 2004

Volume 309: *Soft X-ray Emission from Clusters of Galaxies and Related Phenomena*, edited by R. Lieu and J. Mittaz
Hardbound ISBN 1-4020-2563-7, September 2004

Volume 308: *Supermassive Black Holes in the Distant Universe,* edited by A.J. Barger
Hardbound ISBN 1-4020-2470-3, August 2004

Volume 307: *Polarization in Spectral Lines*, by E. Landi Degl'Innocenti and M. Landolfi
Hardbound ISBN 1-4020-2414-2, August 2004

Volume 306: *Polytropes – Applications in Astrophysics and Related Fields,* by G.P. Horedt
Hardbound ISBN 1-4020-2350-2, September 2004

Volume 305: *Astrobiology: Future Perspectives,* edited by P. Ehrenfreund, W.M. Irvine, T. Owen, L. Becker, J. Blank, J.R. Brucato, L. Colangeli, S. Derenne, A. Dutrey, D. Despois, A. Lazcano, F. Robert
Hardbound ISBN 1-4020-2304-9, July 2004
Paperback ISBN 1-4020-2587-4, July 2004

Volume 304: *Cosmic Gammy-ray Sources,* edited by K.S. Cheng and G.E. Romero
Hardbound ISBN 1-4020-2255-7, September 2004

Volume 303: *Cosmic rays in the Earth's Atmosphere and Underground*, by L.I, Dorman
Hardbound ISBN 1-4020-2071-6, August 2004

Volume 302: *Stellar Collapse,* edited by Chris L. Fryer
Hardbound, ISBN 1-4020-1992-0, April 2004

Volume 301: *Multiwavelength Cosmology*, edited by Manolis Plionis
Hardbound, ISBN 1-4020-1971-8, March 2004

Volume 300: *Scientific Detectors for Astronomy*, edited by Paola Amico, James W. Beletic, Jenna E. Beletic
Hardbound, ISBN 1-4020-1788-X, February 2004

Volume 299: *Open Issues in Local Star Fomation,* edited by Jacques Lépine, Jane Gregorio-Hetem
Hardbound, ISBN 1-4020-1755-3, December 2003

Volume 298: *Stellar Astrophysics - A Tribute to Helmut A. Abt,* edited by K.S. Cheng, Kam Ching Leung, T.P. Li
Hardbound, ISBN 1-4020-1683-2, November 2003

Volume 297: *Radiation Hazard in Space,* by Leonty I. Miroshnichenko
Hardbound, ISBN 1-4020-1538-0, September 2003

Volume 296: *Organizations and Strategies in Astronomy, volume 4,* edited by André Heck
Hardbound, ISBN 1-4020-1526-7, October 2003

Volume 295: *Integrable Problems of Celestial Mechanics in Spaces of Constant Curvature,* by T.G. Vozmischeva
Hardbound, ISBN 1-4020-1521-6, October 2003

Volume 294: *An Introduction to Plasma Astrophysics and Magnetohydrodynamics,* by Marcel Goossens
Hardbound, ISBN 1-4020-1429-5, August 2003
Paperback, ISBN 1-4020-1433-3, August 2003

Volume 293: *Physics of the Solar System,* by Bruno Bertotti, Paolo Farinella, David Vokrouhlický
Hardbound, ISBN 1-4020-1428-7, August 2003
Paperback, ISBN 1-4020-1509-7, August 2003

Volume 292: *Whatever Shines Should Be Observed,* by Susan M.P. McKenna-Lawlor
Hardbound, ISBN 1-4020-1424-4, September 2003

Volume 291: *Dynamical Systems and Cosmology*, by Alan Coley
Hardbound, ISBN 1-4020-1403-1, November 2003

Volume 290: *Astronomy Communication,* edited by André Heck, Claus
Madsen
Hardbound, ISBN 1-4020-1345-0, July 2003

Volume 287/8/9: *The Future of Small Telescopes in the New Millennium,*
edited by Terry D. Oswalt
Hardbound Set only of 3 volumes, ISBN 1-4020-0951-8, July 2003

Volume 286: *Searching the Heavens and the Earth: The History of Jesuit
Observatories,* by Agustín Udías
Hardbound, ISBN 1-4020-1189-X, October 2003

Volume 285: *Information Handling in Astronomy - Historical Vistas,* edited
by André Heck
Hardbound, ISBN 1-4020-1178-4, March 2003

Volume 284: *Light Pollution: The Global View,* edited by Hugo E. Schwarz
Hardbound, ISBN 1-4020-1174-1, April 2003

Volume 283: *Mass-Losing Pulsating Stars and Their Circumstellar Matter,*
edited by Y. Nakada, M. Honma, M. Seki
Hardbound, ISBN 1-4020-1162-8, March 2003

Volume 282: *Radio Recombination Lines,* by M.A. Gordon, R.L. Sorochenko
Hardbound, ISBN 1-4020-1016-8, November 2002

Volume 281: *The IGM/Galaxy Connection,* edited by Jessica L. Rosenberg,
Mary E. Putman
Hardbound, ISBN 1-4020-1289-6, April 2003

Volume 280: *Organizations and Strategies in Astronomy III,* edited by André
Heck
Hardbound, ISBN 1-4020-0812-0, September 2002

Volume 279: *Plasma Astrophysics , Second Edition,* by Arnold O. Benz
Hardbound, ISBN 1-4020-0695-0, July 2002

Volume 278: *Exploring the Secrets of the Aurora,* by Syun-Ichi Akasofu
Hardbound, ISBN 1-4020-0685-3, August 2002

Volume 266: *Organizations and Strategies in Astronomy II,* edited by André Heck
Hardbound, ISBN 0-7923-7172-0, October 2001

Volume 265: *Post-AGB Objects as a Phase of Stellar Evolution,* edited by R. Szczerba, S.K. Górny
Hardbound, ISBN 0-7923-7145-3, July 2001

Volume 264: *The Influence of Binaries on Stellar Population Studies,* edited by Dany Vanbeveren
Hardbound, ISBN 0-7923-7104-6, July 2001

Volume 262: *Whistler Phenomena - Short Impulse Propagation,* by Csaba Ferencz, Orsolya E. Ferencz, Dániel Hamar, János Lichtenberger
Hardbound, ISBN 0-7923-6995-5, June 2001

Volume 261: *Collisional Processes in the Solar System,* edited by Mikhail Ya. Marov, Hans Rickman
Hardbound, ISBN 0-7923-6946-7, May 2001

Volume 260: *Solar Cosmic Rays,* by Leonty I. Miroshnichenko
Hardbound, ISBN 0-7923-6928-9, May 2001

For further information about this book series we refer you to the following web site:
www.springeronline.com

To contact the Publishing Editor for new book proposals:
Dr. Harry (J.J.) Blom: harry.blom@springer-sbm.com
Sonja japenga: sonja.japenga@springer-sbm.com